Thomas Wick

Multiphysics Phase-Field Fracture

Radon Series on Computational and Applied Mathematics

Volume 28

Thomas Wick

Multiphysics Phase-Field Fracture

Modeling, Adaptive Discretizations, and Solvers

DE GRUYTER

Mathematics Subject Classification 2010
35-02, 35Q74, 35Q35, 65-02, 65-04, 65Mxx, 65Nxx, 74-04, 74R10, 76-04, 76S05

Author
Prof. Dr. Thomas Wick
Leibniz Universität Hannover
Institut für Angewandte Mathematik
Welfengarten 1
30167 Hannover
Germany
thomas.wick@ifam.uni-hannover.de

ISBN 978-3-11-049656-7
e-ISBN (PDF) 978-3-11-049739-7
e-ISBN (EPUB) 978-3-11-049419-8
ISSN 1865-3707

Library of Congress Control Number: 2020942738

Bibliographic information published by the Deutsche Nationalbibliothek
The Deutsche Nationalbibliothek lists this publication in the Deutsche Nationalbibliografie; detailed bibliographic data are available on the Internet at http://dnb.dnb.de.

Typesetting: VTeX UAB, Lithuania
Printing and binding: CPI books GmbH, Leck

www.degruyter.com

I dedicate this book to my daily inspiration:
Marie-Cécile, Elisa, Céline, Maxim, Antoine

Preface and personal acknowledgments

Finishing this book was a challenging exercise![1] Let me start differently than usual. First of all, I wish to thank De Gruyter and notably Nadja Schedensack as well as the managing editor of this book series, Ulrich Langer, for this opportunity and their endless patience and professional cooperation. When I proposed the concept in April 2016, I had a tranquille[2] life as a Post doc at the Radon Institute of Computational and Applied Mathematics (RICAM) in Linz, Austria. Originally, this monograph was envisaged as part of the habilitation procedure as final qualification for a university career. I am sure that I would have achieved the goal of completing this book if I had stayed longer at RICAM.

Suddenly, however (luckily as I can say today!), I was overtaken by events: I got an offer as maître de conférences from École Polytechnique in Palaiseau (near Paris), France, and soon after a (permanent) professorship position at the Leibniz University Hannover in Germany. I was incredibly lucky to get involved in so many activities since then. I never wanted to miss all of these activities. But there is one task I did not pursue as I should have: this book! In Spring 2020, I found the most valuable treasure of our times in my profession: time! Time for both my children[3] and this book.[4]

I shall now briefly introduce the main topic, which is to discuss recent developments in mathematical modeling, design of algorithms, and numerical simulations in multiphysics variational phase-field fracture propagation. Variational modeling of fracture has been influential in several communities since 1998 when the first paper by Gilles Francfort and Jean-Jacques Marigo appeared.

The simplified prototype structure of a phase-field model for fracture propagation is: Find vector-valued displacements $u : B \to \mathbb{R}^d$ and a scalar-valued, smoothed, indicator function $\varphi : B \to \mathbb{R}$ such that

$$\begin{aligned} -\nabla \cdot (\varphi^2 \nabla u) &= f, \\ |\nabla u|^2 \varphi - \varepsilon \Delta \varphi - \frac{1}{\varepsilon}(1-\varphi) &\leq 0, \\ \partial_t \varphi &\leq 0, \\ \left[|\nabla u|^2 \varphi - \varepsilon \Delta \varphi - \frac{1}{\varepsilon}(1-\varphi) \right] \partial_t \varphi &= 0. \end{aligned}$$

1 But I should add: When I was in the "flow" of thinking and writing, it has been a great, great pleasure!

2 tranquille (French) $\hat{=}$ carefree.

3 Rather I should say: time for "family" including my wife, but she would not totally agree, I guess.

4 With the virtual summer semester start, my "free time" also had to be shared preparing online lectures, too. Needless to say that I did this also with great pleasure.

https://doi.org/10.1515/9783110497397-201

Therein, ε is a positive regularization parameter and we have appropriate boundary conditions on the outer boundary ∂B, and some initial data. The inequality $\partial_t \varphi \leq 0$ is an irreversibility constraint and states that the fracture cannot heal.

Multiphysics extensions include models in porous media, thermal effects, coupling to pressures and fluid flow inside the fracture, to name a few. Let us denote the information coming from the other physics by g, g_D and g_N. These enter as coupling terms into coefficients, additional operator terms or right-hand side information, e. g., $f := f(g)$. Or data is given on the fracture surface $\partial \mathcal{C} \subset B$ via kinematic and dynamic coupling conditions, namely

$$u = g_D \quad \text{on } \partial \mathcal{C},$$
$$\sigma n = g_N \quad \text{on } \partial \mathcal{C},$$

where $\sigma := \varphi^2 \nabla u$ and n is the outward pointing normal vector on $\partial \mathcal{C}$. Such interface conditions will play a key role throughout this monograph.

As previously mentioned, this book project was originally planned as habilitation thesis (the traditional European university qualification exam) either at Johannes Kepler Universität Linz (Austria) or at École Polytechnique in Palaiseau (France). I got involved in phase-field fracture in February 2013 with Andro Mikelić and Mary F. Wheeler when I was a Post doc at the Institute for Computational Sciences and Engineering (ICES; now Oden Institute) at the University of Texas at Austin (USA). Later, I concentrated more on phase-field fracture multiphysics extensions that include inter alia fluid-structure interaction, a field that I knew from my Master's thesis,[5] PhD[6] and early Post doc time in Heidelberg.

Consequently, this book contains not only interdisciplinary topics because I consider multiphysics that intrinsically connect difference fields, but it is also international in the truest sense of the word by assembling research in four different countries: USA (2012–2014), Austria (2014–2016), France (2016–2017), Germany (2012, 2015, and again since October 2017).

During these times, I am grateful to many people and circumstances, which might all be summarized as: "There is a window of opportunity in life" (Mary F. Wheeler).[7] And I was lucky enough to have several such *windows* (in terms of topics and colleagues/professors) during my Post doc time (Heidelberg/Germany, Austin/USA, Linz/Austria) and early Assistant Professor period (Palaiseau/France). I owe my deep-

5 Master's thesis in 2008 with Franz-Theo Suttmeier, University of Siegen.

6 PhD (Dr. rer. nat.) in 2011 with Rolf Rannacher, Heidelberg University.

7 Thanks to Thomas Carraro, my former office mate in Heidelberg, who took the initiative to call me on my cell phone on Friday, February 17, 2012, for a truly spontaneous first-time meeting with Mary Wheeler at IWR Heidelberg. Very likely without him, I would have missed this "window."

est appreciations to Rolf Rannacher,[8] Thomas Richter,[9] Jeremi Mizerski, Winnifried Wollner, Mary F. Wheeler, Andro Mikelić,[10] Boris Vexler, Ulrich Langer who mentored my Post doc periods. Then becoming assistant professor at École Polytechnique, I want to thank Grégoire Allaire who supported me.

Becoming a professor in October 2017, I got again involved in short time in various projects. These include the excellence cluster PhoenixD (headed by Uwe Morgner), the German-French IRTG proposal, and the French-German Doctoral College Sophisticated Numerical and Testing Approaches (in particular with Udo Nackenhorst and Amélie Fau), DFG-SPP 1748 (including the GAMM article [446] and editing together with Jörg Schröder a special issue in the GAMM-Mitteilungen), DFG-SPP 1962, DAAD-Peru, DAAD-IIT Indore, collaboration with the Institut für Kontinuumsmechanik (Peter Wriggers, Fadi Aldakheel), member as young research scientist in the German excellence application, and the Computing Future course. I just feel extremely flattered about all the opportunities.

There are many others whom I cannot all name in person here, i. e., office mates, co-authors, collaborators in writing joint software, colleagues in joint proposals, and those organizing conferences with me. Various types of collaborations are influential in this monograph and have a fruitful impact for which I am grateful. Special thanks go the Sanghyun Lee with whom I had enriching programming sessions at UT Austin. Likewise pleasure I had with Timo Heister over Thanksgiving in November 2013 in Austin and June 2015 in Linz.

Of course, I want also to thank my current IfAM professor colleagues, Sven Beuchler, Joachim Escher, Marc Steinbach, and Christoph Walker at the Institute of Applied Mathematics (IfAM) and, needless to say, specifically the members, former members, and associated members of my group "Wissenschaftliches Rechnen" for help and joyful discussions. Thanks to: Carmen Gatzen, Roswitha Behrens, Katrin Mang, Philipp Thiele, Sebastian Kinnewig, Denis Khimin, Nima Noii (2018–2020), Meng Fan (exchange, 1 year), Mats Brun (exchange, 3 months), Amirreza Khodadadian, Rajendra Choudhari, Daniel Jodlbauer (RICAM Linz), and Bernhard Endtmayer (RICAM Linz).

I also thank Bernhard Endtmayer, Daniel Jodlbauer, Sebastian Kinnewig, Katrin Mang, Philipp Thiele, and Diva Hurtado for suggestions and proofreading (parts of)

8 Thanks, Rolf, for giving me, as you said literally, the full education (in German: Vollausbildung) in your numerical analysis group in Heidelberg. I learned from you all basic ingredients to pursue a university career. Thank you very much for this wonderful opportunity!

9 Thomas was my actual mentor during my PhD and early Post doc time and he did a great, great job. In 2012, we shared a lecture that finally resulted in the book [381].

10 Meetings with Andro Mikelić have been just incredibly extraordinary, and demonstrated how mathematical PDE analysis and numerical simulations yield fruitful results. During Andro's sabbatical in the first half of 2013 at UT Austin, we had nearly every day discussions resulting in [318] and several follow-up publications. This experience has been influential to me and was an outstanding opportunity being a young Postdoc.

this monograph. Furthermore, I appreciate Katrin's work for doing some of the tikz graphics and for allowing me to take some materials from our joint lecture notes.

Almost finished, I thank my parents-in-law Hélène and Henri for all the support as well as my families in Germany, France, and our (unofficial) Texan grandmother Laura.

Lastly, I thank my wife Marie-Cécile for her dedication to explore together (four!) different countries, cultures, and horizons. It holds true what an African proverb says:

> If you want to go fast, you go alone, but if you want to go far, you go together.

Vielen Dank! Thank you very much! Merci beaucoup!

Austin, USA
Linz, Austria
Versailles, France
Hannover, Germany
June 2020

Thomas Wick

Classes and schools related to this monograph

Parts of this monograph are influenced by presentations and discussions in special classes (Spezialvorlesung) and summer schools that I delivered. For the summer schools and class at JKU (Johannes Kepler University) Linz, the invitations from Sorin Iliu Pop (Hasselt University), Ulrich Langer (JKU Linz), and Marc-André Keip (University of Stuttgart; former Ch. Miehe's group) is gratefully acknowledged. The list of these classes is as follows:

1. Regular class (4 + 2; lecture and theoretical exercises) at my Home University in Hannover in the current summer semester 2020:
   Numerical methods for continuum mechanics
   http://www.thomaswick.org/links/topics_NumContMech_SoSe_2020.txt
2. Special class (2 + 1; lecture and theoretical exercises) at my Home University in Hannover in the winter semester 2019/2020:
   Goal-oriented a posteriori error estimation and adaptive finite elements
   http://www.thomaswick.org/links/topics_online_DWR_WS_19_20.txt
3. Hasselt University (Belgium), June 26, 2019, summer school lecture
   Phase field methods for fractured media
   https://www.uhasselt.be/summer-school-phase-field-modeling-2019
4. Special class (2 + 1 + 1; lecture, theoretical and practical exercises) at my Home University in Hannover in the winter semester 2018/2019:
   Numerical methods for contact problems: application to variational phase-field fracture propagation
   http://www.thomaswick.org/links/ankuendigung_vpff_Wick_Mang_Noii.pdf
5. Johannes Kepler University (JKU) Linz in Austria, March 2018, Course no.: 327.018. This block lecture comprised seven lectures à 90 minutes.
   Numerical methods for variational phase-field fracture problems
   https://www.dk-compmath.jku.at/Courses/2018s/phase-field-fracture-problems
6. University of Stuttgart (Germany), Computational Mechanics of Materials and Structures (COMMAS), Fall 2016, summer school lectures:
   Numerical methods and adaptivity for multiphysics phase-field fracture
   http://www.thomaswick.org/commas_stuttgart_fall_2016.html

https://doi.org/10.1515/9783110497397-202

Research funding acknowledgments

Research funding over the last 8 years, with scientific results that found their way into this monograph, is gratefully acknowledged:

1. German Priority Program 1748 (DFG SPP 1748)
   https://www.uni-due.de/spp1748/
   with the project *Structure Preserving Adaptive Enriched Galerkin Methods for Pressure-Driven 3D Fracture Phase-Field Models*
   https://www.uni-due.de/spp1748/adaptive_enriched_galerkin_methods
2. FWF (Austrian Science Fund) project P-29181 *Goal-oriented error control for phase-field fracture coupled to multiphysics problems*
   http://www.numa.uni-linz.ac.at/Research/Projects/P29181/index.shtml
3. German Priority Program 1962 (DFG SPP 1962)
   https://spp1962.wias-berlin.de/
   with project No. 15 *Optimizing Fracture Propagation Using a Phase-Field Approach*
   https://spp1962.wias-berlin.de/project.php?projectID=15
4. Member in the *Cluster of Excellence PhoenixD (Photonics, Optics, and Engineering – Innovation Across Disciplines)* (EXC 2122)
   https://www.phoenixd.uni-hannover.de/en/
5. DAAD (German Academic Exchange Service) *Indo-German Higher Education Partnership* 2019–2023, Leibniz University Hannover – IIT Indore/Germany – India
6. Alexander von Humboldt Foundation[11] *Feodor Lynen Research Fellowship* (2013–2014)
   https://www.oden.utexas.edu/about/news/228/
7. *ICES*[12] *Post-doctoral fellowship*[13] (2013):
   https://www.oden.utexas.edu/media/uploaded-images/351.jpg

11 https://www.humboldt-foundation.de/web/home.html

12 https://www.oden.utexas.edu/

13 https://www.oden.utexas.edu/programs/postdoc/

https://doi.org/10.1515/9783110497397-203

Own related work

Several parts of this monograph are related to published studies:[14] most of them in peer-reviewed journal papers, some as peer-reviewed conference papers, some non-reviewed work (GAMM Leitartikel or lecture notes), and some under current review. My own contributions to phase-field fracture, multiphysics extensions, and collaborations[15] can be roughly divided into the following categories:

1. *Basic concepts in phase-field fracture numerical modeling:*
   (a) Crack irreversibility constraint with simple penalization [323], augmented Lagrangian [430], primal-dual active methods [221], and complementarity systems [301]
   (b) Higher-order finite elements [301, 249]
   (c) Mixed formulations for (nearly) incompressible fracture [301, 42]
   (d) Monolithic [444, 445], quasi-monolithic [221, 444], and iterative coupling schemes [88, 154, 247] of the displacement/phase-field system
   (e) Nonlinear monolithic solvers [444, 445]
   (f) Parallel multigrid preconditioners [222] (AMG) and [250, 249] (matrix-free GMG)
2. *Adaptive modeling, error control, and local mesh refinement*:
   (a) Predictor-corrector scheme for mesh refinement around the crack [221] (2d), [448, 285] (3d)
   (b) Residual-based error estimators [299, 42]
   (c) Goal-oriented error estimators [442]
   (d) Multiple goal functionals for the perfect slit (not yet phase-field) [153]
   (e) Adaptive nonintrusive global-local approach [341, 9]
   (f) Adaptive time/loading step control [443, 159]
3. *Numerical optimization with phase-field fracture*[16]
   (a) Optimal control [336, 337]
   (b) Parameter estimation [259, 342]
4. *Engineering applications:*
   (a) Fracture and damage in screws [433]
   (b) Fracture and damage in nearly incompressible rubber-like solids [298, 301]
   (c) One-way coupling to a reservoir simulator in porous media [450]
   (d) Fracture initialization with probability maps [287]
   (e) Uniaxial compression test with inclined notch [160]

---

14 Complete publication lists on http://www.thomaswick.org/

15 All of these collaborations are highly appreciated and they are influential in this monograph.

16 Numerical optimization will not be touched in this monograph. For details, I refer immediately to the four listed papers/preprints [336, 337, 259, 342] and in multiphysics fluid-structure interaction to [378, 452] and recent differentiability results for stationary problems [451].

https://doi.org/10.1515/9783110497397-204

5. *Multiphysics modeling and applications:*
   (a) Mathematical and numerical modeling of pressurized fractures in porous media [324, 323, 448, 285, 9]
   (b) Mathematical and numerical modeling of fluid-filled fractures in porous media [321, 322, 285, 286, 14]
   (c) Nonlinear porous media phase-field fracture [425]
   (d) Transport flow in phase-field fractures in porous media [286]
   (e) Two-phase flow in phase-field fractures in porous media [280]
   (f) Nonisothermal modeling of pressurized fractures in porous media [343]
   (g) Coupling to fluid-structure interaction [441, 443]
   (h) IPACS: Integrated Phase-Field Advanced Crack Propagation Simulator. An adaptive, parallel, physics-based-discretization phase-field framework for fracture propagation in porous media [431]
6. *Benchmarking:*
   [449, 222]
7. *Research software development:*
   [222, 431, 207, 183]
8. *GAMM Leitartikel:*
   [446]
9. *Lecture notes as educational advancements:*
   (a) Phase-field fracture lecture notes [300] [partially taken and re-worked in this monograph]
   (b) [447, specifically Chapters 9, 11–14] [few materials taken in this monograph]

Contents

1 Introduction

This book is devoted to multiphysics variational phase-field fracture modeling. For elastic materials, the variational approach to fracture was introduced 20 years ago in [174, 72] (see also [73]). Over the last 10 years, multiphysics extensions of phase-field fracture were addressed by various groups worldwide. In this book, basic concepts, numerical discretizations, solution schemes, and adaptive concepts shall be discussed. A specific focus is on coupled problems, their classification, and their state-of-the art numerical modeling including research software development.

1.1 Multiphysics

In multiphysics problems, several physical phenomena interact with each other, as for instance: solids, fluids, heat transfer, chemical reactions, and electromagnetic forces. The governing equations in this book are based on continuum mechanics. Such models often result in **nonstationary, nonlinear, coupled systems of partial differential equations (PDEs)**. They may be subject to inequality constraints. As we shall work with variational formulations, the resulting system is referred to as a **coupled variational inequality**[1] **system (CVIS)** in this monograph. Characteristic features are:

1. At least two PDEs/VIs are involved;
2. Models in Lagrangian and Eulerian coordinates;
3. Different physical conservation laws such as balance of mass, momentum, angular momentum, and energy;
4. Second law of thermodynamics: entropy inequality principle characterizing the energy transfer direction in order to distinguish irreversible or reversible processes;
5. Information exchange between different PDEs/CVISs via:
   - coefficients,
   - linear or nonlinear coupling terms,
   - right-hand sides,
   - interfaces (internal boundaries).

Remark 1 (Coupled problems versus multiphysics). Phase-field fracture modeling as a key ingredient of multiphysics problems addressed in this book, consists of two interacting problems. However, phase-field fracture is not yet a multiphysics system because only one PDE describes physics, namely discplacements. The second problem is a VI, which is an indicator function that "only" locates a discontinuity, damage, or fracture. Therefore, phase-field fracture is simply a so-called coupled problem.

1 A variational inequality is referred to as VI.

https://doi.org/10.1515/9783110497397-001

1.2 Variational phase-field fracture

Key idea

The main idea is to approximate lower-dimensional phenomena (e. g., fracture or material discontinuities) with a function φ that is obtained by solving an elliptic functional defined in the entire domain. The approximation width is represented through a regularization parameter $\varepsilon > 0$. As shown in Figure 1.1, this smoothed indicator variable φ varies from 0 (fracture) to 1 (intact material).

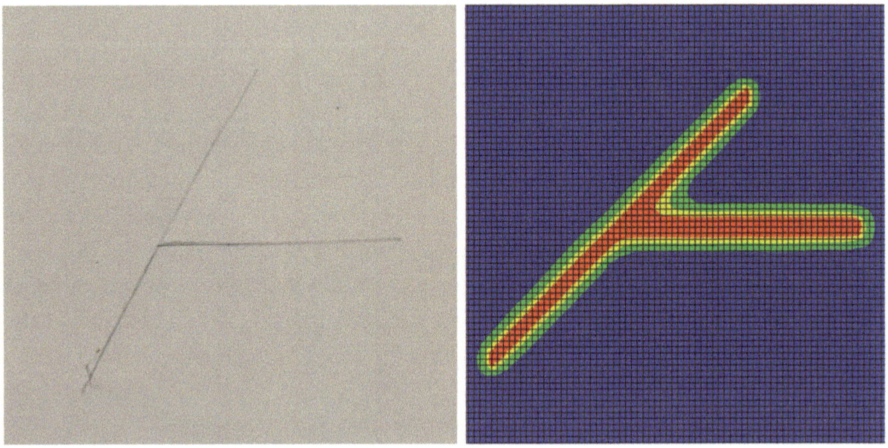

Figure 1.1: Paper cracked with scissors (left) and diffusive fixed-mesh phase-field modeling (right). The transition zone (green) has the half-band width $\varepsilon > 0$. Right subfigure: Reprinted from [430] with permission from Elsevier.

Appealing features

Phase-field fracture is a regularized diffusive interface approach that has (as many numerical methods) advantages and shortcomings. Appealing features are:

1. Phase-field is a variational framework [174, 72];
2. It has a thermodynamical continuum description [317, 313];
3. It is a fixed-mesh approach in which remeshing or updating of basis functions is not required;
4. Phase-field allows for the computation and prediction of complex crack patterns (see Figures 1.2 and 1.3), joining, and branching;
5. Relation to level-set methods, which are themselves very successful with similar methodologies (see for a short paragraph on their relations and references in Section 4.4.3.2);
6. The indicator property of phase-field can be further employed to design indicator functions for modeling multiphysics problems with interfaces (Chapter 11);

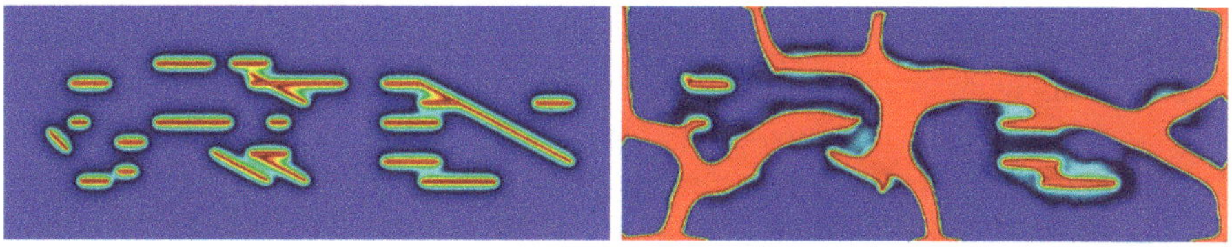

Figure 1.2: Fracture network (initial configuration at left) with growing fractures (at right) using a phase-field model. Red colors indicate the fracture ($\varphi = 0$) and blue colors the unbroken zone ($\varphi = 1$). The transition zone is indicated in yellow/green. This computation was done with the model presented in [323].

Figure 1.3: A 3D setting: two penny-shaped fractures grow, then join and later branch due to the heterogeneity of the solid. This simulation shows that phase-field can cope with complex 3D fracture shapes. The computation was performed using high performance parallel computing (Section 7.10.4) and predictor corrector local mesh adaptivity (Section 9.5.1). The figure is a reworked version from [448].

7. Phase-field is an approach that allows us to compute advanced engineering applications, offers challenges and fruitful contributions in numerical modeling and calculus of variations. Thus, phase-field modeling provides an appealing balance between sufficient sophistication in theory and practical applications.

This last point might be the key for its attractiveness in both applied mathematics and engineering.

Shortcomings and challenges

The biggest challenges are:

1. Interaction of several model and discretization parameters.
2. Smeared interface:
   (a) asking for high-accuracy discretizations;
   (b) complicates prescription of interface conditions;

   (c) complicates crack width computations;
   (d) challenging in computing accurate crack opening displacements, crack tip stresses, crack tip velocities;
   (e) nucleation and crack initiation, which is, despite studies, not yet fully understood.
3. On the energy level, the formulation is nonconvex, constituting a challenge for both theory and design of numerical algorithms.
4. Limit processes toward sharp interface physics require mathematically special function spaces. Moreover, these are challenging in terms of numerical and computational convergence analyses.

As attested in numerous studies (see Section 1.5), ideas have been proposed how to cope with these challenges. Most raise new questions offering new research opportunities.

1.3 Interfaces

A characteristic feature of many coupled problems are interfaces that divide a medium into at least two subproblems. Fracture propagation belongs to this class. However, interfaces may appear in various fashions and it is worth to classify them.

Definition 1 (Types of interfaces). Interfaces are denoted in this monograph by Γ (or Γ_i or $\mathcal{C}$),[2] and they are lower-dimensional manifolds that separate at least two media. They can be divided into three types with increasing level of difficulty:
1. Fixed interfaces that are fixed in space, time, and area (length). Examples are domain decomposition methods and problems in subsurface modeling with over- and underburden;
2. Moving interfaces that vary in space/time and do not (or only slightly) change their area. Examples are two-phase flow problems and fluid-structure interaction;
3. Propagating interfaces that vary in space/time and significantly extend their area (length). Examples are damage and fracture problems.

A precise definition of an interface is provided in Section 3.3.2.

Ω_1 : Domain 1
Ω_2 : Domain 2
$\Gamma_i, i = 1, 2, 3$: interfaces

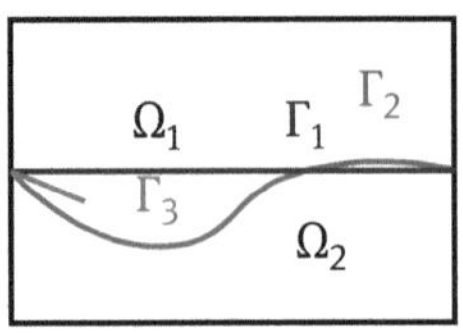

2 Notation is a nightmare. During this book, I was not able to make a decision for only one of the three notations Γ or Γ_i or $\mathcal{C}$, because all three have their justification.

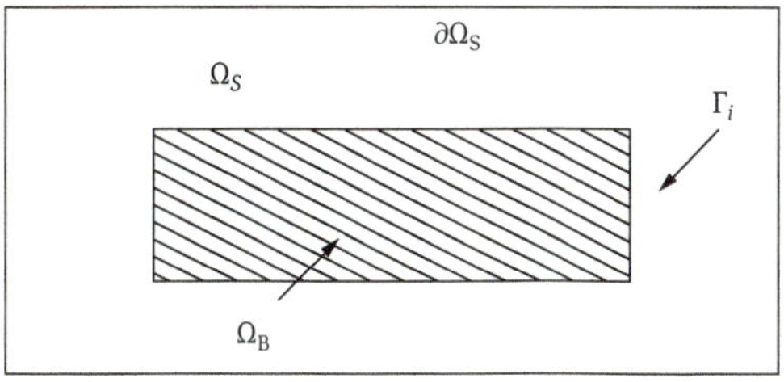

Figure 1.4: Coupling elasticity in Ω_S with poro-elasticity in Ω_B. The interface Γ_i between both subdomains is fixed. For numerical simulations and results, we refer the reader to Section 3.4.

1.3.1 Fixed interfaces

An illustrative example is the overburden example from subsurface modeling in which a porous medium is coupled to elasticity. Here, the interface is static and will never change its position; see Figure 1.4 for a prototype configuration. A concrete realization in terms of the so-called augmented Mandel's problem can be found in Section 3.4.

1.3.2 Moving interfaces

A famous example of type 2 with moving interfaces is fluid-structure interaction (see, for instance, the books [93, 170, 187, 92, 46, 64, 376, 178]) in which the Navier–Stokes equations are coupled to (nonlinear) elasticity. Here, the interface moves and must be tracked or captured (see Section 1.3.4) during a simulation, demanding robust and accurate discretizations. We refer the reader to Figure 1.5 for an illustration.

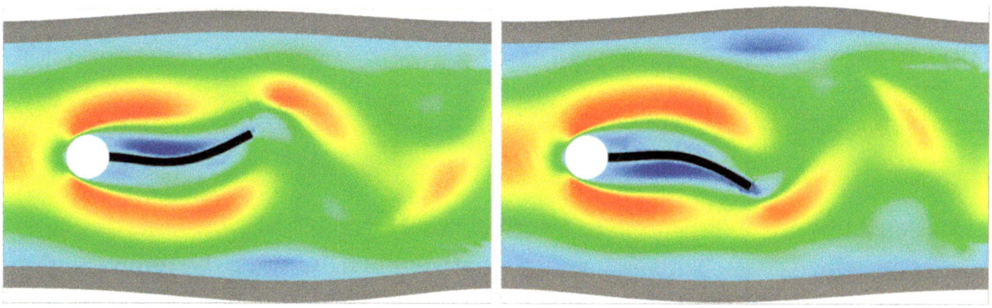

Figure 1.5: Coupling the incompressible Navier–Stokes equations with nonlinear elasticity: fluid-structure interaction. Here, the black and grey interfaces are moving in space and time. Zoom-ins for further discussions are provided in Figure 1.7. Figures reprinted by permission from Springer Nature: Springer, Computational Mechanics, [436], 2013.

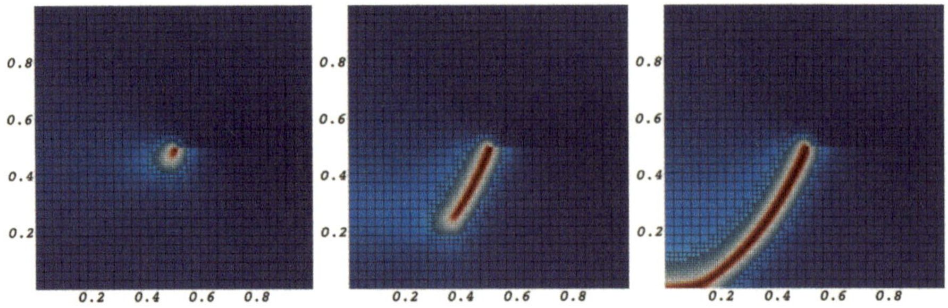

Figure 1.6: Propagating interface in terms of a fracture. Reprinted from [221] with permission from Elsevier.

1.3.3 Propagating interfaces

Third, we consider propagating interfaces as for instance fractures. As before, the interface will change its position, but usually it does also extend its area or length with a moving singularity or discontinuity, i. e., fracture tip or fracture front. This adds an additional level of complexity because the moving front, i. e., the fracture tip, must be resolved with sufficient accuracy. In addition, the future location of fractures are a priori unknown and must be updated at each time instance of a computation. This causes challenges in the mathematical analysis, in the design of numerical methods as well as in reliable software developments. We shall discuss various examples throughout this monograph. An illustration is provided in Figure 1.6.

1.3.4 Realizing interface coupling

To approximate interfaces, such as the solid discontinuity in fracture mechanics, there exist two basic methods:
- Interface-tracking;
- Interface-capturing.

Definition 2 (Interface-tracking). In interface-tracking methods, the interface is explicitly described. Here, we need meshes that enable us to track the interface. Consequently, the mesh needs to be updated as the interface evolves further in time. One shortcoming is that mesh elements may be deformed too much, such that the approach fails, if remeshing is not taken care of in a proper way.

Definition 3 (Interface-capturing). In interface-capturing methods, the computations are done on a fixed spatial domain and the interface is implicitly defined through additional functions. The interface function captures the interface and marks its location.

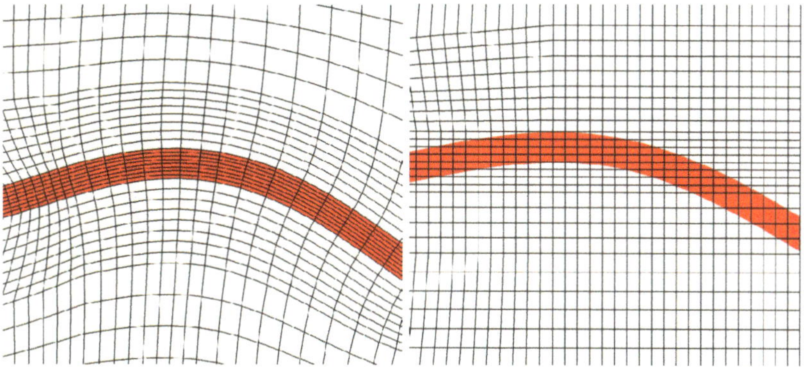

Figure 1.7: Comparison of the Lagrangian perspective with interface-tracking of the red object (left) and an Eulerian fixed-mesh approach with interface-capturing of the red object (right). In the former, the interface is located on mesh edges. In the latter, the interface cuts through mesh elements and is captured with an additional marking function. Both figures are results from numerical simulations done in [435] and [437], respectively.

Mesh degeneration is not a problem, but conserving well-posedness of the interface marker function is important.

Interface-tracking comprises the so-called arbitrary Lagrangian–Eulerian approach [136, 234, 243, 134, 169] or the Deforming-Spatial-Domain/Stabilized Space–Time technique [410, 411]. Diffusive methods for interface-capturing are the famous level-set method [346, 401] or phase-field methods going back to [98, 12]. Some selected nonfracture studies include, for instance, [38, 392, 106, 145, 164, 263, 463, 39]. Hybrid methods in which both techniques are combined or simultaneously employed are known as well, e. g., [394, 44, 5, 412, 436, 439]. The quality of diffusive approaches relies on regularization parameters such as, for instance ε, that determine the thickness of the transition zone and, therefore, the sharpness of the interface (see Figure 1.8).

Furthermore, fitted and unfitted methods are distinguished. Examples for fitted methods are locally modified finite elements [179], interface-fitted subspace projection methods [43] or historical work [31]. Examples for unfitted methods are extended/generalized finite elements [328, 184, 310, 32], cut-cell methods [96], fictious domain methods [204], immersed boundary methods [351], or the finite cell method [349, 358].

The decision whether to work with interface-tracking or interface-capturing techniques is a compromise between algorithmical, implementational, and computational efforts in terms of accuracy of the desired interface approximation and the mathematical rigorous justification. While in general, interface-capturing approaches can deal more easily with moving and evolving interfaces up to topology changes, the accuracy is lower than comparable interface-tracking approaches. The latter ones are, however,

Figure 1.8: Comparison of a sharp interface resolution and a diffusive interface.

more challenging when interfaces are propagating because the mesh must be adapted. This holds in particular true for 3D.[3]

1.4 Engineering problem: fracture and damage in screws

We briefly present a situation in which phase-field fracture was successfully employed to predict fracture and damage in a screw tension test [433]. Some simulation results are shown in Figure 1.9.

Typical material damages are closed wrinkles in the middle of thread flanks, seams, cracks, and hollow-rolled screws. The cause of material damage is often a wrong setup of the rolling machine. In the production process, the thread of the final screw is obtained by rolling an initially unmachined blank. Our developed phase-field model for elastic and elasto-plastic materials was able to predict fracture in those regions in which experimentally the same fracture patterns could be observed. Consequently, these numerical simulations further helped to better understand the mechanisms of failure. For more details, we refer the reader to [433] and [432].

1.5 State-of-the-art literature overview

In this section, lists of developments related to variational approaches to fracture and phase-field fracture methods are provided. Most likely, these lists will not be complete because of various ongoing studies in numerous groups worldwide.

3 3D means three spatial dimensions, i. e., $\mathbb{R}^3$ in space.

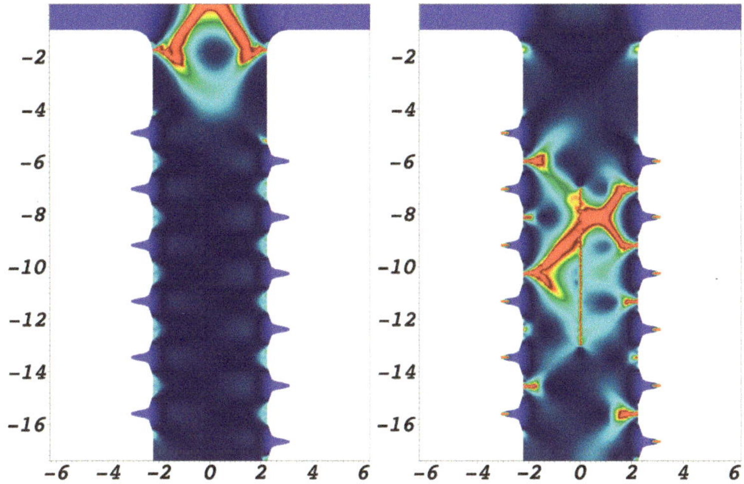

Figure 1.9: Left: Tension test of a screw. At the beginning (not shown here), the screw is completely intact and not damaged. Applying tensile forces on the head, while fixing the bottom, high stresses develop in the region of where the head is attached. Finally, the screw will break (red fracture). Right: material inconsistencies inside the screw yield total damage in the body of the screw rather at the head. Qualitative similar observations have been made in industrial experiments. Details can be found in [433] and [432].

Recent review articles

In the recent review articles [74], [459], and [431], current advancements are summarized. In [74], the pioneering authors review theoretically-oriented aspects [74]. The authors of [459] provide an extensive overall summary. A high-performance parallel-adaptive framework including physics-based-discretizations and various coupling schemes in multiphysics phase-field fracture in porous media is reviewed in [431].

Thematic overview

In this list, we notice upfront that most publications belong to more than one thematic classification.

1. Mathematical modeling and analysis [174, 73, 122, 121, 175, 173, 172, 197, 334].
2. Thermodynamical interpretations [317, 313, 458].
3. Relation to damage mechanics [127, 458, 137, 302, 190].
4. Analysis and computations on crack nucleation [108, 427, 267][theory], [433, 445][screw tests], [17, 311, 444, 192][L-shaped panel].
5. Other discretization techniques: IGA (isogeometric analysis), e. g., [67, 66, 225, 226, 18], special basis functions [265], discontinuous Galerkin [155, 326, 325], virtual finite elements [7, 8].

6. Coupling phase-field fracture with extended finite elements [199, 189].
7. Different formulations of the crack regularization (Ambrosio–Tortorelli) functional [21, 76, 352] and [161][pp. 649–650].
8. Degradation functions [386, 190, 29, 458].
9. Optimal control, parameter estimation, topology/shape optimization, e. g., [336, 337, 259, 258, 457, 11, 383, 342].
10. Spatial mesh adaptivity [30, 221, 442, 35, 285, 94, 95, 462, 350].
11. Time step adaptivity and multirate coupling [468, 443] and [14, 247].
12. Linear solvers, e. g., [161, 222, 250, 468].
13. Nonlinear monolithic solvers [192, 444, 445, 262].
14. Partitioned, staggered, iterative coupling, alternating minimization [72, 70, 311, 94, 313, 88, 157, 288].
15. Dynamic fracture [67, 276, 75, 277, 235, 387, 466, 190].
16. Fracture with thermal/temperature interaction, plasticity, ductile fracture, viscoelastic solids, anisotropic solids, e. g., [76, 316, 405] and [312, 18, 16, 422, 266, 10, 393, 133, 408, 137, 6, 8].
17. Crack width computations and approximation results [338, 286, 112, 460].
18. Pressure-driven and fluid-driven pressure, e. g., [430, 318, 71, 323, 450, 343, 448, 465] and [321, 322, 315, 314, 303, 218, 339, 385, 219, 285, 282, 281, 100, 287, 9, 112, 466, 453, 467].
19. Coupling to fluid-structure interaction [441] and [443].
20. Phase-field global-local and multiscale approaches [194, 341, 9, 189, 350].

Many more references are provided throughout this monograph.

1.6 Outline of this book

Chapter 2

In this chapter, the main notation used throughout this monograph is introduced. Furthermore, some mathematical prerequisites are presented. Then prototype problems are addressed. These serve to introduce and discuss the key aspects of multiphysics phase-field fracture problems.

Chapter 3

This chapter is of general interest. Coupled problems in conjunction with variational inequalities and multiphysics formulations play important roles in today's applications in applied mathematics and engineering. We try to provide a state-of-the-art classification with an emphasis on coupling strategies of such problems and present several illustrating examples.

Chapter 4

In this chapter, a classical quasi-static variational fracture model for brittle materials proposed by Francfort and Marigo is recapitulated. Emphasis is on the approximation with the help of elliptic Ambrosio–Tortorelli functionals including some historic notes. The corresponding Euler–Lagrange system is derived. Finally, some thermodynamic extensions are discussed briefly.

Chapter 5

The main parts of this monograph are partitioned into three numerical chapters and one multiphysics modeling chapter, which are alternated with four simulation chapters.

In this first numerical chapter, the focus is on the regularization of variational inequality constraints. This is then followed by temporal and spatial discretization. While temporal discretization is less important for quasi-static processes, current state-of-the-art time-stepping schemes are introduced nonetheless. The main reason is that multiphysics extensions are most often time-dependent. In the spatial discretization, we recall the basic concepts of the Galerkin finite element method. Then a numerical analysis of simplified, prototype model problems on the relationships of the phase-field regularization parameters and the spatial discretization parameter is established.

Chapter 6

In this simulation chapter, a numerical example is considered to which a manufactured solution was derived by others. The fracture is a steady-state slit. This allows us to study computational convergence in the spatial discretization.

Chapter 7

In this chapter, the emphasis is on the nonlinear solution. Partitioned approaches as well as monolithic techniques are discussed in great detail. Several nonlinear solution methods for fixed-point iterations, functional iterations and Newton's method are presented. Five different Newton-type methods shall be discussed in detail. Finally, the linear solution is briefly discussed. Here, technical details are left out and we refer the reader to published work elsewhere.

Chapter 8

In this second simulation chapter, the so-called single edge notched shear test is considered. In extension to the published literature, comparisons of several nonlinear solution methods are provided. Loading step refinements are presented as well.

Chapter 9

Adaptivity is considered in this chapter. First, we discuss goal-oriented error estimation and spatial mesh adaptivity for steady-state fractures. These presentations are complemented with concepts of multigoal-oriented error estimation that are relevant in multiphysics applications. Then predictor-corrector strategies are discussed. These are used for mesh adaptivity and the determination of local fine-scale problems within a nonintrusive global-local approach. Lastly, a heuristic method for goal-oriented time error control is introduced.

Chapter 10

In this chapter, three numerical examples are presented. First, we continue with the slit configuration from Chapter 6 and present goal-oriented error estimation for a steady-state fracture. Related results have been already published and we conduct some new studies here. Then we consider multigoal-oriented error estimation for a slit domain without phase-field. Of course, the principial problem is much simpler, but allows to compare to (future) phase-field solutions. Lastly, some results of time/loading step control for a screw simulation are shown.

Chapter 11

On the one hand, we have concentrated "only" on coupled problems in the previous chapters—but not yet true multiphysics. On the other hand, most presented algorithms naturally extend to multiphysics problems. In most of them the author has done substantial work since his PhD thesis [434] when studying fluid-structure interaction. Thus, we discussed implicitly multiphysics already. In this chapter, we now consider some specific parts.

First, we start highlighting similarities of phase-field and level-set methods. The level-set concept is key in this chapter and is our red thread throughout this part. Specifically, we define indicator functions that allow us to distinguish different physics in a fixed-mesh domain.

Then we focus on interface-related quantities such as crack width computations. Next, we combine phase-field fracture and multiphysics problems. These are first motiviated by porous media applications and later by fluid-structure interaction. Most results have been published in peer-reviewed journals, but new materials are added as well in this monograph. This holds specifically true for a fully Eulerian description of phase-field fracture, where all equations are defined in Eulerian coordinates on a fixed mesh.

This Eulerian phase-field solid description is finally extended to a concept for fully Eulerian fluid-structure interaction coupled with phase-field fracture. Here, the property of phase-field being a fixed mesh approach is taken into advantage. To determine the fluid and solid subdomains, we use again a level-set function showing the importance of indicator functions for phase-field multiphysics problems.

Chapter 12
In this chapter, some characteristic simulations are presented that demonstrate the concepts from Chapter 11 and our efforts over the last years in multiphysics phase-field fracture applications.

Chapter 13
An important part of the author's work is research software development. Important contributions are summarized in this chapter. Links to open-source programming codes are provided as well.

Chapter 14
We summarize the main contents of this monograph in this chapter. Afterward, some topics for future research are briefly discussed.

Chapter 15
The pleasure of finding things out, summarizing them, and presenting them in written work is a process that requires periods of concentration, periods of creativity, periods of literature research, periods to understand of the why/how/what, scientific disappointments and negative results, and periods of asking the same questions again and again. But also the motivation and hope to achieve these goals. Producing new energy for feeding my motivation and my hope is illustrated in this chapter.[4]

4 Dear Jeremi (Mizerski), even that I knew works by Richard P. Feynman (e. g., the Feynman Lectures on Physics), I was not aware of this beautiful book "The Pleasure of Finding Things Out [167]." Feynman's "pleasure" came to me while writing this monograph. Thanks for recommending this book when I was visiting you in Warsaw and Zamość! Best regards, Thomas W.
M. D. Ph. D. Jeremi Mizerski, Warsaw/Poland, was the scientific supervisor from the Poland-side within the international graduate college IGK 710 Heidelberg-Warsaw (running until December 2009) during the doctoral dissertation studies November 2008–December 2011 of the author.

2 Notation, prerequisites, and prototype model problems

2.1 Notation

We introduce the global notation used in this book. Specific definitions might be added later when necessary.

2.1.1 Spatial domains, boundaries, fracture

Let d be the spatial dimension. In the following, let $B \subset \mathbb{R}^d$, $d = 2, 3$ the total domain wherein $\mathcal{C} := \mathcal{C}(t) \subset \mathbb{R}^{d-1}$ denotes a (propagating) fracture (or multiple fractures) and $\Omega := \Omega(t) \subset \mathbb{R}^d$ is the intact domain (Figure 2.1). A point in B is denoted by

$$x = (x_1, \ldots, x_d).$$

The domain $\Omega_F := \Omega_F(t) \subset \mathbb{R}^d$ is the domain, which arises when $\mathcal{C}$ is approximated with a higher-dimensional object, mainly determined by $\varepsilon > 0$. This introduces the internal boundary (interface) $\partial\mathcal{C} := \partial\mathcal{C}(t) := \overline{\Omega}_F \cap \overline{\Omega}$. For the limit case (no transition zone, phase-field regularization parameter $\varepsilon = 0$), it holds $\Omega_F = \mathcal{C}$. Detailed discussions on how Ω_F is constructed in practice can be found in our recent review paper [431] and also in Section 11.2.

The outer boundary may be decomposed into $\partial B = \partial B_{\mathrm{HD}} \cup \partial B_{\mathrm{ND}} \cup \partial B_{\mathrm{HN}} \cup \partial B_{\mathrm{NN}}$, where ∂B_{HD} denotes homogeneous Dirichlet conditions, ∂B_{ND} denotes nonhomogeneous Dirichlet conditions, ∂B_{HN} denotes homogeneous Neumann conditions, and ∂B_{NN} denotes nonhomogeneous Neumann conditions. Not always do we deal simul-

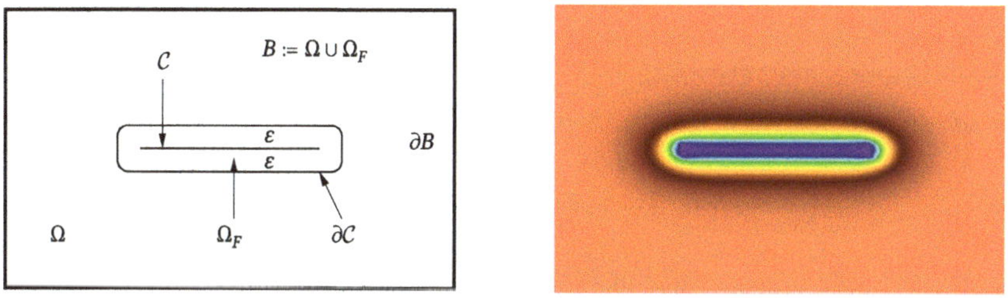

Figure 2.1: Setup of the notation: the unbroken domain is denoted by Ω and $\mathcal{C}$ is the fracture. The latter one is approximated by the domain Ω_F. The half-thickness of Ω_F is ε. The fracture boundary is $\partial\mathcal{C}$ and the outer boundary is ∂B. The corresponding realization using phase-field is shown in the right subfigure. Here, the lower-dimensional fracture ($\varphi = 0$) is approximated with the phase-field variable. The transition zone with $0 < \varphi < 1$ has the thickness of ε on each side of the fracture. Consequently, Ω_F can be represented in terms of φ.

https://doi.org/10.1515/9783110497397-002

taneously with all four situations, but separately all of them will appear within this monograph.

2.1.2 Time/loading intervals

The loading/time-interval[1] is denoted by $I := (0, T)$ with the end time value $T > 0$. Mostly, in this book, we have quasi-static fracture in mind for which I is the interval of incremental loading. This means that at each instance, the body under consideration is in a thermodynamic equilibrium. Nonetheless, the state variables might be time-dependent.

2.1.3 Model, material, and discretization parameters

Regularization and penalization parameters

The most important model parameters are the phase-field regularization parameters $\varepsilon > 0$ and $\kappa > 0$. For some models, we use $\gamma > 0$ for denoting a penalization parameter.

Material parameters

The Lamé parameters are denoted with $\mu > 0$ (shear modulus) and $\lambda > -\frac{2}{3}\mu$ with the unit $\mathrm{Pa} = \mathrm{N/m^2} = \mathrm{kg/m\,s^2}$. We denote with σ_s (often simply σ) the Cauchy solid stress tensor with the unit $\mathrm{Pa/m^2}$. The critical energy release rate is denoted with $G_c > 0$ with the unit $\mathrm{J/m^2} = \mathrm{N/m}$. Poisson's ratio is denoted with $-1 \le \nu_s \le 0.5$ and is dimensionless. Young's modulus $E_Y > 0$ has the unit $\mathrm{Pa} = \mathrm{N/m^2} = \mathrm{kg/m\,s^2}$. Further parameters shall be introduced in the respective sections. For example, in Section 3.4, we discuss in detail a poroelastic medium and provide all poroelastic parameters therein. In Chapter 11, we additionally deal with fluid flow and provide more parameters therein.

Spatial and temporal discretization

The time (loading) step size is defined as $k_n = t_n - t_{n-1}$ with $k := \max_n(k_n)$, where

$$0 =: t_0 < t_1 < \cdots < t_{n-1} < t_n < \cdots < t_N := T.$$

The time points t_n can have nonuniform distances. The spatial discretization parameter, namely the diagonal of a quadrilateral or hexahedral element, is denoted as usually by h. More information on h can be found in Section 5.4.

1 The notation I is also used for the identity matrix/tensor, but the context will make it clear of what we are dealing with.

2.1.4 Solution variables

The unknown solution variables are:
- vector-valued displacements $u := u(x,t) : B \times I \to \mathbb{R}^d$;
- a smoothed scalar-valued indicator phase-field function $\varphi := \varphi(x,t) : B \times I \to [0,1]$. That indeed $0 \leq \varphi \leq 1$ holds true is proven in Theorem 5 (Section 5).

The latter one describes the crack path in a smeared fashion. Specifically:
- $\varphi = 0$ denotes the crack region;
- $\varphi = 1$ characterizes the unbroken material;
- $0 < \varphi < 1$ are intermediate values constituting a smooth transition zone dependent on the regularization parameter ε. In engineering or physics, ε is often a so-called length-scale parameter.[2] This may be justified since this zone weakens the material and is a physical transition zone from the unbroken material to a fully damaged state (some thoughts are provided in Section 5.5.3.6).

In multiphysics fracture (Chapter 11), we denote further
- vector-valued velocities $v := v(x,t) : B \times I \to \mathbb{R}^d$,
- scalar-valued pressure $p := p(x,t) : B \times I \to \mathbb{R}$,
- scalar-valued temperature $\vartheta := \vartheta(x,t) : B \times I \to \mathbb{R}$.

We often consider several solution variables and then use U as joint variable. For instance, $U = (u, \varphi)$ or $U = (v, p, u)$ or as well $U = (v, u, p, \varphi)$ in Section 11.8. Moreover, we notice that the solutions are dependent on the regularization parameter ε, that is,

$$U := U_\varepsilon.$$

For notational brevity, the subscript ε is most often suppressed in this monograph.

2.1.5 Gradient, divergence, trace, Laplace

It is convenient to work with the nabla operator (see [87] for a concise overview) to define spatial derivative expressions. The gradient of a single-valued function $v : \mathbb{R}^d \to \mathbb{R}$ reads:

$$\nabla v = \begin{pmatrix} \partial_1 v \\ \vdots \\ \partial_d v \end{pmatrix},$$

2 In related fields such as phase transitions in materials science, ε is a so-called interaction length. In two-phase flow, ε is known as interfacial thickness. In peridynamics, the range of interactions ε between materials points is known as horizon.

where $\partial_i v := \frac{\partial v}{\partial x_i}$, where x_i is the ith coordinate direction.

The gradient of a vector-valued function $v : \mathbb{R}^d \to \mathbb{R}^m$ is called Jacobian matrix and reads

$$\nabla v = \begin{pmatrix} \partial_1 v_1 & \dots & \partial_d v_1 \\ \vdots & & \vdots \\ \partial_1 v_m & \dots & \partial_d v_m \end{pmatrix}.$$

The divergence is defined for vector-valued functions $v : \mathbb{R}^d \to \mathbb{R}^d$:

$$\operatorname{div} v := \nabla \cdot v := \nabla \cdot \begin{pmatrix} v_1 \\ \vdots \\ v_d \end{pmatrix} = \sum_{j=1}^{d} \partial_j v_j.$$

The divergence for a tensor $\sigma \in \mathbb{R}^{d\times d}$ is defined as

$$\nabla \cdot \sigma = \left(\sum_{j=1}^{d} \frac{\partial \sigma_{ij}}{\partial x_j} \right)_{1\le i\le d}.$$

The trace of a matrix $A \in \mathbb{R}^{d\times d}$ is defined as

$$\operatorname{tr}(A) = \sum_{j=1}^{d} a_{jj}.$$

Definition 4 (Laplace operator). The Laplace operator of a two-times continuously differentiable scalar-valued function $u : \mathbb{R}^d \to \mathbb{R}$ is defined as

$$\nabla \cdot (\nabla u) = \Delta u = \sum_{j=1}^{d} \partial_{jj} u.$$

Definition 5. For a vector-valued function $u : \mathbb{R}^d \to \mathbb{R}^m$, we define the Laplace operator componentwise as

$$\Delta u = \Delta \begin{pmatrix} u_1 \\ \vdots \\ u_m \end{pmatrix} = \begin{pmatrix} \sum_{j=1}^{d} \partial_{jj} u_1 \\ \vdots \\ \sum_{j=1}^{d} \partial_{jj} u_m \end{pmatrix}.$$

2.1.6 Function spaces

2.1.6.1 General notation

We adopt standard notation for the usual Lebesgue and Sobolev spaces [455, 2]. We denote the following: $D := B$ or $D := \Omega$ or $D := \Omega_F$. By $L^p(D), 1 \le p \le \infty$, we indicate the standard Lebesgue space that consists of measurable functions u, which are

Lebesgue-integrable to the p-th power. The set $L^p(D)$ forms a Banach space with the norm $\|u\|_{L^p(D)}$:

$$\|u\|_{L^p(D)} := \left(\int_D |u(x)|^p \, dx\right)^{\frac{1}{p}}, \quad 1 \le p < \infty,$$
$$\|u\|_{L^\infty(D)} := \operatorname*{ess\,sup}_{x\in D} |u(x)|.$$

We obtain the Hilbert space $L^2(D)$ for $p = 2$, equipped with the inner product

$$(u, v)_{L^2(D)} := \int_D u(x)v(x) \, dx.$$

The Sobolev space $W^{m,p}(D), m \in \mathbb{N}, 1 \le p \le \infty$ is the space of functions in $L^p(D)$ that have distributional derivatives of order up to m, which belong to $L^p(D)$. This space is equipped with the norm

$$\|u\|_{W^{m,p}(D)} := \left(\sum_{|\alpha|\le m} \|D^\alpha u\|^p_{L^p(D)}\right)^{\frac{1}{p}}, \quad 1 \le p < \infty,$$
$$\|u\|_{W^{m,\infty}(D)} := \max_{|\alpha|\le m} \|D^\alpha u\|_{L^\infty(D)}.$$

Definition 6. The symbol $\alpha = (\alpha_1, \ldots, \alpha_d) \in \mathbb{N}^d$ denotes a multiindex with the properties

$$|\alpha| := \sum_{j=1}^d \alpha_j, \quad D^\alpha := \frac{\partial^{|\alpha|}}{\partial x_1^{\alpha_1} \cdots \partial x_d^{\alpha_d}}.$$

If $k \in \mathbb{N}_0$, we define the set of all partial derivatives of order k:

$$D^k u(x) := \{D^\alpha u(x) : |\alpha| = k\}.$$

For $p = 2$, $H^m(D) := W^{m,2}(D)$ is a Hilbert space equipped with the inner product

$$(u, v)_{H^m(D)} := \sum_{|\alpha|\le m} (D^\alpha u, D^\alpha v)_{L^2(D)},$$

and the norm $\|\cdot\|_{H^m(D)}$ [455]. Seminorms are defined by

$$|u|_{W^{m,p}(D)} := \left(\sum_{|\alpha|=m} \|D^\alpha u\|^p_{L^p(D)}\right)^{\frac{1}{p}}, \quad 1 \le p < \infty,$$
$$|u|_{W^{m,\infty}(D)} := \max_{|\alpha|=m} \|D^\alpha u\|_{L^\infty(D)}.$$

Finally, we indicate the subspace $W^{m,p}(D)$ of functions with zero trace on ∂X by $W_0^{m,p}(D)$. Specifically, we define $H_0^1(D) = \{u \in H^1(D) : u = 0 \text{ on } \Gamma_D \subset \partial X\}$.

2.1.6.2 Comments on space-time function spaces/almost all

For time-dependent weak formulations, one can work with space-time Sobolev spaces, so-called Bochner spaces (e. g., [124, 455]). This would be the most elegant notation. We rather use fruitful consequences of Lebesgue measures, namely that L^2 integrals in $I = (0, T)$ can be approximated almost everywhere in I on the time-continuous level.

Definition 7 (Almost everywhere, a. e.). Let $\Omega \subset \mathbb{R}^d$ be a Lebesgue-measurable open domain. Measurable functions are defined almost everywhere, in short a. e., in Ω:

$$f(x) = g(x) \quad \text{a. e. } \Omega,$$

when there is $Z \subset \Omega$ such that the Lebesgue measure of Z is zero and

$$f(x) = g(x) \quad \text{for all } x \in \Omega \setminus Z.$$

Short: Almost everywhere means except on a set of Lebesgue measure zero.

Definition 8 (Almost all). On the time-continuous level $I = (0, T)$, rather than working with L^2 integrals, we use the fact from Lebesgue theory that such functions can be approximated for almost all $t \in I$.

Remark 2. Clearly after time-discretization, we evaluate the semidiscrete problem for all $n = 1, 2, 3, \ldots, N$.

2.1.6.3 Notation for phase-field fracture problems

We frequently use the short notation

$$V_B := H^1(B), \quad V_B^0 := H_0^1(B).$$

For vector-valued function spaces (for instance, the displacements $u : B \to \mathbb{R}^d$ are vector-valued), we use the same notation:

$$V_B := H^1(B)^d, \quad V_B^0 := H_0^1(B)^d.$$

The context should make clear whether we work in the scalar-valued or vector-valued case. Our fundamental space for phase-field fracture is the following.

Definition 9 (Governing spaces for displacements and phase-field). The trial space for the displacements is defined as

$$V_u^D := V_u^D(B) := \{u \in V_u \mid u = u_D \text{ on } \partial\Omega_D,\ u = 0 \text{ on } \partial\Omega_D \setminus \partial B\}.$$

The test space for the displacements is the classical $H_0^1(B)$ space and defined in our notation as

$$V_u^0 := V_u^0(B) := \{u \in V_u \mid u = 0 \text{ on } \partial B\}.$$

The trial and test space for the phase-field variable is

$$V_\varphi := V_\varphi(B) = H^1(B).$$

These spaces are sufficient as prototype spaces. Often in applications, homogeneous Neumann conditions ∂B_N will also play a role in the displacements, which is then specifically indicated.

Remark 3. For general algorithmic concepts we simply work with V.

We then seek

$$u \in V_u^D,$$

and

$$\varphi \in V_\varphi.$$

Furthermore, we introduce the convex set

$$K := K(\varphi^{n-1}) := \{\varphi \in V_\varphi : 0 \le \varphi \le \varphi^{n-1} \le 1 \text{ a. e. in } B\},$$

which we need for including the variational inequality constraint in time, the so-called irreversiblity constraint. When useful, we work with the following convention:

Definition 10. We denote the product space by

$$X = V_u^D \times V_\varphi,$$

such that we seek $U \in X$, i. e., $(u, \varphi) \in V_u^D \times V_\varphi$. If necessary, we also use the notation $X^D = X = V_u^D \times V_\varphi$ and $X^0 = V_u^0 \times V_\varphi$.

2.1.7 Notation for inner products, bilinear forms, semilinear forms

We briefly recall our important notation for representing weak variational forms:

1. Let $f, g \in V^d$. Then the scalar product is denoted by

$$(f, g) = \int_B f \cdot g \, dx, \quad f, g \in V^d.$$

2. Let $F, G \in V^{d \times d}$. Then the scalar product for matrices is denoted by

$$(F, G) = \int_B F : G \, dx, \quad F, G \in V^{d \times d}.$$

   Here, : stands for the Frobenius scalar product.

3. Often, we use: Find $u \in V$ such that

$$a(u, \varphi) = l(\varphi) \quad \forall \varphi \in V.$$

   Here, $a(u, \varphi) : V \times V \to \mathbb{R}$ is a bilinear form and $l(\varphi) \in V^*$ is a linear form (linear functional). Here, V^* denotes the dual space $V^* = L(V, \mathbb{R})$.
4. For nonlinear problems, the solution variable $u \in V$ is nonlinear, while the test function is still linear. We use the notation: Find $u \in V$ such that

$$a(u)(\varphi) = l(\varphi) \quad \forall \varphi \in V.$$

   Here, $a(u)(\varphi)$ is a so-called semilinear form, where the first argument $u \in V$ is nonlinear and the second argument $\varphi \in V$ is linear.
5. For (linear) PDE systems, we define the space X, e. g., $X = V_u^D \times V_\varphi$, and the notation is: Find $U = (u, \varphi) \in X$ such that

$$A(U, \Psi) = F(\Psi) \quad \forall \Psi := (w, \psi) \in X^0 := V_u^0 \times V_\varphi.$$

6. For nonlinear PDE systems, the notation is: Find $U \in X$ such that

$$A(U)(\Psi) = F(\Psi) \quad \forall \Psi \in X^0.$$

   We encounter this last notation often throughout this monograph.

2.1.8 Notational conventions for scalars, vectors, tensors

Important to notice is that we do not distinguish between scalar-valued, vector-valued, and tensor-valued variables in this monograph by using bold letters, arrows, or similar notation. All variables are in nonbold fonts. We expect the reader to distinguish them out of the context.

2.1.9 Tables of variables, parameters, and constitutive quantities

In the Tables 2.1 and 2.2, we summarize in a concise way the most important notation.

2.2 Prerequisites in differentiation and integration

Very briefly, we list some fundamental results that are necessary throughout this book.

Table 2.1: Part I: Summary of important notation and their physical units. The units are given for the space $\mathbb{R}^3$.

SYMBOL	*DESCRIPTION*	*UNIT*
d	Spatial dimension	
x	Position	m
t	Time; loading time in quasi-static (incremental) problems	s
B	Total domain: $B = \overline{\Omega} \cup \overline{C}$, i. e., $\overline{\Omega} \cup \overline{\Omega}_F$	m^3
C	Fracture (lower-dimensional)	m^2
I	$I = (0, T)$: Time/loading interval	s
T	End time value	s
h	Spatial discretization parameter (diam of element)	m
k	$k_n = t_n - t_{n-1}$: Temporal discretization parameter	s
U	Total solution vector, often in these notes: $U = (u, \varphi)$	
Ψ	Total test function, often in these notes: $\Psi = (w, \psi)$	
V_u^D	Trial function space for displacements	
V_u^0	Test function space for displacements	
V_φ	Function space for phase-field	
K	Convex set for phase-field with inequality constraint	
X	$X = V_u^D \times V_\varphi$: Trial product function space	
X^0	$X = V_u^0 \times V_\varphi$: Test product function space	
$a(\cdot,\cdot)$	Bilinear form to express weak/variational forms	
$A(\cdot)(\cdot)$	Semilinear form to express weak/variational forms	
(f, g)	Scalar product: $\int_B f \cdot g \, dx$ with $x, y \in \mathbb{R}^d$	
(F, G)	Scalar product: $\int_B F : G \, dx$ with $F, G \in \mathbb{R}^{d\times d}$	
κ	Phase-field regularization parameter in the bulk	
ε	Phase-field regularization parameter	m
γ	Penalization parameter	
$g(\varphi)$	$g(\varphi) = (1-\kappa)\varphi^2 + \kappa$: Degradation function	
u	Displacements	m
φ	Phase-field variable	
∇u	Gradient of displacements	
Δu	$\Delta u = \nabla \cdot \nabla u$: Laplacian	1/m

2.2.1 Gauss–Green theorem/divergence theorem

The Gauss–Green theorem or often known as divergence theorem. Let $\Omega \subset \mathbb{R}^d$ a bounded, open domain and $\partial\Omega$ of class C^1.

Theorem 1 (Gauss–Green theorem/divergence theorem). *Suppose that* $u := u(x) \in C^1(\bar{\Omega})$ *with* $x = (x_1, \ldots, x_d)$. *Then*

$$\int_\Omega u_{x_i} \, dx = \int_{\partial\Omega} u n_i \, ds, \quad \text{for } i = 1, \ldots, d.$$

Table 2.2: Part II: Summary of important notation and their physical units. The units are given for the space $\mathbb{R}^3$.

SYMBOL	*DESCRIPTION*	*UNIT*
$e(u)$	$e = e(u) = \frac{1}{2}(\nabla u + \nabla u^T)$, linearized strain tensor	
$\partial_t u$	$\partial_t u = v$: velocity (time-derivative displacements)	$\mathrm{m/s}$
$\partial_t v$	Acceleration ($\partial_t v = \partial_{tt}^2 u$)	$\mathrm{m/s^2}$
$\partial_t \varphi$	Time-derivative phase-field	$1/\mathrm{s}$
$E(\cdot)$	Energy functional	$\mathrm{J} = \mathrm{kgm^2/s^2}$
v	Velocity	$\mathrm{m/s}$
∇v	Gradient of the velocity/inner friction of fluids	$1/\mathrm{s}$
p	In general: pressure	$\mathrm{Pa} = \mathrm{N/m^2} = \mathrm{kg/ms^2}$
∇p	Gradient of pressure	$\mathrm{Pa/m} = \mathrm{N/m^3} = \mathrm{kg/m^2s^2}$
$\hat{F}$	Deformation gradient	
$\hat{J}$	Determinant of $\hat{F}$	
n	Normal vector	m
$\hat{E}$	Green–Lagrange/Green–St. Venant strain tensor	
I	Identity matrix/tensor	
σ_f	Cauchy fluid stress tensor	$\mathrm{Pa} = \mathrm{N/m^2} = \mathrm{kg/m\,s^2}$
σn	Traction force	$\mathrm{Pa\,m} = \mathrm{N/m}$
ρ	Density	$\mathrm{kg/m^3}$
ν	Kinematic fluid viscosity	$\mathrm{m^2/s}$
f	Volume force (e.g., gravity)	$\mathrm{N/kg} = \mathrm{m/s^2}$
σ	Cauchy solid stress tensor	$\mathrm{Pa/m^2}$
σ^+	Cauchy solid stress tensor: tensile forces	$\mathrm{Pa/m^2}$
σ^-	Cauchy solid stress tensor: compressive forces	$\mathrm{Pa/m^2}$
G_c	Critical energy release rate	$\mathrm{J/m^2} = \mathrm{N/m}$
μ	Lamé parameter/shear modulus	$\mathrm{Pa} = \mathrm{N/m^2} = \mathrm{kg/ms^2}$
λ	Lamé parameter	$\mathrm{Pa} = \mathrm{N/m^2} = \mathrm{kg/ms^2}$
ν_s	Poisson's ratio	
E_Y	Young's modulus	$\mathrm{Pa} = \mathrm{N/m^2} = \mathrm{kg/ms^2}$
$\hat{\Sigma}$	Second Piola–Kirchhoff stress tensor	$\mathrm{Pa/m^2}$
J	Goal functional	
η	Error estimator	
TOL	Solver tolerance, e. g., $\mathrm{TOL} = 10^{-10}$	
TOL_N	or $\mathrm{TOL}_{\mathrm{New}}$ same as TOL	

Here, n_i is the ith component of the normal vector n. In compact notation, we have

$$\int_{\Omega} \operatorname{div} u \, dx = \int_{\partial\Omega} u \cdot n \, ds$$

for each vector field $u \in C^1(\bar{\Omega}; \mathbb{R}^d)$.

Proof. Please see, for example, [271]. □

2.2.2 Integration by parts and Green's formula

From the divergence theorem, we obtain immediately the following

Proposition 1 (Integration by parts). *Let $u, v \in C^1(\bar{\Omega})$. Then*

$$\int_{\Omega} u_{x_i} v\,dx = -\int_{\Omega} u v_{x_i}\,dx + \int_{\partial\Omega} u v n_i\,ds, \quad \text{for } i = 1, \dots, d.$$

In compact notation:

$$\int_{\Omega} \nabla u v\,dx = -\int_{\Omega} u \nabla v\,dx + \int_{\partial\Omega} u v n\,ds.$$

Proof. Follows from the divergence theorem to uv. □

We obtain some further results, which are useful, but are all based directly on integration by parts.

Proposition 2 (Green's formulae). *Let $u, v \in C^2(\bar{\Omega})$. Then it holds:*

$$\int_{\Omega} \Delta u\,dx = \int_{\partial\Omega} \partial_n u\,ds,$$

$$\int_{\Omega} \nabla u \cdot \nabla v\,dx = -\int_{\Omega} \Delta u\, v\,dx + \int_{\partial\Omega} v\,\partial_n u\,ds.$$

Proof. Follows from integration by parts. □

2.2.3 Change of variables in Lebesgue integrals

Theorem 2 (Integration by substitution/change of variables). *Let $\hat{B} \subset \mathbb{R}^d$ be open. Let $\hat{T} : \hat{B} \to \mathbb{R}^d$ be a diffeomorphism in $\mathbb{R}^d$. Then, the function $f : \hat{T}(\hat{B}) \to \mathbb{R}$ is Lebesgue-integrable on $\hat{T}(\hat{B})$ if and only if the function $\hat{x} \in \hat{B} \to f(\hat{T}(\hat{x}))|\det(\hat{\nabla}\hat{T}(\hat{x}))| \in \mathbb{R}$ is Lebesgue-integrable on $\hat{B}$. Then we have*

$$\int_{\hat{T}(\hat{B})} f(x)\,dx = \int_{\hat{B}} f(\hat{T}(\hat{x}))|\det(\hat{\nabla}\hat{T}(\hat{x}))|\,d\hat{x}.$$

Theorem 3. *Let $B \subset \mathbb{R}^m$ be measurable and $A \subset \mathbb{R}^d$ be open. Let the function $f : A \times B \to \mathbb{R}$ be Lebesgue integrable in B for all fixed $x \in A$. Moreover, let f be continuously differentiable for $x \in A$ for almost all $y \in B$. We assume the bound $g : B \to \mathbb{R}$ with $\|\nabla_x f(x, y)\| \le g(y)$ for all $x \in A$ and almost all $y \in B$. Then:*

1. $\nabla_x f(x,y)$ *is Lebesgue integrable on* B *for all fixed* $x \in A$;
2. *The parameter integral* $F(x) = \int_B f(x,y)\,dy$ *is continuously differentiable with the derivative*

$$\nabla F(x) = \int_B \nabla_x f(x,y)\,dy.$$

Both theorems are important in continuum mechanics in which Lagrangian and Eulerian coordinates play a role, such as phase-field fluid-structure interaction in Chapter 11.

2.2.4 Differentiation in Banach spaces

At several places, we need differentiation in Banach spaces. From the literature, e. g., [419, 116, 429], we cite some well-known results in this section.

Definition 11 (Directional derivative). Let V and W be normed vector spaces and let $U \subset V$ be nonempty. Let $f : U \to W$ be a given mapping. If the limit

$$f'(u)(\phi) := \lim_{h\to 0} \frac{f(u+h\phi) - f(u)}{h}, \quad u \in U, \phi \in V,$$

exists, then $f'(u)(\phi)$ is called the directional derivative of the mapping f at u in direction ϕ. If the directional derivative exists for all $\phi \in V$, then f is called directionally differentiable at u.

Remark 4. Often, the direction ϕ is denoted by δu in order to highlight that the direction is related with the variable u. This notation is useful, when several solution variables exist and several directional derivatives need to be computed.

Remark 5. The definition of the directional derivative in Banach spaces is in agreement with the definition of derivatives in $\mathbb{R}$ at $x \in \mathbb{R}$ (see [270]):

$$f'(x) := \lim_{h\to 0} \frac{f(x+h) - f(x)}{h},$$

and in $\mathbb{R}^d$ we have (see [271]) into the direction $\phi \in \mathbb{R}^d$:

$$f'(x)(\phi) := \lim_{h\to 0} \frac{f(x+h\phi) - f(x)}{h}.$$

A function is called differentiable when all directional derivatives exist in the point $x \in \mathbb{R}^d$. The derivatives in the directions $e_i, i = 1,\dots,d$ of the standard basis are the well-known partial derivatives.

Definition 12 (Gâteaux derivative). Let the assumptions hold as in Definition 11. A directional-differentiable mapping as defined in Definition 11, is called Gâteaux-differentiable at $u \in U$, if there exists a linear continuous mapping $A : U \to W$ such that

$$f'(u)(\phi) = A(\phi),$$

holds true for all $\phi \in U$. Then A is the Gâteaux derivative of f at u and we write $A = f'(u)$.

Remark 6. The Gâteaux derivative is computed with the help of directional derivatives and it holds $f'(u) \in L(U, W)$. If $W = \mathbb{R}$, then $f'(u) \in U^*$.

Definition 13 (Fréchet derivative). A mapping $f : U \to W$ is Fréchet-differentiable at $u \in U$ if there exists an operator $A \in L(U, W)$ and a mapping $r(u, \cdot) : U \to W$ such that for each $\phi \in U$ with $u + \phi \in U$, it holds:

$$f(u + \phi) = f(u) + A(\phi) + r(u, \phi),$$

with

$$\frac{\|r(u, \phi)\|_W}{\|\phi\|_U} \to 0 \quad \text{for } \|\phi\|_U \to 0.$$

The operator $A(\cdot)$ is the Fréchet derivative of f at u and we write $A = f'(u)$.

Definition 14 (Equivalent formulation denoting derivatives). In the literature and above, we have (at least) three common notations and ways to compute directional derivatives:

$$\begin{aligned} f'(u)(\phi) &= \lim_{h \to 0} \frac{f(u + h\phi) - f(u)}{h} \\ &= \frac{d}{dh} f(u + h\phi)|_{h=0} \\ &= f(u + \phi) - f(u) - r(u, \phi). \end{aligned}$$

Example 1. The bilinear form $a(u, \varphi) = (\nabla u, \nabla \varphi)$ is Fréchet-differentiable in the first argument u (of course also in the second argument! But u is the variable we are interested in):

$$a(u + \phi, \varphi) = (\nabla(u + \phi), \nabla\varphi) = \underbrace{(\nabla u, \nabla\varphi)}_{=a(u,\varphi)} + \underbrace{(\nabla\phi, \nabla\varphi)}_{=a'_u(u,\varphi)(\phi)}.$$

Here, the remainder term is zero, i. e., $r(u, \phi) = 0$, because the bilinear form is linear in u. Thus, the Fréchet derivative of $a(u, \varphi) = (\nabla u, \nabla\varphi)$ is $a'_u(u, \varphi)(\phi) = (\nabla\phi, \nabla\varphi)$.

2.2.5 Chain rules

The chain rule in $\mathbb{R}^{d+1}$ is as follows.

Definition 15. Let the functions $g : (a,b) \to \mathbb{R}^{d+1}$ and $f : \mathbb{R}^{d+1} \to \mathbb{R}$ and its composition $h = f \circ g \in \mathbb{R}$ be given and specifically $g(t,x) := (t,x) := (t,x_1,x_2,\dots,x_d)$:

$$\begin{aligned}
\frac{d}{dt}h(t,x) &= \frac{d}{dt}f(g(t,x)) = \frac{d}{dt}f(t,x) \\
&= \sum_{j=0}^{d} \partial_j f(g(x)) \cdot \partial_t g_j \\
&= \sum_{j=0}^{d} \partial_j f(t,x_1,\dots,x_d) \cdot \partial_t x_j, \quad \text{where } x_0 := t \\
&= \partial_t f \cdot \partial_t t + \sum_{j=1}^{d} \partial_j f(t,x_1,\dots,x_d) \cdot \partial_t x_k \\
&= \partial_t f + \nabla f \cdot (\partial_t x_1,\dots,\partial_t x_d)^T \\
&= \partial_t f + \nabla f \cdot v.
\end{aligned}$$

For instance $d = 3$ and time t, this means that we deal with a four-dimensional setting (t,x,y,z).

Remark 7. See also [271, p. 54 and p. 93] for definitions of the chain rule.

The chain rule in Banach spaces is [419, 116, 429]:

Definition 16. Let U, V, W be Banach spaces. Let U and V be open sets. We have the mappings $F : U \to V$ and $G : V \to W$. Let F be differentiable at $u \in U$ and G be differentiable at $F(u) \in V$. Then the composition $H = G \circ F$ defined by $H(u) = G(F(u))$ with $H : U \to W$ is Fréchet differentiable at $u \in U$ and it holds

$$H'(u) = G'(F(u))F'(u).$$

More specifically, in a direction $\delta u \in U$, it holds

$$H'(u)(\delta u) = G'(F(u))F'(u)(\delta u).$$

Example 2. We apply the chain rule to an operator $T : U \to W$:

$$T(u) = u^3$$

then

$$T(u+\phi) = (u+\phi)^3 = u^3 + 3u^2\phi + 3u\phi^2 + \phi^3.$$

Here, we now have four terms to describe $T(u)$, $A(\phi)$ and $r(u,\phi)$.

- The identification of $T(u) = u^3$ is obvious.
- According to the theory $A(\phi)$ is a linear operator in ϕ. Therefore, $A(\phi) = 3u^2\phi$.
- The rest goes into $r(u, \phi) = 3u\phi^2 + \phi^3$.

To check that we deal with a Fréchet derivative, we need to verify

$$\frac{\|r(u,\phi)\|_W}{\|\phi\|_U} \to 0 \quad \text{for } \|\phi\|_U \to 0.$$

Here, it is important to remark that u is fixed and we only vary in ϕ. Since we have ϕ^2 and ϕ^3 in the nominator, but only ϕ in the denominator, one can indeed check that $T(u) = u^3$ is Fréchet-differentiable.

Example 3. Let $v \in V_1$ and $u \in V_2$ be solution variables. For brevity, we neglect the governing equation for v. Let u be determined by the semilinear form

$$A(v,u)(\varphi) := (v\nabla u, \nabla\varphi) + (u^3, \varphi).$$

Then the directional derivative in direction $(\delta v, \delta u) \in V_1 \times V_2$ is given by

$$A'(v,u)((\delta v, \delta u), \varphi) = (\delta v \nabla u + v\nabla \delta u, \nabla\varphi) + (3u^2\delta u, \varphi),$$

using the product and chain rules, respectively. If $v \in V_1$ would be given and not an unknown, then

$$A'(v,u)((\delta v, \delta u), \varphi) = (v\nabla \delta u, \nabla\varphi) + (3u^2\delta u, \varphi).$$

2.3 A prototype phase-field formulation

A prototype model problem is introduced in this section. This model allows us to explain the characteristic features of quasi-brittle phase-field fracture. To determine both solution variables u and φ, we need two equations. Later, displacements are computed with the help of elasticity. For the moment, however, we work with the Laplacian to make it as simple as possible. Let B a domain with boundary ∂B with classical assumptions on their smoothness (e. g., [79]). Find $u : B \to \mathbb{R}$ such that

$$-\nabla \cdot (a\nabla u) = f \quad \text{in } B, \quad \text{plus bc on } \partial B, \tag{2.1}$$

where "bc" = "boundary conditions" and with a material coefficient $a > 0$. Soon, we shall see that in phase-field fracture, the parameter a depends on $x \in B$ as well as on a second solution variable.

The phase-field function φ describes the crack path ($\varphi = 1$ for an unbroken solid and $\varphi = 0$ for a fully damaged material) and is in the limit a lower-dimensional manifold. Here, [72] proposed to employ an Ambrosio–Tortorelli elliptic functional

(1990/1992) [20, 21]. Further details will follow in Chapter 4. For now, we are satisfied to know that we work with an elliptic functional, whose corresponding PDE is very familiar to us (again a Laplacian with a reaction term and a parameter $\varepsilon > 0$): Find $\varphi : B \to \mathbb{R}$ such that

$$-\varepsilon\Delta\varphi - \frac{1}{\varepsilon}(1-\varphi) = g \quad \text{in } B, \quad \text{plus bc on } \partial B, \tag{2.2}$$

where $g : B \to \mathbb{R}$ is a right-hand side force. We later prove in Theorem 5, Section 5.5.2, that (2.2) is bounded in the H^1 norm, which is the natural norm for this equation. Moreover, it can be shown that $0 \leq \varphi \leq 1$; see again Theorem 5 (and also the literature [336, 173]). Together with (2.2), this yields the following.

Proposition 3. *Given equation* (2.2) *and assuming the later result from Theorem* 5 *that* φ *is indeed well-defined in the* H^1 *setting and* $0 \leq \varphi \leq 1$, *it holds:*
1. *The variable* φ *is an indicator function with values between* 0 *and* 1.
2. *The gradient* $\nabla\varphi$ *smoothes the transition from* $\varphi = 0$ *to* $\varphi = 1$. *For large* ε, *we have a smoother (larger) transition zone. For small* ε, *the transition zone becomes smaller and the gradients* $\nabla\varphi$ *steeper, i. e., values* $\varphi = 0$ *and* $\varphi = 1$ *are accented and we lose smoothness in the limit for* $\varepsilon \to 0$.

Proof. We refer to Theorem 5 for the mathematical statements. The descriptive illustration follows trivially. See also Figure 2.2. □

However, an additional constraint is the crack irreversibility (the crack cannot heal in time), which is imposed as

$$\partial_t\varphi \leq 0. \tag{2.3}$$

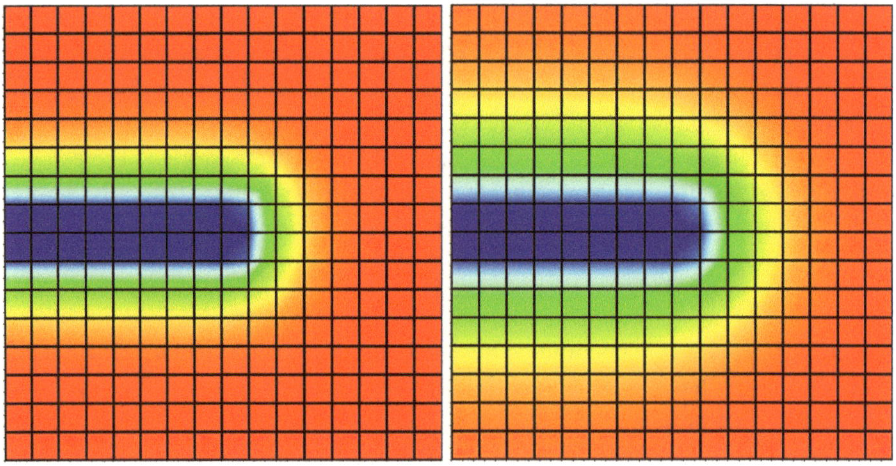

Figure 2.2: Approximation of a fracture with phase-field. At left, the regularization parameter is chosen as $\varepsilon = h$. At right, the regularization parameter is chosen as $\varepsilon = 2h$.

Consequently, the equation determining φ becomes an inequality and is linked to the constraint (2.3) via a complementarity condition. Therefore, the system is very close to the well-known and well-studied obstacle problem; see e. g., the classical books [141, 260, 261, 418, 417]. However, the two main PDEs do not have any time derivatives, but only the constraint does. This constitutes a difference to the standard obstacle problem.

With the help of these definitions, a simplified phase-field fracture problem can be formulated. This allows us to identify two important characteristic aspects of the full problem. To this end, we study

Formulation 1 (A simplified strong problem formulation). Let B be a domain with the boundary $\partial B := \partial\Omega_{\mathrm{HD}} \cup \partial\Omega_{\mathrm{ND}} \cup \partial\Omega_{\mathrm{HN}}$. Find a displacement function $u: B \times I \to \mathbb{R}$ and a phase-field indicator function $\varphi: B \times I \to [0,1]$, such that

$$-\nabla\cdot(\varphi^2\nabla u) = f \quad \text{in } B \times I, \quad (u\text{-equation}) \tag{2.4}$$

$$\varphi|\nabla u|^2 - \varepsilon\Delta\varphi - \frac{1}{\varepsilon}(1-\varphi) \leq 0 \quad \text{in } B \times I, \quad (\varphi\text{-equation}) \tag{2.5}$$

$$\partial_t\varphi \leq 0 \quad \text{in } B \times I, \quad \text{(crack irreversibility)} \tag{2.6}$$

$$\left[\varphi|\nabla u|^2 - \varepsilon\Delta\varphi - \frac{1}{\varepsilon}(1-\varphi)\right]\cdot\partial_t\varphi = 0 \quad \text{in } B \times I, \quad \text{(complementarity condition).} \tag{2.7}$$

These equations are complemented with the following boundary and initial conditions:

$$\begin{aligned}
u(x,t) &= u_D(x,t) && \text{on } \partial\Omega_{\mathrm{ND}} \times I,\\
u(x,t) &= 0 && \text{on } \partial\Omega_{\mathrm{HD}} \times I,\\
\varphi^2\nabla u\cdot n &= 0 && \text{on } \partial\Omega_{\mathrm{HN}} \times I,\\
\varepsilon\partial_n\varphi &= 0 && \text{on } \partial B \times I,\\
\varphi(x,0) &= \varphi_0 && \text{on } B \times \{0\},
\end{aligned}$$

with an initial fracture φ_0 and with $\varepsilon > 0$ as the so-called phase-field regularization parameter.

Remark 8 (Coupling terms). Observing (2.1) and (2.2), we obtain Formulation 1 by setting $a = \varphi^2$ and $g = -\varphi|\nabla u|^2$.

Remark 9. The PDE (2.4) and VI (2.5) are not time-dependent, but the constraint (2.6) depends on time. This constraint can be discretized using a backward difference quotient:

$$\partial_t\varphi \approx \frac{\varphi^n - \varphi^{n-1}}{k} \leq 0,$$

where $\varphi^n := \varphi(t_n)$ and the time stepsize $k = t_n - t_{n-1}$. This implies $\varphi^n \leq \varphi^{n-1}$. Further discussions on how to impose this constraint are provided in Chapter 5.

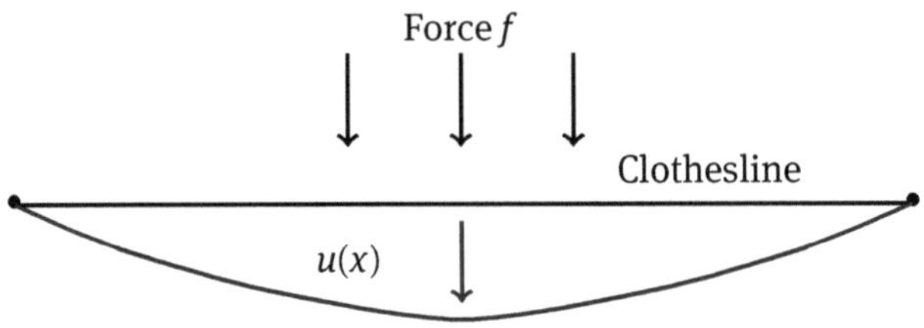

Figure 2.3: Clothesline problem (Poisson problem in one dimension). Displacement $u(x)$ induced by a force f.

2.4 Comments on the prototype phase-field problem

2.4.1 Relation to the obstacle problem

There are several striking similarities between the obstacle problem [141, 417, 260, 261, 209, 418] and the previously introduced model. For this reason, we recapitulate known models and results for an illustrative understanding; see also Figure 2.3.

The classical Poisson problem reads: Find $u : B \to \mathbb{R}$ such that

$$\begin{aligned} -\Delta u &= f \quad \text{in } B \\ u &= 0 \quad \text{on } \partial B. \end{aligned}$$

The obstacle problem is defined as

$$\begin{aligned} -\Delta u &\geq f \quad \text{in } B, && \text{(PDE)}, \\ u &\geq g \quad \text{in } B, && \text{(Inequality constraint)}, \\ (f + \Delta u)(u - g) &= 0 \quad \text{in } B, && \text{(Complimentarity condition)}, \\ u &= 0 \quad \text{on } \partial B, && \text{(Boundary condition)}. \end{aligned}$$

The obstacle problem (see Figure 2.4) is a free boundary problem that splits the domain B into two parts $\mathcal{N}$ (inactive set) and $\mathcal{A}$ (active set).

- If $f + \Delta u = 0$, then we solve the PDE and the constraint is fulfilled (inactive), i. e., $u > g$ in the so-called inactive set $\mathcal{N}$. In other words, the obstacle condition is not active.
- If $f + \Delta u < 0$, we are “sitting” on the obstacle, i. e., $u = g$, and we are in the active set $\mathcal{A}$.

2.4.2 Brief comments on the classification

The fully coupled system (2.4) to (2.7) is nonlinear (to be specific: quasi-linear; see Section 3.1). If we decouple the system, the single expressions (2.4) and (2.5) become

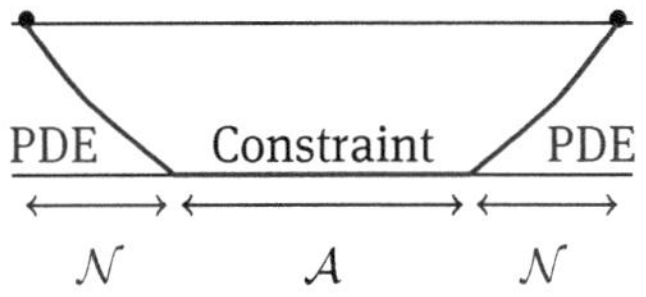

Figure 2.4: Obstacle problem: a clothesline touches the ground in the active set $\mathcal{A}$ in which the obstacle condition is active. In the inactive set $\mathcal{N}$, the PDE is solved.

linear. However, (2.5)–(2.7) forms an inequality and, therefore, we nevertheless deal with a nonlinear problem since the underlying space is only a convex set and not anymore a linear function space.

2.4.3 Energy formulations

The PDE $-\nabla \cdot (\varphi^2 \nabla u) = f$ with homogeneous Dirichlet conditions reads in a variational formulation: Given $\varphi^2 \in L^\infty(B)$, find $u \in H_0^1(B)$ such that

$$(\varphi^2 \nabla u, \nabla w) = (f, w), \quad \text{for all } w \in H^1(B), \tag{2.8}$$

respectively boundary conditions of Dirichlet type. We introduce the notation

$$E_b'(u, \varphi)(w) := (\varphi^2 \nabla u, \nabla w) - (f, w) = \int_B \varphi^2 \nabla u \nabla w \, dx - \int_B f \cdot w \, dx.$$

Then (2.8) reads: Given $\varphi^2 \in L^\infty$, find $u \in H_0^1(B)$ such that

$$E_b'(u, \varphi)(w) = 0.$$

Integrating yields the energy of the system:

$$E_b(u, \varphi) = \frac{1}{2} \int_B \varphi^2 |\nabla u|^2 \, dx - \int_B f \cdot u \, dx. \tag{2.9}$$

The resulting minimization problem reads: Given $\varphi^2 \in L^\infty$, solve

$$\min_{u \in H_0^1(B)} E_b(u, \varphi).$$

Considering also (2.5) with φ as a solution variable yields a fourth-order expression

$$\min_{u \in H_0^1(B), \varphi \in K} E_b(u, \varphi).$$

Specifically, the solution of E_b is not unique and the corresponding Hessian E'' is indefinite. For an inspection of a finite-dimensional prototype problem, we refer to [445][Section 4]. These properties cause challenges for the nonlinear solution. However, as we easily observe, fixing one variable, yields a standard second-order problem. This has led to the so-called alternating minimization algorithm in phase-field fracture [70].

2.5 A simplified phase-field model for dynamic fracture

We now introduce a prototype dynamic fracture model in which time derivatives are added to both the principle PDE and the VI. The displacement equation is augmented with a second order in time acceleration term, which yields a wave equation and is of hyperbolic type. The phase-field inequality is extended to a parabolic-type problem. Such parabolic phase-field equations are known in Allen–Cahn models [12]; see e. g., [163, 463, 41].

Formulation 2. Let $B \subset \mathbb{R}^d$ be open and let $I := (0, T)$ with $T > 0$ being the end time value. Find $u : B \times I \to \mathbb{R}$ and $\varphi : B \times I \to \mathbb{R}$ such that

$$\begin{aligned}
\rho_s \partial_t^2 u - \nabla \cdot (\varphi^2 \nabla u) &= f \quad \text{in } B \times I, \\
\partial_t \varphi + \varphi |\nabla u|^2 - \varepsilon \Delta \varphi - \frac{1}{\varepsilon}(1 - \varphi) &\leq 0 \quad \text{in } B \times I, \\
\partial_t \varphi &\leq 0 \quad \text{in } B \times I, \\
\left[\partial_t \varphi + \varphi |\nabla u|^2 - \varepsilon \Delta \varphi - \frac{1}{\varepsilon}(1 - \varphi) \right] \cdot \partial_t \varphi &= 0 \quad \text{in } B \times I, \\
u &= 0 \quad \text{on } \partial B \times I, \\
\varepsilon \partial_n \varphi &= 0 \quad \text{on } \partial B \times I, \\
u &= u_0 \quad \text{in } B \times \{0\}, \\
\partial_t u &= v_0 \quad \text{in } B \times \{0\}, \\
\varphi &= \varphi_0 \quad \text{in } B \times \{0\}.
\end{aligned}$$

Remark 10. Dynamic phase-field fracture models will be studied less in the following chapters. We briefly comment on them in Section 5.6. However, in Chapter 11 on multiphysics extensions, the dynamic model becomes more important, for instance, in fully time-dependent problems with large deformations and fluid-structure interaction.

3 Classifications of partial differential equations and variational inequalities

In this chapter, we classify CVISs, namely nonstationary, nonlinear, coupled differential equations subject to inequality constraints.

Definition 17 (Classification of differential equations). The following properties are distinguished:

1. Dependent variables $(x, y, z, t, \dots)$; if only one, then ODE (ordinary differential equation), otherwise PDE;
2. Steady-state problems or time-dependent PDEs;
3. Conservation laws such as mass, momentum, angular momentum, energy, resulting in diffusion, transport, convection, waves, reaction systems, and their combinations;
4. Order of a PDE (in time, space, or both);
5. Single equations and PDE systems;
6. Nonlinear problems:
   - Nonlinearities in the PDE through state variables,
   - Nonlinear constitutive laws,
   - Inequality constraints;
7. Coupled PDE systems and CVIS:
   - coupling through model or material parameters,
   - linear or nonlinear coupling terms,
   - right-hand sides,
   - interfaces.

We briefly focus on three of these classifications in more detail. First, we recall from Evans [158] the general definition of a PDE and four different cases of linear/nonlinearities. Second, we recapitulate variational inequalities from the literature [260]. Third, we provide our own definitions of coupled problems in Section 3.3.

3.1 Linear versus nonlinear PDEs

3.1.1 General definition of a differential equation

For a general description of differential equations, the multiindex notation introduced in Definition 6 (Section 2.1.6) is commonly used. Next, we cite some results from Evans [158].

Definition 18. Let $\Omega \subset \mathbb{R}^n$ be a space-time domain in this section. Here, n denotes the total dimension including time, i. e., $n = d+1$, thus $\Omega = B \times I$. Furthermore, let $k \geq 1$ an

https://doi.org/10.1515/9783110497397-003

integer that denotes the order of the differential equation. Then a differential equation in implicit form can be expressed as: Find $u : \Omega \to \mathbb{R}$ such that

$$F(D^k u, D^{k-1}u, \dots, D^2 u, Du, u, x) = 0 \quad x \in \Omega,$$

where

$$F : \mathbb{R}^{n^k} \times \mathbb{R}^{n^{k-1}} \times \dots \times \mathbb{R}^{n^2} \times \mathbb{R}^n \times \mathbb{R} \times \Omega \to \mathbb{R}.$$

Definition 19 (PDE system). For F_i PDEs and the solution variables u_i with $i = 1, \dots, s$, we deal with a PDE system with s equations.

3.1.2 Classifications into linear and nonlinear PDEs

Based on the previous notation, we now provide classifications of linear and nonlinear differential equations. Simply speaking: each differential equation, which is not linear is called nonlinear. However, in the nonlinear case a further refined classification can be undertaken.

Definition 20 (Evans [158]). Differential equations are divided into linear and nonlinear classes as follows:

1. A differential equation is called linear if it is of the form:

$$\sum_{|\alpha| \le k} a_\alpha(x) D^\alpha u - f(x) = 0.$$

2. A differential equation is called semilinear if it is of the form:

$$\sum_{|\alpha| = k} a_\alpha(x) D^\alpha u + a_0(D^{k-1}u, \dots, D^2 u, Du, u, x) = 0.$$

   Here, nonlinearities may appear in all terms of order $|\alpha| < k$, but the highest order derivative $|\alpha| = k$ is fully linear.
3. A differential equation is called quasi-linear if it is of the form:

$$\sum_{|\alpha| = k} a_\alpha(D^{k-1}u, \dots, D^2 u, Du, u, x) D^\alpha u + a_0(D^{k-1}u, \dots, D^2 u, Du, u, x) = 0.$$

   Here, full nonlinearities may appear in all terms of order $|\alpha| < k$, in the highest order $|\alpha| = k$, nonlinear terms appear up to order $|\alpha| < k$.
4. If none of the previous cases applies, a differential equation is called (fully) nonlinear.

Remark 11. In theory, as well as practice, it holds: the higher the nonlinearity of an equation is, the more difficult the analysis becomes and the more challenging the design of good numerical methods will be.

3.1.3 Examples

Some examples from fluid mechanics are provided in the following:

1. Euler equations (fluid dynamics, special case of Navier–Stokes with zero viscosity). Here, $n = 2+1 = 3$ (two spatial dimensions and one time dimension) in which we have 2 velocity components $v = (v_x, v_y)$, and 1 pressure variable p. Let us consider the momentum part of the Euler equations:

$$\partial_t v + v \cdot \nabla v + \nabla p = f.$$

Here, the highest order is $k = 1$ (in the temporal variable as well as the spatial variable). But in front of the spatial derivative, we multiply with the zero-order term v. Consequently, the Euler equations are quasi-linear because a lower-order term of the solution variable is multiplied with the highest derivative.

2. Navier–Stokes momentum equation:

$$\partial_t v - \rho_f \nu_f \Delta v + v \cdot \nabla v + \nabla p = f.$$

Here, $k = 2$. But the coefficients in front of the highest order term, the Laplacian, do not depend on v. Consequently, the Navier–Stokes equations are neither fully nonlinear nor quasi-linear. We have again the nonlinearity of the first-order convection term $v \cdot \nabla v$. Thus the Navier–Stokes equations are in principle semilinear. However, if $\nu_f \to 0$ (high Reynolds number), then we obtain the Euler equations in the limit, and thus a quasi-linear PDE.

3. A fully nonlinear situation would be:[1]

$$\partial_t v - \rho_f \nu_f (\Delta v)^2 + v \cdot \nabla v + \nabla p = f.$$

3.2 Variational equations versus variational inequalities

Variational inequalities generalize weak forms, such that constraints on the solution variables can be incorporated.

3.2.1 Variational inequalities

Let $(V, \|\cdot\|)$ be a real Hilbert space and K a nonempty, closed, convex subset of V. We define the mapping $A : K \to V^*$, where V^* is as usual the dual space of V. Then we define the abstract problem:

1 This last equation is purely academic without specific application.

Formulation 3 (Abstract VI). Find $u \in K$ such that

$$A(u)(\varphi - u) \geq 0 \quad \forall \varphi \in K.$$

If K is a linear subspace (and not only a set!) of V, then we have linearity, which allows us to take

$$\varphi := u \pm w \in K,$$

for each $w \in K$. Since the semilinear form $A(\cdot)(\cdot)$ is linear in the second argument, we obtain:

$$A(u)(w) \geq 0 \quad \text{and} \quad A(u)(-w) \geq 0,$$

from which follows:

$$A(u)(w) = 0 \quad \forall w \in K.$$

This shows that weak equations are contained in the more general setting of variational inequalities.

In the literature and later in the Sections 5.2.3 and 7.9.5, the notation of the duality pairing $\langle \cdot, \cdot \rangle$ is often found for describing weak forms and variational inequalities. This notation better highlights the spaces we are working in. Here, the previous statements read as follows:

Formulation 4 (Abstract VI – duality pairing notation). Find $u \in K$ such that

$$\langle A(u), \varphi - u \rangle \geq 0 \quad \forall \varphi \in K.$$

Here, we see more easily that $A(u) \in V^*$ and $\varphi - u \in V$.

If K is a linear subspace (and not only a convex set!) of V, then

$$\varphi := u \pm w \in K,$$

for each $w \in K$. Since the semilinear form $A(\cdot)(\cdot)$ is linear in the second argument, we obtain

$$\langle A(u), w \rangle \geq 0 \quad \text{and} \quad \langle A(u), -w \rangle \geq 0,$$

from which follows:

$$\langle A(u), w \rangle = 0 \quad \forall w \in K.$$

3.2.2 Minimization of functionals

For energy formulations, the following results are useful, which we briefly summarize from [260]. Let $(V, \|\cdot\|)$ be a reflexive, real Hilbert space, K be a nonempty closed convex subset of V and $F : K \to \mathbb{R}$ be a real functional defined on K. In the following, we formulate everything in terms of F. A natural candidate is the energy, i. e., $F := E$.

Formulation 5 (Minimization problem). Find $u \in K$ such that

$$F(u) = \inf_{v \in K} F(v)$$

or equivalently

$$F(u) \leq F(v) \quad \forall v \subset K.$$

Definition 21 (Convexity). A functional $F : K \to \mathbb{R}$ is convex if and only if

$$F(\theta u + (1-\theta)v) \leq \theta F(u) + (1-\theta)F(v) \quad u, v \in K, \quad 0 \leq \theta \leq 1.$$

The functional is strictly convex when $<$ holds true.

Definition 22. As before, let

$$\lim_{h \to 0} \frac{d}{dh} F(u + hv) = F'(u)(v) = \langle F'(u), v \rangle$$

be the Gâteaux derivative of F.

Theorem 4. *Let K be a subset of a normed linear space V and let $F : K \to \mathbb{R}$ be Gâteaux differentiable. If u is a minimizer of F in K, then u can be characterized as follows:*

1. *If K is a nonempty closed convex subset of V, then*

$$F'(u)(v - u) \geq 0 \quad \forall v \in K.$$

2. *If K is a nonempty closed convex subset of V and $u \in \operatorname{int}(K)$, then*

$$F'(u)(v) \geq 0 \quad \forall v \in K.$$

3. *If K is a linear variety with respect to w (i. e., K is a linear subspace translated by a factor w with respect to the origin), then*

$$F'(u)(v) = 0 \quad \forall v \in V,$$

   and $v - w \in K$.
4. *If K is a linear subspace of V, then*

$$F'(u)(v) = 0 \quad \forall v \in K.$$

Proof. Sketch. See also [260].

1. The set K is convex. Thus $w := u + \theta(v - u) \in K$ for $\theta \in [0, 1]$ and for every $u, v \in K$. If u is a minimizer of K, then we have

$$F(u) \leq F(w) = F\big(u + \theta(v - u)\big) \quad \forall v \in K.$$

Differentiation yields:

$$\frac{d}{d\theta}[F(u + \theta(v - u)) - F(v)]|_{\theta=0} = F'(u)(v - u) \geq 0.$$

2. If $u \in \operatorname{int}(K)$ and any $v \in K$. Then there exists $\delta \geq 0$ such that $u + \delta v \in K$. Therefore, we insert an element in into the previous characterization and obtain

$$F'(u)(u + \delta v - u) = F'(u)(\delta v) \geq 0.$$

Since δ is arbitrary, this implies the assertion.

3. Let $K = \{w\} + M$, where M is a linear subspace of V. Then $w - v \in K$ implies $v \in M$. The same holds for $-v \in M$. Then

$$F'(u)(v) = 0 \quad v \in V \quad \text{such that } v - w \in K.$$

4. Take $w = 0$. Then

$$F'(u)(v) = 0 \quad \forall v \in K.$$

□

3.3 Coupled problems and multiphysics PDEs

Coupled problems represent an important field in applied mathematics thanks to advancements in computing resources, numerical analysis techniques, mathematical modeling, and analysis. Coupled problems are key in understanding multiphysics PDEs and CVISs. We provide basic definitions in this section, which shall be complemented by solution algorithms in Chapter 7.

Definition 23 (Coupled PDEs). A coupled PDE (system) or CVIS consists of at least two PDEs/VIs that interact with each other via

1. model or material parameters,
2. linear or nonlinear coupling terms,
3. right-hand sides,
4. interfaces.

The solution of one PDE will influence the solution of the other PDE. This definition extends to $S \in \mathbb{N}$ interacting PDEs.

Remark 12. The previous definition also holds for ODEs. However, the last condition of interface coupling is significantly more difficult in the PDE case.

The understanding of coupling terms is important for the mathematical analysis and the design of numerical solution algorithms. Often, coupling terms introduce additional nonlinearities. Explicit treatments (e. g., linearizations, time-lagging) might result in linear subproblems.

Definition 24 (Multiphysics PDEs/CVISs). A multiphysics problem is a coupled PDE system in which at least two PDEs, representing different physics, are interacting. A multiphysics problem is always a coupled problem, but not vice versa. If inequality constraints are taken into consideration, it is referred to as a multiphysics CVIS.

Proposition 4. *We distinguish three cases:*
1. *coupled problems as the most general definition;*
2. *internally physics-coupled problems;*
3. *multiphysics PDEs/CVISs.*

The following examples hold:
1. *Phase-field fracture (Section 4.4.4) is a coupled problem, but not a multiphysics PDE. Furthermore, it is a CVIS because a temporal inequality constraint is considered;*
2. *The incompressible Navier–Stokes equations* [409, 364, 201] *are an internally coupled problem (pressure and velocities interact), but not a multiphysics system;*
3. *The Maxwell equations (e.g.,* [168]*) for describing electromagnetics are an internally coupled problem.*
4. *The Biot equations* [58, 59, 119, 414] *for porous media modeling (displacements and pressure are interacting) is a multiphysics problem and thus also a coupled problem;*
5. *Thermo-poroelasticity* [119, 424, 426] *is a multiphysics problem in which solids, fluid flow and temperatures interact;*
6. *Fluid-structure interaction* [376, 93, 187, 46] *is a multiphysics problem. Here, velocities/pressure interact with solid displacements.*

Proof.
1. In phase-field fracture, two problems couple. The phase-field variational inequality, however, does not represent any physics, but does only mark the fracture location. The only physics are displacements.
2. In the Navier–Stokes equations, two state variables couple internally, namely pressure and velocities, but they describe the same physical phenomenon: fluid flow. Therefore, it is an internally coupled problem, but not multiphysics.
3. In the Maxwell equations, we have two interacting variables: the electric field and the magnetic field. Together they describe electromagnetic phenomena.
4. The Biot equations represent two physics, i. e., multiphysics: solid displacements in a porous medium and fluid flow modeled with the help of Darcy's law. In the

classical formulation (see also Section 3.4), this problem has the specific property that it is a linear multiphysics problem.
5. In thermo-poroelasticity, the Biot equations couple in addition with temperature. Here, three physics couple.
6. In fluid-structure interaction two physical phenomena couple with in total three solution variables (v, p, u): solid deformations u and fluid flow (v, p). □

Definition 25 (One-way coupling versus two-way coupling). Let PDE 1 and PDE 2 be given. In one-way coupling information transport is one-sided, i. e., solution variables of PDE 1 enter into PDE 2, but not vice versa. In two-way coupling, both PDEs interact with each other: solution variables of PDE 1 enter into PDE 2 and solution variables of PDE 2 enter into PDE 1.

Example 4. An example for one-way coupling is

$$\partial_t u - \nabla \cdot (\varphi^2 \nabla u) = f(\varphi) \quad \text{(PDE 1)},$$
$$\partial_t \varphi - \Delta \varphi = g \quad \text{(PDE 2)}.$$

The second PDE enters into the first PDE, but not vice versa. An example for two-way coupling is the prototype phase-field model from Section 2.3.

We investigate in more detail the four coupling situations from Definition 23 in the following and further identify two situations.

Definition 26 (Domain coupling versus interface coupling). We define:
1. Domain-coupling in which the solution of one PDE enters as a domain variable (defined in the entire domain B) into the other PDE(s).
2. Interface-coupling in which information is exchanged on an interface Γ.

In general, interface coupling is more challenging because the interface is a lower-dimensional manifold for which mathematical regularity as well as numerical discretizations require often more sophisticated tools than in domain coupling.

Remark 13 (Simultaneous domain and interface coupling). Both coupling types can arise simultaneously, as for instance in arbitrary Lagrangian-Eulerian fluid-structure interaction, in which the problem is formulated in the reference domain [139, 434, 376, 178] (see also Chapter 11) or the augmented Mandel's problem [200] for coupling elasticity and poroelasticity (see Section 3.4). Moreover, in coupled fluid-structure interaction with phase-field fracture, both types arise (see again Chapter 11).

Remark 14 (Volume coupling). Domain coupling is also known as volume coupling.

3.3.1 Domain coupling

We discuss both coupling types in more detail and consider first domain coupling. To this end, we observe the following system.

Formulation 6. Let V_1 and V_2 be Hilbert or Sobolev spaces to describe each single PDE such that these are well-posed. Let two weak formulations be given:

$$\begin{aligned}
&\text{Given } u_2 \in V_2, \text{ find } u_1 \in V_1: \quad A_1((u_1,u_2))(\varphi_1) = F_1(\varphi_1) \quad \forall \varphi_1 \in V_1,\\
&\text{Given } u_1 \in V_1, \text{ find } u_2 \in V_2: \quad A_2((u_1,u_2))(\varphi_2) = F_2(\varphi_2) \quad \forall \varphi_2 \in V_2,
\end{aligned}$$

where A_1 and A_2 are semilinear forms and F_1 and F_2 right-hand side functionals.

We refine the first notation from Definition 23.

Definition 27 (Coupling via model or material parameters). Let us consider a specific realization of Laplace-type:

$$A_1((u_1,u_2))(\varphi_1) = (\mu(u_2)\nabla u_1, \nabla \varphi_1)$$

and some nonspecified form of $A_2((u_1,u_2))(\varphi_2)$. Here, the solution variable u_2 determined in A_2 enters as coefficient $\mu(u_2)$ into A_1.

Examples are phase-field fracture modeling as in this book or non-Newtonian fluid flow, e. g., p-Stokes [233].

Definition 28 (Coupling via linear or nonlinear coupling terms). We consider specifically

$$A_1((u_1,u_2))(\varphi_1) = (\nabla u_2 + \nabla u_1, \nabla \varphi_1),$$

or

$$A_1((u_1,u_2))(\varphi_1) = (\nabla u_2 \cdot \nabla u_1, \nabla \varphi_1),$$

and again some nonspecified form of $A_2((u_1,u_2))(\varphi_2)$. We assume that the dimensions match such that $(\nabla u_2 \cdot \nabla u_1, \nabla\varphi_1)$ can be evaluated as L^2 inner product and is well-defined. This definition is close to the previous one and the difference is that u_2 does not enter via a coefficient, but directly into A_1.

Examples are the linear Biot equations for porous media modeling, e. g., [58, 59, 119, 414] (see Section 3.4) or the Stokes equations [409, 364, 201] (see Section 9.3.1 and Chapter 11).

Definition 29 (Coupling via right-hand sides). Let us consider

$$\text{Find } u_1 \in V_1: \quad A_1((u_1,u_2))(\varphi_1) = F_1(u_2)(\varphi_1),$$

where u_2 enters into the right-hand side F_1. Hence, F_1 becomes solution dependent and is not a right-hand side in the strict sense any more. Rather we must bring all terms on the left-hand side:

$$\text{Find } u_1 \in V_1: \quad A_1((u_1, u_2))(\varphi_1) - F_1(u_2)(\varphi_1) = 0.$$

As before, the specific form of A_2 is not important. To this end, this definition reduces to the previous definition as a linear or nonlinear domain coupling.

3.3.2 Interface coupling

We now address the fourth coupling from Definition 23 in more detail. Here, both (or multiple) PDEs/CVISs live in different subdomains.

Definition 30 (Interface). Let $\Omega_1 := \Omega_1(t) \subset \mathbb{R}^d$ and $\Omega_2 := \Omega_2(t) \subset \mathbb{R}^d$ be open domains that may vary in time. Let the total domain be defined as $B := \Omega_1 \cup \Omega_2$. The interface is then defined by

$$\Gamma(t) := \overline{\Omega}_1(t) \cap \overline{\Omega}_2(t).$$

Often, we use the short hand notation $\Gamma := \Gamma(t)$. Furthermore, we assume that both the outer boundary ∂B and the interface Γ are sufficiently smooth such that normal vectors n are well-defined.

Remark 15. We refer the reader back to Section 1.3 for qualitative descriptions and illustrations of three different types of interfaces.

Definition 31 (Interface coupling conditions; scalar-valued case). Let $u_1 : \Omega_1 \to \mathbb{R}$ and $u_2 : \Omega_2 \to \mathbb{R}$. For second-order-in-space equations (for instance, two Poisson problems), we need to impose two interface conditions on Γ:

$$\begin{aligned} u_1 &= u_2, \\ \sigma_1(u_1)n &= \sigma_2(u_2)n, \end{aligned}$$

with the normal vector n, e. g., $n := n_1$ and the relation $n_1 = -n_2$. The first condition is a kinematic condition and the second is the continuity of normal stresses, also known as balance of contact forces, the so-called dynamic (lateral) condition. The first condition is a Dirichlet-like condition, which needs to be imposed in the function space. The second one is a Neumann-type condition and appears through integration by parts in the PDE.

Example 5. The dynamic coupling condition reads for two Poisson operators $\nabla \cdot (\alpha \nabla u)$ and $\nabla \cdot (\beta \nabla u)$:

$$\alpha \nabla u_1 n = \beta \nabla u_2 n.$$

The previous definition extends naturally to vector-valued solution variables.

Definition 32 (Interface coupling conditions; vector-valued case). Let $u_1 : \Omega_1 \to \mathbb{R}^d$ and $u_2 : \Omega_2 \to \mathbb{R}^d$ and consider as before two second-order-in-space equations. Then we need to impose two interface conditions on Γ:

$$u_1 = u_2,$$
$$\sigma_1(u_1)n = \sigma_2(u_2)n.$$

Here, $\sigma_1(u_1), \sigma_2(u_2) \in \mathbb{R}^{d\times d}$.

Example 6. Let $d = 3$. Then, the componentwise notation is given by

$$u_1 = (u_{1,(1)}, u_{1,(2)}, u_{1,(3)})^T \in R^3, \quad u_2 = (u_{2,(1)}, u_{2,(2)}, u_{2,(3)})^T \in \mathbb{R}^3$$

and

$$\begin{pmatrix} \sigma_{1,(11)} & \sigma_{1,(12)} & \sigma_{1,(13)} \\ \sigma_{1,(21)} & \sigma_{1,(22)} & \sigma_{1,(23)} \\ \sigma_{1,(31)} & \sigma_{1,(32)} & \sigma_{1,(33)} \end{pmatrix} \in \mathbb{R}^{3\times 3}, \quad \begin{pmatrix} \sigma_{2,(11)} & \sigma_{2,(12)} & \sigma_{2,(13)} \\ \sigma_{2,(21)} & \sigma_{2,(22)} & \sigma_{2,(23)} \\ \sigma_{2,(31)} & \sigma_{2,(32)} & \sigma_{2,(33)} \end{pmatrix} \in \mathbb{R}^{3\times 3},$$

and

$$n = (n_{(1)}, n_{(2)}, n_{(3)})^T.$$

Then $\sigma_1(u_1)n, \sigma_2(u_2)n \in \mathbb{R}^3$.

3.3.3 Coupling strategies

There exist several ways for realizing numerically the coupling. We describe two of them conceptionally in this section and provide concrete algorithms in Chapter 7. Two principles algorithms exist:

- Iterative coupling, also known as partitioned or staggered approaches: iterate between the given PDEs/CVISs while only solving for the actual solution variable in a specific PDE/CVIS and fixing all others.
- Monolithic approaches: solving equations simultaneously for the given unknowns.

A flow chart is provided in Figure 7.1 at the beginning of Chapter 7.

Remark 16. When more than two PDEs/CVISs are considered, combinations of both coupling strategies can be employed.

In the following, we briefly describe a partitioned scheme and in more detail specific forms of monolithic coupling.

Definition 33 (Variational-monolithic coupling). In variational-monolithic coupling the coupling conditions are realized in an exact fashion in the weak (i. e., variational) formulation.

In the following, we consider: Find $U = (u_1, u_2) \in V_1 \times V_2$ such that

$$A_1(U)(\psi_1) = F(\psi_1) \quad \forall \psi_1 \in V_1, \tag{3.1}$$

$$A_2(U)(\psi_2) = F(\psi_2) \quad \forall \psi_2 \in V_2. \tag{3.2}$$

3.3.3.1 Partitioned schemes for coupled problems

For coupled problems, often partitioned schemes are employed. The idea is to iterate between the different equations per time step.

Algorithm 1. Let initial iteration guesses u_1^0 and u_2^0 be given. For $j = 1, 2, 3, \ldots$ iterate

$$\text{Given } u_2^{j-1}, \text{find } u_1^j: \quad A_1(u_1^j, u_2^{j-1})(\psi_1) = F(\psi_1),$$

$$\text{Given } u_1^j, \text{find } u_2^j: \quad A_2(u_1^j, u_2^j)(\psi_2) = F(\psi_2).$$

Check a stopping criterium:

$$\max(\|u_1^j - u_1^{j-1}\|, \|u_2^j - u_2^{j-1}\|) < \text{TOL}$$

If the stopping criterion is fulfilled, stop. If not, increment $j \to j + 1$.

Remark 17. Alternatively, a residual-based stopping criterion can be employed:

$$\max(\|A_1(u_1^j, u_2^j)(\psi_1) - F(\psi_1)\|, \|A_2(u_1^j, u_2^j)(\psi_2) - F(\psi_2)\|) < \text{TOL}.$$

The previous algorithm extends naturally to time-dependent problems for a sequence $n = 1, 2, 3, \ldots, N$. Here, we set for instance $u_1^{n,0} := u_1^{n-1}$ and $u_2^{n,0} := u_2^{n-1}$ as initial guesses for the iteration procedure at a given time point t_n.

3.3.3.2 Variational-monolithic domain coupling

We now describe a strategy in which the governing system is exact on the variational level and treated all-at-once. To this end, the design of a common function space is the first task:

$$X = V_1 \times V_2.$$

This allows us to define a compact semilinear form with both solution variables.

Formulation 7. Let $U := (u_1, u_2) \in X$. Sum-up both equations (3.1) and (3.2) such that

$$\text{Find } U \in X: \quad A(U)(\Psi) = F(\Psi) \quad \forall \Psi \in X,$$

where $\Psi = (\psi_1, \psi_2) \in X$ and

$$A(U)(\Psi) := A_1((u_1, u_2))(\psi_1) + A_2((u_1, u_2))(\psi_2).$$

Numerically, various options exist depending on the classification of the above problem. In the more rare case that $A(U)(\Psi)$ is linear, direct solvers, conjugate gradients (CG), generalized minimal residuals (GMRES), domain decomposition (DD), or multigrid methods can be employed to name a few. Therein, appropriate preconditioners must be constructed. In the more likely situation that $A(U)(\Psi)$ is nonlinear, either functional/operator-splitting iterations, fixed-point iterations, linearizations in form of time-lagging or extrapolation, or Newton-type methods can be employed, which are topics in Chapter 7. In the latter, however, linear systems must be solved again at each Newton step. Therein, the design of well-performing preconditioners and efficient parallel schemes are often the most challenging parts.

3.3.3.3 Example for variational-monolithic domain coupling

We couple two elliptic PDEs. The first describes some displacements while the second is a heat transfer problem. Find $u : \Omega \to \mathbb{R}$ such that

$$\begin{aligned} -\nabla \cdot (\vartheta \nabla u) &= f \quad \text{in } \Omega, \\ u &= 0 \quad \text{on } \partial\Omega, \end{aligned}$$

and for a given $\alpha > 0$, find $\vartheta : \Omega \to \mathbb{R}$ such that

$$\begin{aligned} -\nabla \cdot (\alpha \nabla \vartheta) &= g(u) \quad \text{in } \Omega, \\ \vartheta &= 0 \quad \text{on } \partial\Omega. \end{aligned}$$

According to Definition 23, in the first PDE we have a nonlinear coupling via the solution variables and in the second PDE we have a coupling via the right-hand side.

The function spaces for the single PDEs read

$$\begin{aligned} V_1 &:= H_0^1(\Omega), \\ V_2 &:= H_0^1(\Omega). \end{aligned}$$

The product space reads:

$$X := V_1 \times V_2.$$

This allows us to define $U := (u, \vartheta) \in X$ and we obtain the following compact semilinear form:

$$A(U)(\Psi) := (\vartheta \nabla u, \nabla \psi_1) + (\alpha \nabla \vartheta, \nabla \psi_2) - (g(u), \psi_2), \tag{3.3}$$

and the right-hand side:

$$F(\Psi) := (f, \psi_1). \tag{3.4}$$

Then we obtain in abstract form

Formulation 8. Find $U \in X$ such that

$$A(U)(\Psi) = F(\Psi) \quad \forall \Psi \in X,$$

with $A(U)(\Psi)$ defined in (3.3) and $F(\Psi)$ given by (3.4).

3.3.3.4 Variational-monolithic interface coupling

We now address interface coupling in more detail. Here, both (or multiple) PDEs/CVISs live in different subdomains. One possibility is variational-monolithic interface coupling. Here, we first define the space

$$X = \{v \in H_0^1(\Omega) |\ v_1 = v|_{\Omega_1}, v_2 = v|_{\Omega_2}, v_1 = v_2 \text{ on } \Gamma\}.$$

Here, the function v is defined over the entire domain Ω. Moreover, the kinematic coupling condition from Definition 32 is built into X. This allows us to define a compact semilinear form with both solution variables. To this end, let $U := (u_1, u_2) \in X$. Summing-up both equations yield

Proposition 5. *With the previous notation, an interface-coupled problem reads in a variational-monolithic formulation:*

$$\textit{Find } U \in X: \quad A(U)(\Psi) = F(\Psi) \quad \forall \Psi \in X,$$

where $\Psi = (\psi_1, \psi_2) \in X$ *and*

$$A(U)(\Psi) := A_1((u_1, u_2))(\psi_1) + A_2((u_1, u_2))(\psi_2).$$

3.3.3.5 Example for variational-monolithic interface coupling

Let $u_1 : \Omega_1 \to \mathbb{R}$ and $u_2 : \Omega_2 \to \mathbb{R}$. Consider two Poisson-type problems:

$$\begin{aligned} -\nabla \cdot (\alpha \nabla u_1) &= f_1 \quad \text{in } \Omega_1 \\ u_1 &= 0 \quad \text{on } \partial\Omega_1 \setminus \Gamma \\ -\nabla \cdot (\beta \nabla u_2) &= f_2 \quad \text{in } \Omega_2 \\ u_2 &= 0 \quad \text{on } \partial\Omega_2 \setminus \Gamma. \end{aligned}$$

We need to impose two interface conditions on Γ (see Definition 32):

$$\begin{aligned} u_1 &= u_2 && \text{(Kinematic condition)}, \\ \alpha \nabla u_1 n &= \beta \nabla u_2 n && \text{(Dynamic condition)}. \end{aligned}$$

Again, let us recall the two-step procedure for deriving weak formulations:

Step 1: Design a function space X that includes the correct boundary conditions and interface data.

Step 2: Multiply the strong form with a test function from X and integrate.

Here, the global function space is defined as

$$X = \{v \in H^1(\Omega) \mid v_1 = v_2 \text{ on } \Gamma,\ v_1 = 0 \text{ on } \partial\Omega_1 \setminus \Gamma,\ v_2 = 0 \text{ on } \partial\Omega_2 \setminus \Gamma\}$$

with $\Omega := \Omega_1 \cup \Omega_2$. Working with X, the function v is defined over the entire domain Ω, with

$$v_1 = v|_{\Omega_1}, \quad v_2 = v|_{\Omega_2}.$$

The kinematic coupling condition is built into X. The Dirichlet outer boundary conditions are build into X as well. Then we obtain

Proposition 6. *Find $U = (u_1, u_2) \in X$ such that*

$$A(U, \Psi) = F(\Psi) \quad \forall \Psi \in X,$$

where $\Psi = (\psi_1, \psi_2) \in X$ and $F(\Psi) := (f_1, \psi_1) + (f_2, \psi_2)$ and

$$A(U, \Psi) := A_1(u_1, \psi_1) + A_2(u_2, \psi_2)$$

with

$$A_1(u_1, \psi_1) = (\alpha \nabla u_1, \nabla \psi_1), \quad A_2(u_2, \psi_2) = (\beta \nabla u_2, \nabla \psi_2).$$

Proof. It holds after partial integration of the strong forms:

$$(\alpha \nabla u_1, \nabla \psi_1) - \int_\Gamma \alpha \partial_{n_1} u_1 \psi_1 \, ds, \quad (\beta \nabla u_2, \nabla \psi_2) - \int_\Gamma \beta \partial_{n_2} u_2 \psi_2 \, ds.$$

We sum-up:

$$\begin{aligned}&(\alpha \nabla u_1, \nabla \psi_1) - \int_\Gamma \alpha \partial_{n_1} u_1 \psi_1 \, ds + (\beta \nabla u_2, \nabla \psi_2) - \int_\Gamma \beta \partial_{n_2} u_2 \psi_2 \, ds \\ &= A_1(u_1, \psi_1) + A_2(u_2, \psi_2) - \int_\Gamma (\alpha \partial_{n_1} u_1 \psi_1 + \beta \partial_{n_2} u_2 \psi_2) \, ds.\end{aligned}$$

We argue that $\Psi = (\psi_1, \psi_2) \in X$. In particular, therein $\psi_1 = \psi_2$ on Γ (kinematic condition). Also we use the fact that $n_1 = -n_2$. Then

$$\int_\Gamma (\alpha \partial_{n_1} u_1 \psi_1 + \beta \partial_{n_2} u_2 \psi_1) \, ds = \int_\Gamma (\alpha \partial_{n_1} u_1 - \beta \partial_{n_1} u_2) \psi_1 \, ds.$$

Our second coupling condition $\alpha \partial_{n_1} u_1 - \beta \partial_{n_1} u_2 = 0$ comes into play now and, therefore, the integral on Γ vanishes

$$\int_\Gamma (\alpha \partial_{n_1} u_1 - \beta \partial_{n_1} u_2) \psi_1 \, ds = 0.$$

Hence, the sum of both semilinear forms does not contain any explicit terms on the interface Γ, and consequently, we have

$$A(U,\Psi) := A_1(u_1,\psi_1) + A_2(u_2,\psi_2). \qquad \square$$

3.4 Domain- and interface-coupling: Biot's equations coupled to elasticity

In this section, we study an example that is on the one hand simple enough to explain the previous concepts. On the other hand, this problem is widely used in porous media applications and is a prototype for problem classes in which an enriched model is coupled to a simpler model. The reason is two-fold: first, boundary conditions of the detailed model are often unknown, and second, addressing large-scale practical field problems.

In the following, a poroelastic medium is modeled with the help of Biot's equations. A well-known benchmark is Mandel's problem [297]. A possible surrounding medium (the non-pay zone) is considered as a static elastic solid. This setting is motivated from [200]. In fact, the Biot system [58, 59, 414] is a homogenized problem, which is identified on the microscale as a fluid-structure interaction problem (details on the interface law are found in Mikelić and Wheeler [320]). Through homogenization, the Biot system is derived on the macroscale level. For recent algorithmic and computational advances in solving the Biot problem, we refer to [319, 45, 237, 68, 105]. The objective of this section is now to formulate a variational monolithically-coupled framework.

3.4.1 Notation and equations

We begin by describing the setting for a pure poroelastic setting, the so-called pay-zone. Let Ω_B be the domain of interest and $\partial\Omega_B$ its boundary with the partition:

$$\partial\Omega_B = \Gamma_u \cup \Gamma_t = \Gamma_p \cup \Gamma_f,$$

where Γ_u denotes the displacement boundary (Dirichlet), Γ_t the total stress or traction boundary (Neumann), Γ_p the pore pressure boundary (Dirichlet), and Γ_f the fluid flux boundary (Neumann). Concretely, we have for a material with the displacement variable u and its Cauchy stress tensor σ:

$$\begin{aligned} u &= \bar{u} \quad \text{on } \Gamma_u, \\ \sigma(u)n &= \bar{t} \quad \text{on } \Gamma_t, \end{aligned}$$

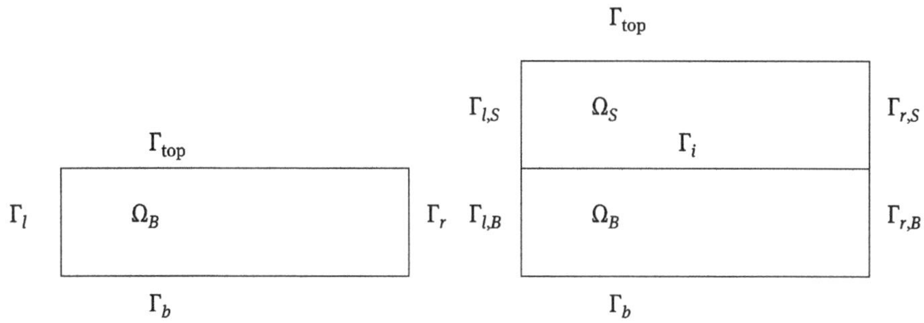

Figure 3.1: Configuration of the classical and augmented Mandel's problem. Biot equations in the pay zone Ω_B and linear elasticity in the nonpay zone Ω_S.

for given $\bar{u}$ and $\bar{t}$, and the normal vector n. For the pressure with the positive definite permeability tensor $K \in L^\infty(\Omega_B, \mathbb{R}^{d\times d})$ and fluid viscosity η_f, we have the kinematic and normal stress balance conditions:

$$\begin{aligned} p &= \bar{p} \quad \text{on } \Gamma_p, \\ -\frac{K}{\eta_f}(\nabla p - \rho_f g) \cdot n &= \bar{q} \quad \text{on } \Gamma_f, \end{aligned}$$

for given $\bar{p}$ and $\bar{q}$; and the density ρ_f and the gravity g. For the initial conditions at time $\tau = 0$, we prescribe

$$\begin{aligned} p(\tau = 0) &= p_0 \quad \text{in } \Omega_B, \\ \sigma(u)(\tau = 0) &= \sigma_0 \quad \text{in } \Omega_B. \end{aligned}$$

The extension of the previous setting with some non-pay zone Ω_S (e. g., rock modeled as an elastic medium) is displayed in Figures 1.4 and 3.1 (right subfigure). In this case (if Ω_B is totally embedded in Ω_S), the boundary conditions on $\partial\Omega_B$ reduce to interface conditions $\partial\Omega_B := \Gamma_i = \overline{\Omega}_B \cap \overline{\Omega}_S$.

3.4.2 Governing equations

The underlying Biot system reads: Find the pressure $p_B : \Omega \times I \to \mathbb{R}$ and displacement $u_B : \Omega \times I \to \mathbb{R}^d$ with the initial data $p_B(0) = p_B^0$ and $\operatorname{div} u_B(0) = \operatorname{div} u_B^0$ such that

$$\begin{aligned} \partial_t(c_B p_B + \alpha_B \operatorname{div} u_B) - \frac{1}{\eta_f}\operatorname{div} K(\nabla p_B - \rho_f g) &= q \quad \text{in } \Omega_B \times I, \\ -\operatorname{div}(\sigma_B(u)) + \alpha_B \nabla p_B &= f_B \quad \text{in } \Omega_B \times I, \end{aligned}$$

with

$$\sigma_B(u_B) := \mu_B(\nabla u_B + \nabla u_B^T) + \lambda_B \operatorname{div} u_B I,$$

and the compressibility coefficient $c_B \geq 0$, the Biot–Willis constant $\alpha_B \in [0,1]$, (this constant relates to the amount of coupling between the flow part and the elastic part) and the permeability tensor K, fluid viscosity and its density η_f and ρ_f, gravity g and a volume source term q. In the second equation, the Lamé coefficients are denoted by $\lambda_B > -\frac{2}{3}\mu_B$ and $\mu_B > 0$ and f_{B} is a volume force.

The velocity v_{B} in the porous medium is obtained with the help of Darcy's law [123] and the Darcy equations which are obtained through homogenization of Stokes's equations [54]. It holds:

$$v_{\mathrm{B}} = -\frac{1}{\eta_f} K(\nabla p_{\mathrm{B}} - \rho_f g).$$

Usually the nonpay zone is described in terms of linear elasticity: Find a displacement $u_S : \Omega_S \times I \to \mathbb{R}^d$ such that

$$-\mathrm{div}(\sigma_S(u_S)) = f_S \quad \text{in } \Omega_S \times I,$$

with[2]

$$\sigma_S(u_S) := \mu_S(\nabla u_S + \nabla u_S^T) + \lambda_S \mathrm{div}\, u_S I,$$

with the Lamé coefficients μ_S and λ_S and a volume force f_S. Examples of boundary conditions are provided below.

3.4.2.1 Interface conditions for the augmented Mandel problem

It remains to describe the interface conditions on Γ_i between the two subsystems. We have two subdomains and in one subdomain two PDEs (elasticity and pressure) and in the other subdomain a pressure equation only. Therefore, we arrive at three coupling conditions on Γ_i: two for the displacements u and one for the pressure p:

$$\begin{aligned} u_{\mathrm{B}} &= u_S, \\ \sigma_{\mathrm{B}}(u_{\mathrm{B}})n_{\mathrm{B}} - \sigma_S(u_S)n_S &= \alpha p_{\mathrm{B}} n_B, \\ -\frac{1}{\eta_f} K(\nabla p_{\mathrm{B}} - \rho_f g) \cdot n_S &= 0. \end{aligned} \tag{3.5}$$

It is important to notice that the second condition in (3.5) requires a careful application of integration by parts and a corresponding algorithmic realization for which we refer to subsection 3.4.5 below.

2 We remind the reader that I stands first for the time interval and afterward for the identity matrix.

3.4.3 Examples of typical setups

Typical configurations and parameters can be taken from [186, 211]. For instance, as domain (0 m, 100 m) × (0 m, 20 m). For the classical Mandel problem, the boundary conditions can be chosen as

$$
\begin{aligned}
\sigma_B n = \bar{t} - \alpha_B p_B n \quad &\text{on } \Gamma_{\text{top}},\\
p = 0 \quad &\text{on } \Gamma_r,\\
u_y = 0 \quad &\text{on } \Gamma_b,\\
u_x = 0 \quad &\text{on } \Gamma_l.
\end{aligned}
$$

As parameters, the following choices are found in the literature: $M_B = 2.5 * 10^{12}$ Pa, $c_B = 1/M_B\,\text{Pa}^{-1}$, $\alpha_B = 1.0$, $\nu_F = 1.0 * 10^{-3}\,\text{m}^2/\text{s}$, $K_B = 100\,\text{md} = 10^{-13}\,\text{m}^2$, $\rho_F = 1.0\,\text{kg/m}^3$, $\bar{t} = F = 1.0 * 10^7\,\text{Pa}\cdot\text{m}$. As elasticity parameters in Biot's model, we use $\mu_S = 10^8$ Pa, $\nu_S = 0.2$. The time step is chosen as $k = 1000$ s. The final time is $5 * 10^6$ s (corresponds to computing 5000 time steps).

For the augmented Mandel's problem, the domain of choice is (0 m, 100 m) × (0 m, 40 m).

$$
\begin{aligned}
\sigma_B n = \bar{t} \quad &\text{on } \Gamma_{\text{top}},\\
\sigma_B n = 0 \quad &\text{on } \Gamma_{r,S},\\
p = 0 \quad &\text{on } \Gamma_{r,B},\\
u_y = 0 \quad &\text{on } \Gamma_b,\\
u_x = 0 \quad &\text{on } \Gamma_{l,B} \cup \Gamma_{l,S}.
\end{aligned}
$$

In addition to the previous parameters, we may use in the pure elastic zone the same Lamé coefficients, such that $\mu_B = \mu_S$ and $\lambda_B = \lambda_S$.

3.4.4 Variational-monolithic formulation

The coupled system of equations is formulated in terms of a variational monolithically-coupled system:

- Domain-coupling: solution variables enter as linear terms into the other problem. Here, coupling both equations yields a linear overall problem.
- Interface-coupling on Γ_i between the poroelastic model and the linearized elastic problem.

We first define the function spaces using the boundary conditions from Section 3.4.3:

$$
V_P := \{p \in H^1(\Omega_B) \mid p = 0 \text{ on } \Gamma_r\},
$$

$$V_S := \{u \in H^1(\Omega_B) \mid u_y = 0 \text{ on } \Gamma_b,\ u_x = 0 \text{ on } \Gamma_l\}.$$

The definition of a weak formulation in a variational-monolithic setting is

Formulation 9 (Biot Problem in Ω_B). Find $(p,u) \in V_P \times V_S$, with $p(0) = p^0$, such that for almost all times $t \in I$, it holds

$$\begin{aligned} c_B(\partial_t p, \phi^p) + \alpha_B(\nabla \cdot u, \phi^p) + \frac{K}{\nu_F}(\nabla p, \nabla \phi^p) \\ - \rho_F(g, \phi^p) - (q, \phi^p) = 0 \quad \forall \phi^p \in V_P, \\ (\sigma_B, \nabla \phi^u) - \alpha_B(pI, \nabla \phi^u) - \int_{\Gamma_{\text{top}}} (\bar{t}_S - \alpha p n)\phi^u \, ds - (f_B, \phi^u) = 0 \quad \forall \phi^u \in V_S. \end{aligned}$$

By augmenting the domain with an outer elastic part we must account for an additional elastic equation. To this end, we redefine

$$V_S := \{u \in H^1(\Omega_B \cup \Omega_S) \mid u_y = 0 \text{ on } \Gamma_b,\ u_x = 0 \text{ on } \Gamma_{l,B} \cup \Gamma_{l,S}\},$$

in which we now implicitly have built-in the kinematic condition $u_B = u_S$ on Γ_i. The dynamic coupling condition from (3.5) is implicitly contained in the coupled PDE system.

Formulation 10 (Biot–Lamé–Navier problem in $\Omega_B \cup \Omega_S$). Find $(p,u) \in V_P \times V_S$, with the initial data $p(0) = p^0$, such that for almost all times $t \in I$, it holds

$$\begin{aligned} c_B(\partial_t p, \phi^p) + \alpha_B(\nabla \cdot u, \phi^p) + \frac{K}{\nu_F}(\nabla p, \nabla \phi^p) \\ - \rho_F(g, \phi^p) - (q, \phi^p) = 0 \quad \forall \phi^p \in V_P, \\ (\sigma_B, \nabla \phi^u) - \alpha_B(pI, \nabla \phi^u) - (f_B, \phi^u) = 0 \quad \forall \phi^u \in V_S(\Omega_B), \\ (\sigma_S, \nabla \phi^u) - \int_{\Gamma_{\text{top}}} \bar{t}_S \phi^u \, ds - (f_S, \phi^u) = 0 \quad \forall \phi^u \in V_S(\Omega_S). \end{aligned}$$

3.4.5 Discretization, implementation, results

We notice that the previous formulation is implemented and available via [207, 413] www.dopelib.net in `dopelib/Examples/PDE/InstatPDE/Example6`. The reader is also referred to Chapter 13 for some further details on software developments. Graphical results of the solutions at the time index $N = 100$ with their range of values are shown in Figure 3.2. Moreover, observing the pressure over time yields a nonmonotone behavior, well known as so-called Mandel–Cryer effect [297]. Our results are published in [207, Section 4].

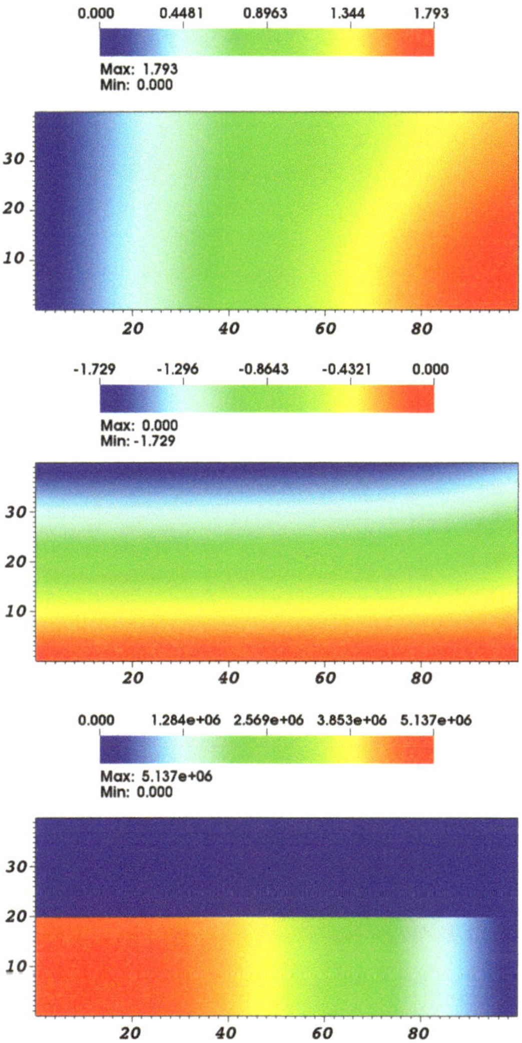

Figure 3.2: Graphical output and range of values of augmented Mandel's problem: u_x and u_y of the vector-valued displacements. At the bottom, the pressure distribution. The displacement fields are continuous in both subdomains. The pressure is discontinuous at the interface Γ_i.

3.5 Classification of the prototype quasi-static model problem

After defining and illustrating classifications of PDEs and CVISs, we finally address our previous simplified model problem from Formulation 1 (Section 2.3) in detail. To this end, we work through Definition 17 step-by-step:

1. We have $d + 1$ independent variables.

2. It is a quasi-static problem with no time-derivatives in the displacement PDE and the phase-field variational inequality. However, the solution variables might be loading-dependent. In the thermodynamic context, a quasi-static problem is at each instance in a thermo-dynamic equilibrium.
3. The main problems are of elliptic type and satisfy energy conservation. The inequality constraint is an irreversibility condition, i. e., an entropy condition.
4. In space, both problems are of second order.
5. The entire system is a CVIS with $d + 1$ equations: d equations for the solid and 1 equation for the phase-field variable.
6. Nonlinearities:
   (a) Through the coupling, the problem is nonlinear. More specifically, the first equation is quasi-linear (see Definition 20) because φ enters as coefficient into the higher-order operator in the displacement equation. The second equation is semilinear because the highest-order operator in φ is linear and the nonlinearity enters as quadratic coefficient into a low-order term.
   (b) The constitutive laws are linear in the proposed model.
   (c) The function set K is not a vector space because we deal with a VI.
7. The coupling is of domain-type via coefficients and solution variables. In a partitioned approach assuming that one variable is given, the first PDE becomes linear. However, the overall setting remains nonlinear because of the inequality constraint in the phase-field part.

3.6 Classification of the prototype dynamic model problem

In this section, we classify the previous simplified model problem in Formulation 2 (Section 2.5) in detail. To this end, we again work through Definition 17 step-by-step:
1. We have $d + 1$ independent variables.
2. Both problems are now time-dependent.
3. The first PDE is a wave equation of hyperbolic type and satisfies the law of energy conservation including kinetic energy. The VI is of parabolic type. The inequality constraint is an irreversibility condition, i. e., an entropy condition.
4. The first PDE is second order in time and second order in space. The second problem is first order in time and second order in space.
5. The entire setting is a CVIS with $d + 1$ equations: d equations for the solid and 1 equation for the phase-field variable.
6. Nonlinearities:
   - Through the coupling, the problem is nonlinear. More specifically it is quasi-linear because φ enters as coefficient into the higher-order operator in the displacement equation.
   - The constitutive laws are linear in the proposed model.
   - The function set K is not a vector space because we deal with a VI.

7. The coupling is of domain-type via coefficients and solution variables. In a partitioned approach assuming that one variable is given, the displacement PDE becomes linear. However, the overall setting remains nonlinear because of the inequality constraint.

4 Modeling of variational phase-field fractures

In this chapter, we concentrate on the modeling of quasi-static variational phase-field fractures in brittle materials. The original work, forming the basis, can be found in [174, 72, 73] and [317, 313]. We aim to provide a compact summary with all tools that are important to us. Many original references are listed in which all details can be found.

4.1 Ingredients from continuum mechanics and some constitutive laws

In this section, we first introduce basic ingredients from continuum mechanics.

Definition 34 (Displacements). Let $B \subset \mathbb{R}^d$ and $I \subset \mathbb{R}$. The displacement variable is denoted by $u := u(x,t) : B \times I \to \mathbb{R}^d$. In detail,

$$u := u(x,t) = (u_1(x,t), \dots, u_d(x,t)) : B \times I \to \mathbb{R}^d.$$

A precise definition in terms of continuum mechanics is provided in Definition 78.

Remark 18. Usually in the phase-field literature, infinitesmal small displacements are considered such that Lagrangian and Eulerian domains are identified (Proposition 108 in Section 11.6.1). For these reasons, the concepts of transformations between coordinate systems are not required. An exception is the work [441, 443] in which a damaged solid inside a fluid-structure framework was considered, where Lagrangian and Eulerian coordinate systems were distinguished (Section 11.7 and Section 11.8).

Definition 35. The linearized strain tensor is denoted by

$$e = e(u) = \frac{1}{2}(\nabla u + \nabla u^T).$$

Example 7 (Gradient of displacements in $\mathbb{R}^3$). In $\mathbb{R}^3$ the gradient of displacements is a $\mathbb{R}^{3\times 3}$ matrix:

$$\nabla u = \begin{pmatrix} \partial_x u_x & \partial_y u_x & \partial_z u_x \\ \partial_x u_y & \partial_y u_y & \partial_z u_y \\ \partial_x u_z & \partial_y u_z & \partial_z u_z \end{pmatrix}.$$

Definition 36 (Linear-stress relationship). A well-known linear stress-strain relationship is given by, e. g., [114, 364]:

$$\sigma := \sigma(u) = Ce(u).$$

The tensor $C = (C_{ijkl})_{i,j,k,l=1}^3$ is a fourth-order material tensor that is positive definite, i. e., there exists a positive constant $\alpha > 0$ such that for any symmetric matrix ξ with

https://doi.org/10.1515/9783110497397-004

the entries ξ_{ij} and any point $x \in B$,

$$\alpha|\xi|^2 \le C(x)\xi : \xi = \sum_{i,j,k,l=1}^{d} C_{ijkl}\xi_{ij}\xi_{kl}.$$

This reduces under the assumption of isotropy, homogeneity, and symmetry (see, e. g., [364, p. 51]) to the following well-known law:

$$\sigma = Ce(u) = 2\mu e(u) + \lambda \operatorname{tr}(e(u))I, \tag{4.1}$$

where I is the identity tensor and $\operatorname{tr}(\cdot)$ the trace operator.

Definition 37. We denote by $W(e(u))$ the elastic (strain) energy density.

Example 8. Two examples of energy density functions are provided.
1. In the case of the Laplacian or Poisson equation, the energy density function is
$$W(e(u)) = \frac{\mu}{2}|\nabla u|^2,$$
where $\mu > 0$ is a material parameter. The energy is defined as
$$E_b(u) = \frac{1}{2}\int_B \mu|\nabla u|^2\,dx - \int_B fu\,dx. \tag{4.2}$$
The PDEs which we have to solve, are the Euler–Lagrange equations:
$$E_b'(u)(\psi) = \int_B \mu\nabla u \cdot \nabla\psi\,dx - \int_B f\psi\,dx. \tag{4.3}$$
2. For a general elastic model, we have
$$W(e(u)) = Ce(u) : e(u),$$
which is the energy density using the Frobenius scalar product. Further details on $W(e(u))$ can be found in the literature, e. g., [236].

Remark 19. Often, and in this monograph as well, C reduces to a few independent components as in (4.1). As an example that sometimes several components of C are of interest, we point to our own work [424] in which we computed 36 components (out of 81).

Definition 38. Solid material properties are often defined in terms of Young's modulus E_Y and the Poisson ratio ν_s. The relationship to the Lamé parameters μ_s and λ_s is as follows:

$$\nu_s = \frac{\lambda_s}{2(\lambda_s + \mu_s)}, \quad E_Y = \frac{\mu_s(\lambda_s + 2\mu_s)}{\lambda_s + \mu_s},$$
$$\mu_s = \frac{E_Y}{2(1 + \nu_s)}, \quad \lambda_s = \frac{\nu_s E_Y}{(1 + \nu_s)(1 - 2\nu_s)}.$$

Remark 20 (Signs of the solid material parameters). It can be shown that

$$E_Y, \mu_s, \lambda_s > 0 \quad \text{and} \quad -1 \leq \nu_s \leq 0.5.$$

Further calculations (e. g., [361]), yield the refined choice:

$$-1 \leq \nu_s \leq 0.5, \quad E_Y, \mu_s > 0, \quad \lambda_s > -\frac{2}{3}\mu_s. \tag{4.4}$$

As background, we notice that Poisson's ratio describes the ratio that a compressed material expands into the two directions orthogonal to the compression direction. For $\nu_s \to 0.5$, an elastic material becomes incompressible. For most materials, it holds $0.5 \geq \nu_s > 0$. Some polymers have negative ν_s, which means that they become thicker when they are stretched. Moreover, the bulk modulus is often used, which is defined by $k = \lambda_s + \frac{2}{3}\mu_s$.

Definition 39 (Degradation function). For a given small parameter $\kappa > 0$, let us define the degradation function in terms of the phase-field function $\varphi : B \times I \to [0,1]$ as

$$g(\varphi) := (1-\kappa)\varphi^2 + \kappa.$$

4.2 Modeling brittle fracture in a nutshell

We summarize the necessary notation to understand the following sections. A fundamental quantity is the critical surface energy density G_c (e. g., [461]), which can be related to the so-called fracture toughness.

Definition 40 (Critical energy release rate). The critical energy release rate (related to the fracture toughness) G_c is a fundamental quantity that describes the resistance of a material to break. In other words, G_c is the energy required to create an infinitesimal crack at a point $x \in \bar{B}$, where B is the domain occupied by the material under consideration [461, Chapter 4].

Definition 41 (Current energy release rate). The current energy release rate is denoted by G and defined as

$$G := G(x) = -\frac{\partial(U-V)}{\partial A}, \quad x \in B, \tag{4.5}$$

where U is the potential energy available in the specimen $B \subset \mathbb{R}^d$ to create new fractures, V is the work associated to external forces and A is the fractured area. The current energy release rate has the unit $[G] = \mathrm{J/m^2}$.

The propagation of a fracture can be divided into three cases, e. g., [461, 73]:
- $G < G_c$: current surface energy density smaller than critical;
- $G = G_c$: propagation, (stable) growth;
- $G > G_c$: unstable growth.

4.3 Griffith's model

According to Griffith [208], a fracture propagates if the rate of elastic energy G decrease per unit surface area of the incremental step is equal to the quasi-static critical energy release rate G_c. The surface energy is a macroscopic description of the lattice debonding observed on the microscopic level.

Prescription of crack growth is based on a competition between bulk (elastic) and crack energy as modeled for brittle fracture:

Proposition 7. *Griffith's theory postulates:*
- *Surface energy is associated with surfaces where the deformation is discontinuous;*
- *A propagation criterion for these surfaces;*
- *An irreversibility condition for the fracture process.*

Remark 21. Only recently, strong minimizers of Griffith's model was rigorously established in [117] for a 2D configuration.

In Griffith's postulates, the fracture process is quasi-static and the laws hold true for brittle fractures, e. g., glass, concrete (in contrast to ductile fracture, e. g., steel), with a preexisting and well-defined crack path.

Definition 42 (Crack). A crack or fracture is defined as follows:

$$\mathcal{C}(l) = \mathcal{C}_0 \cup \{x(s); 0 \leq s \leq l\}.$$

An illustration is provided in Figure 4.1. The length of a sufficiently smooth curve can be computed via the line integral $L_0 = \int_{x_0}^{x_l} ds$.

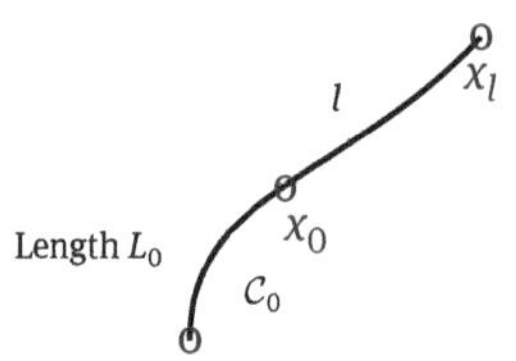

Figure 4.1: Old $\mathcal{C}_0$ and new part $\{x(s)\}$ of a crack resulting in $\mathcal{C}(l)$. The new crack contains the old crack.

Definition 43 (Bulk and surface energies). We define the potential and surface energies, respectively:

$$\text{Potential energy: } P(t,l) := E_b(\mathcal{C}(l), u_0) \qquad \rightarrow \text{monotonically decreasing}$$
$$\text{Surface energy: } Q(t,l) := E_s(\mathcal{C}(l)) - E_s(\mathcal{C}_0) \qquad \rightarrow \text{monotonically increasing}$$

Here, u_0 is a given initial displacement field. We shall investigate the trajectory of a crack along its path:

$$t \mapsto l(t),$$

where t is the time or loading step. The (current) energy release rate is defined as

$$G := -\frac{\partial P}{\partial l}.$$

Proposition 8 (Griffith's laws). *Let $l(t)$ be absolutely continuous in t, then Griffith's law of crack evolution is satisfied:*
(a) $l(0) = 0$ *(initial crack);*
(b) $\partial_t l(t) \geq 0$ *(crack can only grow; irreversibility);*
(c) $G \leq G_c$ *(energy release rate is bounded by the critical energy release rate);*
(d) $(-G + G_c)\partial_t l(t) = 0$ *(complementarity condition; crack can only grow when energy release rate is critical).*

The proof can be found in Proposition 4.8 in [174].

4.4 Francfort and Marigo's variational model for brittle fracture

In the year 1998, Francfort and Marigo [174] generalized Griffith's idea within a variational framework and for which we briefly recapitulate the main ideas.

4.4.1 Surface and bulk energies

In the following, we introduce the (crack) surface energy E_s, which connects G_c with the energy of current fracture.

Definition 44 (Surface energy). The surface energy of the crack $\mathcal{C} \subset \bar{B}$ is defined as

$$E_s(\mathcal{C}) = \int_{\mathcal{C}} G_c(s)\, ds = G_c \mathcal{H}^{d-1}(\mathcal{C}),$$

where $\mathcal{H}$ is the Hausdorff measure [217]. For smooth surfaces, the Hausdorff measure corresponds to the length of a crack $\mathcal{C}$ (1D) in a 2D setting and the crack area (2D) in a 3D setting.

Remark 22. The interpretation of the previous definition is as follows. The higher $E_s(\mathcal{C})$, the more material is fractured. Moreover, the larger $G_c > 0$, the more energy is required to create new fractures. For $G_c = \infty$, the material cannot break.

Next, we define a second energy, the bulk energy. The interplay of surface and bulk energies are key for a variational formulation of fracture.

Definition 45 (Bulk energy). The bulk energy is defined as

$$E_b(\mathcal{C}, u) = \int_\Omega W(e(u))\, dx - \int_\Omega fu\, dx.$$

In fracture settings, volume forces are often of minor importance and, therefore, $f \equiv 0$. Then

$$E_b(\mathcal{C}, u) = \int_\Omega W(e(u))\,dx.$$

In the following, elementary properties of the energy integrals are listed.

Proposition 9. *The bulk and surface energies shall satisfy the following properties:*
(a) *E_s is strictly monotonically increasing in $\mathcal{C}$;*
(b) *E_b is monotonically decreasing in $\mathcal{C}$ for any fixed $u \in V_u^D$.*

Proposition 10 (Total energy in B). *Let $\mathcal{C}$ be a given crack and u_D a given loading on $\partial\Omega_D$. Then the total energy $E_T(\mathcal{C}, u)$ is defined as*

$$E_T(\mathcal{C}, u) = E_b(\mathcal{C}, u) + E_s(\mathcal{C}).$$

Formally, we obtain $u \in V_u^D$ by solving

$$\min_{u \in V_u^D} E_T(\mathcal{C}, u). \tag{4.6}$$

The functional $E_T(\mathcal{C}, u)$ is nonlinear. Discussions, technical details and further information can be found in [73].

4.4.2 Evolution laws

Proposition 11 (Law 2.9 from Francfort and Marigo [174]). *Let $u(x,t)$ be a monotonically increasing load on $\partial\Omega_D$ with $u(x,t) = t\bar{u}(x)$, given $\bar{u}$, for $t \geq 0$. Then $\mathcal{C}(t)$ for $t \in I$ should satisfy:*
1. *Irreversibility (entropy inequality principle): $\mathcal{C}(t)$ is an increasing function in t, where $\mathcal{C}(t) = \mathcal{C}_0$ is the initial fracture for $t = 0$.*
2. *Energy minimization: For $t \in I$ and $\tau \in I$ it holds*

$$E_T(\mathcal{C}(t), u(t)) \leq E_T(\mathcal{C}, u(t)) \quad \forall \bigcup_{\tau<t} \mathcal{C}(\tau) \subset \mathcal{C}.$$

3. *Constraint on the set of the solution:*

$$E_T(\mathcal{C}(t), u(t)) \leq E_T(\mathcal{C}(\tau), u(t)) \quad \forall \tau < t.$$

This last law avoids admissibility of too many fracture solutions. For all details, we refer the reader to [174].

With these laws, irreversible quasi-static fracture propagation with minimal energy can be defined as a mapping $t \mapsto (u(t), \mathcal{C}(t))$ satisfying three conditions:

1. global stability: for every time $t \in I$, the solution $(u(t), \mathcal{C}(t))$ yields minimal energy at time t;
2. irreversibility: $\mathcal{C}(\tau) \subset \mathcal{C}(t)$ for $0 \le \tau < t \le T$;
3. energy balance: increment in stored energy plus crack energy increase equals the work performed by external forces (first law of thermodynamics).

Remark 23. For $t \to \infty$ with a monotonically increasing load, a mechanical failure (thus total cracking) of the specimen arises. Thermodynamic interpretations of the previous laws $t \to \infty$ are discussed in detail in [313].

Remark 24 (Entropy). Entropy characterizes the direction of energy transfer. In general, mechanical energy can be dissipated into other forms of energy. For instance, in fracture processes as in this monograph, we consider dissipation of mechanical energy into crack energy. The contrary is in general not true: once cracks (at least these cracks that we have in mind) are created, they will not heal. Physically speaking, when entropy holds in the strict sense, crack energy cannot be transformed back into mechanical energy. These considerations can be illustrated with a daily life example: take your scissors and cut a sheet of paper as shown in Figure 1.1 (Section 1.2). The cut in the paper will always remain and the paper (if not cheating with adhesive tape) cannot be repaired.

4.4.3 Specific forms of the total energy and Euler-Lagrange systems

Employing Proposition 10, we provide two specific examples:

1. Given the displacement $u : \Omega \to \mathbb{R}$ as a scalar valued function, we deal with

$$E_T(\mathcal{C}, u) = \frac{1}{2} \int_{\Omega} |\nabla u|^2 \, dx + \int_{\mathcal{C}} G_c(x) \, ds.$$

2. Given $u : \Omega \to \mathbb{R}^d$ as the vector-valued elasticity, the energy functional is given as

$$E_T(\mathcal{C}, u) = \frac{1}{2} \int_{\Omega} \sigma(u) : e(u) \, dx + \int_{\mathcal{C}} G_c(x) \, ds.$$

In the following, we discuss some properties, their consequences and treatments.

4.4.3.1 Methods for evaluating the crack surface integral

In principle, we could now compute fracture problems that are covered by the previous physical laws. However, E_b is defined on $\Omega \subset \mathbb{R}^d$ and E_s is defined in $\mathcal{C} \subset \mathbb{R}^{d-1}$. This causes numerical and mathematical problems in terms of both discretization and well-posedness, because the fracture $\mathcal{C}$ is an interface itself (see Section 1.2). In fact, extended/generalized finite elements [328, 184, 310, 32] or boundary element methods [110, 347, 382, 304] represent the interface as a sharp line or area. In phase-field, the main idea is to approximate E_S via domain integrals that live on the entire domain B.

4.4.3.2 Treatment of the surface energy: interface-capturing with elliptic functionals

Our problem belongs to the class of free discontinuity problems. Here, we seek an unknown pair: (u, Γ), where Γ is varying in a class of closed subsets of a fixed open set $B \subset \mathbb{R}^d$. Then we seek $u : B\backslash\Gamma \to \mathbb{R}^d$ by minimizing $E_T(\mathcal{C}, u)$. That the limit case for $\varepsilon = 0$ is well-posed requires the space of special functions of bounded variations, going back to [198] based mainly on ideas of di Giorgi.

The idea of using elliptic functionals for approximation goes back to Modica and Mortola [327] who approximated the perimeter functional with such a procedure. Ambrosio and Tortorelli [20, 21] introduced an approximation procedure with an auxiliary variable φ. Beside fracture, other applications of this approximation procedure is image segmentation in which contours from input images shall be reconstructed [330, 107, 109, 81] or in materials sciences [356]. A concise overview is provided in [19, Introduction, Part I, p. 3]. Further chapters in that book provide more information, for instance from Alberti (in particular, pp. 99–101) and Braides, pp. 171–180.

Proposition 12 (Elliptic approximation functional). *To approximate the surface energy on the interface $\mathcal{C}$, in* [20, 21], *the following elliptic functional is proposed:*

$$\frac{1}{2}\int_B G_C\Big(\frac{1}{\varepsilon}(1-\varphi)^2 + \varepsilon|\nabla\varphi|^2\Big)\,dx \quad \to \quad \int_{\mathcal{C}} G_c\,ds \quad \text{for } \varepsilon \to 0.$$

Remark 25 (AT1 vs. AT2). The term $(1-\varphi)^\alpha$ with $\alpha = 2$ is also known as part of a so-called AT2 model.[1] In some works, the AT1 model with $\alpha = 1$ is of interest, e. g., [76, 161].

Let us give some qualitative explanations of the choice of

$$\frac{1}{2}\int_B \Big(\frac{1}{\varepsilon}(1-\varphi)^2 + \varepsilon|\nabla\varphi|^2\Big)\,dx.$$

The first term is actually the term we are interested in, namely determining values that are 0 (crack) or 1 (unbroken solid). In fact, for $\varepsilon \to 0$ we enforce the behavior of sharp changes between 0 and 1 since the first term dominates. For regularity purposes, we add the gradient penalty term smoothing the transition from 0 to 1. The larger ε, the smoother the transition zone. Proofs are provided later in Section 5.5.2 and also in [336, 173].

We now apply the previous proposition to $E_T(u, C)$:

1 AT = Ambrosio–Tortorelli.

Proposition 13 (Regularized phase-field for Laplacian). *The nonregularized energy functional reads*

$$E_T(u, C) = \frac{1}{2}\int_\Omega |\nabla u|^2\, dx + G_c H^{d-1}(C).$$

Using AmbrosioTortorelli elliptic functionals, we replace the sharp lower-dimensional crack C by a smoothed indicator function φ and we obtain with the help of Definition 39 the following regularized version:

$$E_{T,\varepsilon}(u, \varphi) = \frac{1}{2}\int_B g(\varphi)|\nabla u|^2\, dx + \frac{1}{2}\int_B G_c\Big(\frac{1}{\varepsilon}(1-\varphi)^2 + \varepsilon|\nabla\varphi|^2\Big)\, dx$$

with $u \in H^1(B)$, $u\varphi \in H^1(B)$, and $0 \le \varphi \le 1$.

Remark 26 (Regularization parameter κ). While using the elliptic surface approximation, we also modified the bulk energy term by inserting the degradation function $g(\varphi)$ with the regularization parameter κ in order to extend the bulk energy from Ω to B. This parameter is numerically useful when $\varphi \to 0$ such that the discrete system matrix remains regular:

$$[(1-\kappa)\varphi^2 + \kappa]|\nabla u|^2 \quad \to \quad \kappa|\nabla u|^2 \quad \text{for } \varphi = 0.$$

Mathematically, $\varphi^2|\nabla u|^2$ (thus $\kappa = 0$) works as well; see Braides 1998 [80] or [21]. In the latter reference on page 107, there is however a discussion what might happen in the theory when $\kappa = 0$; see further in [22]. For dynamic fracture, Borden et al. [67] also worked with $\kappa = 0$ because here, the acceleration term $\partial_t^2 u$ is not weighted with $g(\varphi)$ and thus provides nonzero matrix entries in the discrete system.

Remark 27 (κ and ε). More discussions on the phase-field regularization parameters shall follow in Section 5.5.

Proposition 14. *Using Γ-convergence (e.g., again [80]) for $\varepsilon \to 0$:*

$$E_{T,\varepsilon}(u, \varphi) = \begin{cases} E_T(u, C) & \text{if } \varphi = 1 \text{ a.e. in } B, \\ +\infty, & \text{otherwise.} \end{cases}$$

Proof. See [72, Section 2.1]. □

In a similar way, we obtain for linearized elasticity.

Proposition 15 (Regularized phase-field for linearized elasticity). *The total energy functional reads*

$$E_{T,\varepsilon}(u, \varphi) = \frac{1}{2}\int_B g(\varphi)\sigma(u) : e(u)\, dx + \frac{1}{2}\int_B G_c\Big(\frac{1}{\varepsilon}(1-\varphi)^2 + \varepsilon|\nabla\varphi|^2\Big)\, dx$$

with $u \in H^1(B)^d$, $\varphi u \in H^1(B)^d$, and $0 \le \varphi \le 1$.

4.4.3.3 First variations: Euler–Lagrange equations without inequality constraint

For solving $\min E_{T,\varepsilon}(u,\varphi)$, we derive the first variation, resulting in the Euler–Lagrange system. Of course, we obtain stationary points rather than local minima. Starting with energy minimization

$$\min_{u\in V_u^D,\varphi\in K} E_{T,\varepsilon}(u,\varphi),$$

we obtain from differentiation the first variation:

$$E'_{T,\varepsilon}(u,\varphi)(w,\psi-\varphi)\geq 0.$$

We deal with a VI in φ because of the irreversibility constraint $0\leq\varphi^n\leq\varphi^{n-1}\leq 1$. Let us assume for the moment that we are interested in

$$E'_{T,\varepsilon}(u,\varphi)(w,\psi)=0.$$

We then obtain

Lemma 1. *The first variations for* $(u,\varphi)\in V_u^D\times H^1(B)$ *are given by*

$$E'_{T,\varepsilon,u}(u,\varphi)(w)=\int_B g(\varphi)\nabla u\cdot\nabla w\,dx,$$

$$E'_{T,\varepsilon,\varphi}(u,\varphi)(\psi)=\int_B (1-\kappa)\varphi|\nabla u|^2\psi\,dx+\int_B G_c\Big(\frac{1}{\varepsilon}(1-\varphi)(-\psi)+\varepsilon\nabla\varphi\cdot\nabla\psi\Big)\,dx.$$

Proof. We apply the differentiation rules from Section 2.2.4. First, we notice that the directional derivative of $f(u):=|\nabla u|^2$ into the direction w is formally obtained with the help of the chain rule (Section 2.2.5) and given by

$$\begin{aligned}f'(u)(w)&=\lim_{s\to 0}\frac{f(u+sw)-f(u)}{s}\\&=[|\nabla u|^2]'(w)=\Big(\sum_{i=1}^d(\partial_i u)^2\Big)'(w)=2\sum_{i=1}^d(\partial_i u)(\partial_i w)\\&=2\nabla u\cdot\nabla w.\end{aligned}$$

For the energy functional itself, we simply evaluate for u,

$$E'_{T,\varepsilon,u}(u,\varphi)(w)=\lim_{s\to 0}\frac{E_{T,\varepsilon}(u+sw,\varphi)-E_{T,\varepsilon}(u,\varphi)}{s}.$$

To this end, we obtain for variations in u, while fixing φ:

$$\begin{aligned}&\lim_{s\to 0}\frac{E_{T,\varepsilon}(u+sw,\varphi)-E_{T,\varepsilon}(u,\varphi)}{s}\\&=\lim_{s\to 0}\Big(\frac{1}{2}\int_B g(\varphi)|\nabla(u+sw)|^2\,dx-\frac{1}{2}\int_B g(\varphi)|\nabla u|^2\,dx\Big)s^{-1}\end{aligned}$$

$$
\begin{aligned}
&= \lim_{s\to 0}\left(\frac{1}{2}\int_B g(\varphi)(|\nabla u|^2 + 2s\nabla u\cdot\nabla w + s^2|\nabla w|^2)\,dx - \frac{1}{2}\int_B g(\varphi)|\nabla u|^2\,dx\right)s^{-1}\\
&= \lim_{s\to 0}\left(\frac{1}{2}\int_B g(\varphi)(2s\nabla u\cdot\nabla w + s^2|\nabla w|^2)\,dx\right)s^{-1}\\
&= \int_B g(\varphi)\nabla u\cdot\nabla w\,dx.
\end{aligned}
$$

In the last step, we utilized

$$\lim_{s\to 0}\frac{1}{2}\int_B g(\varphi)(s|\nabla w|^2)\,dx = 0.$$

For the directional derivative into φ, while fixing u, we notice first

$$g'(\varphi)(\psi) = 2(1-\kappa)\varphi\psi.$$

Thus

$$E'_{T,\varepsilon,\varphi}(u,\varphi)(\psi) = \lim_{s\to 0}\frac{E_{T,\varepsilon}(u,\varphi+s\psi) - E_{T,\varepsilon}(u,\varphi)}{s}.$$

Then we obtain with the same differentiation rules as before

$$
\begin{aligned}
E'_{T,\varepsilon,\varphi}(u,\varphi)(\psi) &= \frac{1}{2}\int_B g'(\varphi)(\psi)|\nabla u|^2\,dx\\
&\quad + \frac{1}{2}\int_B G_c\left(2\frac{1}{\varepsilon}(1-\varphi)(-\psi) + 2\varepsilon\nabla\varphi\cdot\nabla\psi\right)dx\\
&= \int_B (1-\kappa)\varphi\psi|\nabla u|^2\,dx + \int_B G_c\left(-\frac{1}{\varepsilon}(1-\varphi)\psi + \varepsilon\nabla\varphi\cdot\nabla\psi\right)dx.
\end{aligned}
$$ □

Writing the same contents in the usual notation, we obtain the following

Formulation 11. Introducing a semilinear form $A(\cdot)(\cdot)$ in which the first argument may be nonlinear and the second argument is linear, we write the previous statements in the following notation:

$$
\begin{aligned}
A_1(U)(\Psi) &= E'_{T,\varepsilon,u}(u,\varphi)(w) = (g(\varphi)\nabla u, \nabla w)\\
A_2(U)(\Psi) &= E'_{T,\varepsilon,\varphi}(u,\varphi)(\psi) = ((1-\kappa)\varphi|\nabla u|^2, \psi) - \frac{1}{\varepsilon}(G_c(1-\varphi),\psi) + \varepsilon(G_c\nabla\varphi,\nabla\psi).
\end{aligned}
$$

These formulations are similar in their structure as earlier discussed in (2.4) and (2.5).

4.4.3.4 First variations with inequality constraint. Demonstration for the obstacle problem

For demonstration reasons, let us consider the following obstacle problem

Formulation 12. Let $g \in L^2(\Omega)$. The obstacle problem as energy minimization problem reads

$$\min_{u\in V, u\geq g} E(u).$$

The admissible function space is defined as follows:

$$K := \{v \in H^1(\Omega) \mid v = u_D \text{ on } \partial\Omega, v \geq g \text{ a.e. in } \Omega\},$$

such that we can write

$$\min_{u\in K} E(u).$$

Definition 46 (Convex set). We say K is a convex set if

$$u, v \in K: \quad su + (1-s)v \in K,$$

for $0 \leq s \leq 1$.

Definition 47 (Convex functional). A functional $E : K \to \mathbb{R}$ is convex if and only if

$$E(su + (1-s)v) \leq sE(u) + (1-s)E(v)$$

for all $u, v \in K$ and $0 \leq s \leq 1$.

Using these definitions, we obtain as minimization problem the following.

Formulation 13.

$$E(u) = \min_{v\in K} E(v),$$

where $u \in K$ yields the minimum of the energy functional.

We derive the first variation in the following. Let $v, u \in K$ and $sv + (1-s)u = u + s(v-u) \in K$ with $s \in [0,1]$. Then, by using the previously defined convex set properties, it can be inferred that

$$E(u + s(v-u)) \geq E(u).$$

This yields

$$\frac{d}{ds}E(u + s(v-u))|_{s=0} \geq \frac{d}{ds}E(u) \quad \Leftrightarrow \quad \frac{d}{ds}E(u + s(v-u))|_{s=0} \geq 0.$$

In more detail, we have

$$\frac{d}{ds}\int_\Omega \left(\mu\frac{1}{2}|\nabla(u+s(v-u))|^2 - f(v-u)\right) dx|_{s=0} \geq 0$$

$$\Rightarrow \int_\Omega (\mu\nabla(u+s(v-u))\cdot\nabla(v-u) - f(v-u))\, dx|_{s=0} \geq 0$$

$$\Rightarrow \int_\Omega (\mu\nabla u\cdot\nabla(v-u) - f(v-u))\, dx \geq 0$$

$$\Leftrightarrow \int_\Omega \mu\nabla u\cdot\nabla(v-u)\, dx - \int_\Omega f(v-u)\, dx \geq 0$$

$$\Leftrightarrow \int_\Omega \mu\nabla u\cdot\nabla(v-u)\, dx - \int_\Omega f(v-u)\, dx \geq 0.$$

In shorthand notation (see Section 2.1.9), the last line is written as

$$(\mu\nabla u, \nabla(v-u)) \geq (f, v-u).$$

With these calculations, we obtain the following variational form.

Proposition 16 (Variational inequality of the obstacle problem). *The obstacle problem written as VI is given by: Find $u \in K$ such that*

$$(\mu\nabla u, \nabla(v-u)) \geq (f, v-u) \quad \forall v \in K,$$

with the convex set

$$K = \{v \in H^1 \mid v = u_D \text{ on } \partial\Omega,\ v \geq g \text{ a.e. in } \Omega\}.$$

4.4.3.5 First variations: Euler–Lagrange system with crack irreversibility

With the help of the previous subsection, we now consider

$$E'_{T,\varepsilon}(u,\varphi)(w, \psi-\varphi) \geq 0.$$

Then, we obtain

Proposition 17. *The first variations for $(u,\varphi) \in V_u^D \times K$ are given by*

$$E'_{T,\varepsilon,u}(u,\varphi)(w) = \int_B g(\varphi)\nabla u\cdot\nabla w\, dx,$$

$$E'_{T,\varepsilon,\varphi}(u,\varphi)(\psi-\varphi) - \int_B (1-\kappa)\varphi|\nabla u|^2(\psi-\varphi)\, dx$$

$$+ \int_B G_c\left(-\frac{1}{\varepsilon}(1-\varphi)(\psi-\varphi) + \varepsilon\nabla\varphi\cdot(\nabla(\psi-\varphi))\right) dx.$$

Proof. Follows from Lemma 1 by considering the convex set K and the steps from Section 4.4.3.4. □

4.4.3.6 First variations: Euler–Lagrange system for elasticity with crack irreversibility

In the previous sections, we have only computed the first variations for the scalar-valued case. The starting point now is Proposition 15. Then, we obtain

Proposition 18. *We consider vector-valued displacements $u : B \times I \to \mathbb{R}^d$, namely elasticity. The first variations for $(u, \varphi) \in V_u^D \times K$ are given by*

$$E'_{T,\varepsilon,u}(u,\varphi)(w) = \int_B g(\varphi)\sigma(u) : \nabla w \, dx,$$

$$E'_{T,\varepsilon,\varphi}(u,\varphi)(\psi - \varphi) = \int_B (1-\kappa)\varphi\sigma(u) : e(u)(\psi - \varphi)\, dx + \int_B G_c\Big(-\frac{1}{\varepsilon}(1-\varphi)(\psi-\varphi) + \varepsilon\nabla\varphi \cdot \nabla(\psi - \varphi)\Big)\, dx.$$

We notice since $\sigma(u)$ is symmetric, it holds $\sigma(u) : e(w) = \sigma(u) : \nabla w$.

Proof. The proof follows the same steps as Proposition 17. Specifically, the elliptic Ambrosio–Tortorelli functional is the same. For the first integral, we notice that

$$\sigma(u) : e(u) = \sum_{i,j=1}^{d} \sigma_{ij}(u)e_{ij}(u).$$

The directional derivative into the direction $w = (w_1, \ldots, w_d)$ (now vector-valued) is obtained from applying the product rule to $\sigma(u) : e(u)$:

$$(\sigma(u) : e(u))'(w) = \sigma'(u)(w) : e(u) + \sigma(u) : e(u)'(w).$$

Here, $e(u)'(w) = e(w)$ because $e(\cdot)$ is linear. Due to symmetry and linearity, we have

$$(\sigma(u) : e(u))'(w) = 2\sigma(u) : e(w).$$

As example, we obtain for $d = 3$, for instance,

$$e(u)'(w) = \frac{1}{2}(\nabla w + \nabla w^T),$$

with

$$\nabla w = \begin{pmatrix} \partial_x w_x & \partial_y w_x & \partial_z w_x \\ \partial_x w_y & \partial_y w_y & \partial_z w_y \\ \partial_x w_z & \partial_y w_z & \partial_z w_z \end{pmatrix}. \qquad \square$$

4.4.4 A weak formulation of quasi-static brittle phase-field fracture

Collecting the previous pieces, allows us to formulate the first key model for phase-field fracture, serving as basic model throughout this book. Applying Proposition 18 to a loading sequence, we obtain as CVIS

Proposition 19 (CVIS for phase-field fracture). *For the loading steps $n = 1, 2, 3, \dots, N$, given φ^{n-1}, set up $K := K(\varphi^{n-1})$, find $U = (u, \varphi) := U^n = (u^n, \varphi^n) \in V_u^D \times K$ with $\varphi(0) = \varphi_0$ such that*

$$A(U)(\Psi - \Phi) \geq 0 \quad \forall \Psi = (w, \psi) \in V_u^0 \times K,$$

with $\Phi = (0, \psi)$ and the decomposition into single semilinear forms

$$A(U)(\Psi - \Phi) - A_1(U)(\Psi) + A_2(U)(\Psi - \Phi),$$

and

$$\begin{aligned} A_1(U)(\Psi) &= (g(\varphi)\sigma(u), \nabla w), \\ A_2(U)(\Psi - \Phi) &= ((1-\kappa)\varphi\sigma(u) : e(u), \psi - \varphi) \\ &\quad + \left(-\frac{G_c}{\varepsilon}(1-\varphi), \psi - \varphi \right) + (G_c \varepsilon \nabla\varphi, \nabla(\psi - \varphi)). \end{aligned}$$

The corresponding energy formulation is provided in Proposition 15.

Proof. We employ Proposition 18 for the loading sequence $n = 1, 2, 3, \dots, N$ while definining the convex set $K := K(\varphi^{n-1})$, and then obtain for each step t_n the corresponding formulation. □

4.5 Thermodynamical extensions and interpretations

In this section, we briefly discuss two extensions based on thermodynamical arguments:

1. Distinguishing fracture propagation due to tension and compression;
2. Replacing the inequality constraint by observing the maximal strain energy [313], yielding a history function.

4.5.1 Energy splitting for fracture under tension and compression

Proposition 19 has appeared to be impractical for fracture under compressive forces. Based on thermodynamical arguments, the authors in [23] and [313] (see also Section 4.5.1.3) suggested some extensions. For detailed comparisons of both models, we refer to [17] and other comments can be found in [67, p. 79]. Following Miehe et al. [313], the stress tensor (or directly the energy) is additively decomposed into

- a tensile part σ^+ and
- a compressive part σ^-.

From the physical/engineering point of view, energy degradation, yielding new fractures should be only caused by tensile stresses. This motivates the following modification of the solid stress tensor:

$$\sigma(u) = \sigma^+(u) + \sigma^-(u) \quad \to \quad g(\varphi)\sigma^+(u) + \sigma^-(u),$$

in which the degradation function only acts on σ^+, but not on σ^-. For numerical simulations and further discussions, we refer to [317]. A sketch of what might happen is provided in Figure 4.2. In this book, we also use this test as representative numerical example starting in Chapter 8.

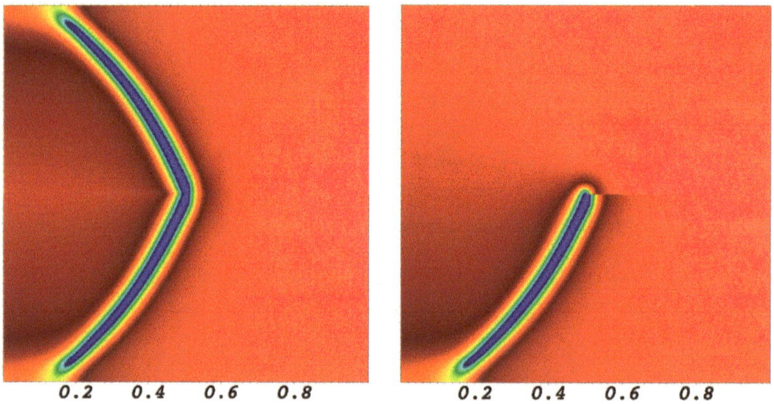

Figure 4.2: Single edge notched shear test (see Chapter 8 for details). At left, a result without any splitting. At right, correct crack path when stress splitting is used. Here, Miehe's et al. approach [317] is adopted.

4.5.1.1 Strain energy splitting à la Miehe et al. [317]

First, the energy can be split into tensile σ^+ and compressive parts σ^-, where damage occurs due to tensile stresses. The term

$$\frac{1}{2}\int_B g(\varphi)\sigma(u) : e(u)\,dx,$$

is split into two parts:

$$\frac{1}{2}\int_B g(\varphi)\sigma^+(u) : e(u)\,dx + \frac{1}{2}\int_B \sigma^-(u) : e(u)\,dx.$$

The tensile and compressive stresses are defined as

$$\sigma^+(u) := 2\mu e^+ + \lambda[\mathrm{tr}(e)]^+ I,$$
$$\sigma^-(u) := 2\mu(e - e^+) + \lambda(\mathrm{tr}(e) - [\mathrm{tr}(e)]^+)I,$$

with the strains

$$e := P\Lambda P^T, \quad e^+ := P\Lambda^+ P^T, \quad e^- := P\Lambda^- P^T.$$

Furthermore, we introduce the cut-off function[2]

$$[x]^+ := \begin{cases} x, & \text{if } x > 0, \\ 0, & \text{if } x \leq 0. \end{cases}$$

The matrices P and Λ form the spectral decomposition with the help of the eigenvalues $\lambda_i := \lambda_i(u)$, $i = 1, 2, 3$ of the strain tensor $e(u)$. In more detail:

$$\Lambda = \begin{pmatrix} \lambda_1 & & \\ & \lambda_2 & \\ & & \lambda_3 \end{pmatrix}, \quad \Lambda^+ = \begin{pmatrix} [\lambda_1]^+ & & \\ & [\lambda_2]^+ & \\ & & [\lambda_3]^+ \end{pmatrix}, \quad \Lambda^- = \Lambda - \Lambda^+.$$

4.5.1.2 Strain energy splitting à la Amor et al. [23]

The other frequently used model seems simpler in terms of implementation, but not necessarily in terms of numerical simulations and mechanical interpretations. Here, the two stress contributions are given by an additive split:

$$\sigma^+(u) := k\,\mathrm{tr}^+(e)I + 2\mu e_D,$$
$$\sigma^-(u) := k\,\mathrm{tr}^-(e)I,$$

with the bulk modulus $k = \frac{2}{d}\mu + \lambda$ and the deviatoric part of the strain tensor $e(u)$ defined as

$$e_D := e - \frac{1}{d}\mathrm{tr}(e)I, \quad d = 2, 3.$$

As before, we need to deal with cut-off terms:

$$\mathrm{tr}^+(e) = [\mathrm{tr}(e)]^+, \quad \mathrm{tr}^-(e) = \mathrm{tr}(e) - \mathrm{tr}^+(e).$$

4.5.1.3 Further constitutive stress-splitting models

Numerous other models for energy (or strain, stress) splitting have been proposed in the literature. We are currently aware of Amor et al. [23], Miehe et al. [317], Zhang et al. [464], Strobl/Seelig [399], Steinke/Kaliske [398], Bryant/Sun [89], Freddi/Royer-Carfagni [176], Bilgen/Homberger/Weinberg [57], and finally a model that we recently developed in our group [160].

---

2 For notational reasons, we use sometimes $[x]_+ := [x]^+$.

4.5.2 A weak phase-field fracture formulation using stress splitting

We summarize the previous developments into our second main formulation describing phase-field fracture propagation in brittle materials.

Proposition 20 (CVIS for phase-field fracture using stress splitting). *For the loading steps $n = 1, \dots, N$, given φ^{n-1}, define $K := K(\varphi^{n-1})$, find $U := U^n := (u^n, \varphi^n) \in V_u^D \times K$ with $\varphi(0) = \varphi_0$ such that*

$$A(U)(\Psi - \Phi) \geq 0 \quad \forall \Psi = (w, \psi) \in V_u^0 \times K,$$

with the decomposition into single semilinear forms

$$A(U)(\Psi - \Phi) = A_1(U)(\Psi) + A_2(U)(\Psi - \Phi),$$

and

$$\begin{aligned} A_1(U)(\Psi) &= (g(\varphi)\sigma^+(u), \nabla w) + (\sigma^-(u), \nabla w), \\ A_2(U)(\Psi - \Phi) &= ((1-\kappa)\varphi\sigma^+(u) : e(u), \psi - \varphi) \\ &\quad + \left(-\frac{G_c}{\varepsilon}(1-\varphi), \psi - \varphi\right) + (G_c\varepsilon\nabla\varphi, \nabla(\psi - \varphi)). \end{aligned}$$

Proof. Use Proposition 19 with the stress splitting terms σ^+ and σ^-. □

4.5.3 Derivation of a strain history field/replacing the irreversibility constraint

Further thermodynamical interpretations discussed by Miehe et al. (2010) [313] are concerned with the fracture energy, resulting in a derivation of a strain history field. The surface integral $E_s(\varphi)$ is interpreted from a thermodynamic viewpoint and allows for a relaxation of the inequality constraint, but destroys the variational energy structure.

Proposition 21 (Thermodynamical properties of $E_s(\varphi)$). *Given as before*

$$E_s(\varphi) = \frac{1}{2}\int_B G_c\left(\frac{1}{\varepsilon}(1-\varphi)^2 + \varepsilon|\nabla\varphi|^2\right) dx.$$

The following postulates are made:
1. *Cracking is a dissipative procedure.*
2. *The system loses the ability to perform mechanical work (entropy).*
3. *Dissipation is expressed through crack irreversibility, which is the work to create a diffusive crack topology.*

Proposition 22 (Crack dissipation). *We have*

$$\frac{d}{dt}E_S(\varphi) = \int_B G_c\Big(-\frac{1}{\varepsilon}(1-\varphi)\partial_t\varphi + \varepsilon\nabla\varphi\cdot\nabla\partial_t\varphi\Big)\,dx$$
$$= \int_B G_c\Big(-\frac{1}{\varepsilon}(1-\varphi) - \varepsilon\Delta\varphi\Big)\partial_t\varphi\,dx.$$

Proof. This follows from change of differentiation and integration, then differentiation using the chain rule. Next, integration by parts in the second term and considering homogeneous Neumann conditions on ∂B for φ, yields the assertion. □

The second law of thermodynamics (see, e. g., [236, Chapter 4.6] or [144, Chapter 5.7]), is related to entropy with a positive dissipation (irreversibility):

$$\frac{d}{dt}E_S(\varphi) \ge 0.$$

In detail,

$$\int_B G_c\underbrace{\Big(\frac{1}{\varepsilon}(1-\varphi) + \varepsilon\Delta\varphi\Big)}_{\ge 0}\underbrace{(-\partial_t\varphi)}_{\ge 0}\,dx \ge 0.$$

The first term is the so-called variational derivative [317]. The second term is well known to us already, noting that $\partial_t\varphi \le 0$. Consequently, changing the sign yields a positive dissipation.

Let us introduce a so-called driving force field β according to Miehe et al. (2010) [313]. We can then formulate the following inequality constraint of the regularized crack energy:

$$\frac{1}{\varepsilon}(1-\varphi) + \varepsilon\Delta\varphi \ge \beta \ge 0.$$

We shall see that the variable β corresponds to the φ-derivative in the u-equation of the prototype Formulation 1, Proposition 19 or 20, respectively:

$$\begin{aligned}
&\text{Laplacian:} && \beta := (1-\kappa)\varphi\,|\nabla u|^2 && \\
&\text{Elasticity:} && \beta := (1-\kappa)\varphi\,\sigma(u) : e(u) && \text{(without stress splitting)}\\
&\text{Elasticity:} && \beta := (1-\kappa)\varphi\,\sigma^+(u) : e(u) && \text{(with stress splitting).}
\end{aligned}$$

Thus, working with the third option, we obtain

$$G_c\Big(\frac{1}{\varepsilon}(1-\varphi) + \varepsilon\Delta\varphi\Big) \ge \beta := (1-\kappa)\varphi\sigma^+(u) : e(u). \tag{4.7}$$

For the next results, we first need

Proposition 23 (Complementarity condition). *In the Euler–Lagrange system in Proposition 20 (similar for Proposition 19 without stress splitting), the following complementarity condition holds true:*

$$\underbrace{\left[(1-\kappa)\varphi\sigma^+(u):e(u)+G_c\left(-\frac{1}{\varepsilon}(1-\varphi)-\varepsilon\Delta\varphi\right)\right]}_{(\mathrm{I})}\cdot\underbrace{\partial_t\varphi}_{(\mathrm{II})}=0. \tag{4.8}$$

Proof. Follows immediately from the strong formulation. We refer the reader also to the obstacle problem in Section 2.4.1; see also the prototype Formulation 1. In the presence of fractures the complementarity condition is also known as Rice condition [372]. □

For a growing crack, we combine (4.7) and (4.8). In (4.8), we can have a variation in the crack length, i. e., $\partial_t\varphi < 0$ (strictly), when the first expression (I) is zero, i. e., when $(1-\kappa)\varphi\sigma^+(u):e(u)$ is maximal:

$$(1-\kappa)\varphi\sigma^+(u):e(u)+G_c\left(-\frac{1}{\varepsilon}(1-\varphi)-\varepsilon\Delta\varphi\right)=0.$$

But this also means that in (4.7), we have now

$$G_c\left(\frac{1}{\varepsilon}(1-\varphi)+\varepsilon\Delta\varphi\right)=\beta=(1-\kappa)\varphi\,\sigma^+(u):e(u).$$

On the other hand, when $\partial_t\varphi=0$, then in (4.8), we obtain

$$\begin{aligned}&(I)<0\\ \Leftrightarrow\quad &(1-\kappa)\varphi\,\sigma^+(u):e(u)-G_c\left(\frac{1}{\varepsilon}(1-\varphi)+\varepsilon\Delta\varphi\right)<0\\ \Leftrightarrow\quad &(1-\kappa)\varphi\,\sigma^+(u):e(u)<G_c\left(\frac{1}{\varepsilon}(1-\varphi)+\varepsilon\Delta\varphi\right).\end{aligned}$$

With these considerations, we obtain the following results.

Definition 48. The following strain energy storage function is defined:

$$\psi_e^+ := \sigma^+(u):e(u).$$

Proposition 24. *Let $\varepsilon>0$ be fixed. For $\psi_e^+\gg 0$, then $\varphi\approx 0$.*

Proof. We provide a heuristic sketch without details on the function spaces. We start with

$$\frac{1}{\varepsilon}(1-\varphi)+\varepsilon\Delta\varphi\ge\beta\ge 0.$$

We previously obtained

$$\beta=(1-\kappa)\varphi\sigma^+(u):e(u)=(1-\kappa)\varphi\psi_e^+.$$

For $\varphi > 0$, but small, we obtain

$$\frac{G_c(\frac{1}{\varepsilon}(1-\varphi) + \varepsilon\Delta\varphi)}{(1-\kappa)\varphi} \geq \psi_e^+.$$

We notice that according to Theorem 5 (therein for a prototype problem, but the generalization holds true as well; see [336]), the nominator is bounded for fixed $\varepsilon > 0$. When we take the maximum in u, we obtain

$$\frac{G_c(\frac{1}{\varepsilon}(1-\varphi) + \varepsilon\Delta\varphi)}{(1-\kappa)\varphi} = \max_u \psi_e^+. \tag{4.9}$$

For large $\max_u \psi_e^+ \gg 0$, the phase-field function on the left-hand side must tend to zero with $\varphi \approx 0$. □

Corollary 1. *The previous proposition also shows that we can enforce crack irreversibility by considering the maximal strain energy ψ_e^+. In this case in (4.7) the equality holds true. Then (I) holds true in the complementarity condition (4.8). Thus, we have $\partial_t\varphi < 0$ strictly for all times t. To this end, the crack irreversibility condition can be eliminated.*

Proof. It holds

$$G_c\left(\frac{1}{\varepsilon}(1-\varphi) + \varepsilon\Delta\varphi\right) \geq (1-\kappa)\varphi\sigma^+(u) : e(u).$$

Taking the maximum in u over space and time yields:

$$G_c\left(\frac{1}{\varepsilon}(1-\varphi) + \varepsilon\Delta\varphi\right) = (1-\kappa)\varphi \max_{s\in[0,t]} \psi_e^+(s). \tag{4.10}$$

Since we have an equality at all times $s \in [0,t]$ in equation (4.10), the complementarity condition (4.8) yields $\partial_t\varphi < 0$. □

Definition 49 (Strain history function). The right-hand side in the previous proof is defined as

$$\mathcal{H} := \max_{s\in[0,t]} \psi_e^+(u(s)).$$

The function $\mathcal{H}$ is a so-called strain history function [313].

4.5.4 A weak phase-field fracture formulation using strain history

The previous section yields our third main formulation for describing phase-field fracture inspired by Miehe et al. (2010) [313].

Proposition 25 (Weak formulation for phase-field fracture with strain history). *For the loading steps $n = 1, 2, 3, \dots, N$, given $\mathcal{H}(u^{n-1})$, find $U := U^n = (u^n, \varphi^n) \in V_u^D \times V_\varphi$ with $\varphi(0) = \varphi_0$, such that*

$$A(U)(\Psi) = 0 \quad \forall \Psi = (w, \psi) \in V_u^0 \times V_\varphi,$$

where

$$A(U)(\Psi) = A_1(U)(\Psi) + A_2(U)(\Psi),$$

and

$$A_1(U)(\Psi) = (g(\varphi)\sigma^+(u), \nabla w) + (\sigma^-(u), \nabla w),$$
$$A_2(U)(\Psi) = ((1-\kappa)\varphi\mathcal{H}, \psi) + \left(-\frac{G_c}{\varepsilon}(1-\varphi), \psi\right) + (G_c \varepsilon \nabla\varphi, \nabla\psi).$$

Remark 28. The alternative formulation offers a PDE system rather than a CVIS, because we immediately deal with a phase-field equation rather than a VI. But the implementation of the strain-history function $\mathcal{H}$ requires some care because the history must be stored at each quadrature point of the FEM program. This is somewhat similar to materials with memory effects. In our implementation (Chapter 8), we took ideas from `step-18` in deal.II [26].

4.6 Comments on variational energy formulations versus Euler–Lagrange systems

As we discussed in this chapter, modeling phase-field fracture in elasticity starts from an energy formulation and seeking local minima, e. g., [73], i. e.,

$$\min_{U \in X} E_T(U).$$

By differentiation, we obtain a CVIS, the so-called Euler–Lagrange system (Section 4.4.3.6), i. e.,

$$A_1(U)(w) = E_u'(U)(w) = 0,$$
$$A_2(U)(\psi) = E_\varphi'(U)(\psi - \varphi) \geq 0.$$

Solving this CVIS yields not only minima, but all stationary points. Specifically, it can happen that our solution is a saddle-point. Therefore, energy formulations yield a smaller, but stronger, class of solutions and are possibly more desirable from a physical viewpoint.

On the other hand, PDE formulations allow us to incorporate a much broader class of models. For instance, the Stokes equations (being symmetric) can be formulated with the help of an energy functional. But the extension to the incompressible Navier–Stokes equations does not have a corresponding energy form. Also, simple equations of parabolic type (such as the heat equation) do not allow for an energy concept. The same holds for the Biot equations describing flow and deformation of porous media. For detailed further reasons and formulating phase-field fracture models within such frameworks, we refer the reader to [322, explanations in Section 2.4] or [441] or [425, explanations in Section 3] or [324, Section 2.3].

For some models under certain assumptions, one can still arrive at an energy formulation. For instance in quasi-static fluid-filled fracture in porous media, some authors interpret the given system as an incremental (time-discretized) problem and analyze their formulations on a given time step. On the time-discretized level, we can probably again formulate elliptic, symmetric PDEs for which a corresponding energy formulation exists, e. g., [315, 314]. In the time-continuous case, specifically in porous media, this procedure does not apply anymore. Also for extensions to phase-field fracture fluid-structure interaction energy formulations are not known.

5 Numerical modeling part I: regularization and discretization

In Chapter 4, we have derived variational phase-field fracture models on the space/time continuous level:
1. Proposition 19 using an inequality constraint;
2. Proposition 20 using an inequality constraint and stress splitting into tensile and compressive parts;
3. Proposition 25 using a strain history function and again stress splitting.

In the Chapters 5–10, we shall discuss regularizations and discretizations (Chapter 5), nonlinear and linear solutions (Chapter 7), and finally adaptive schemes (Chapter 9). These concepts are supplemented in alternating chapters with some prototype numerical examples. Some of them are published and just newly interpreted for this monograph. Other simulations are new and have not been published elsewhere.

5.1 Paradigm of numerical modeling: physics-based discretizations and solvers

A crucial driving force in our work are problem-specific discretizations and solvers that respect the governing physical conservation laws. Ideally, such techniques are enhanced with adaptive routines in combination with parallel computing. Two well-known examples are:
- Low dissipative time discretization schemes for wave-type equations such as elastodynamics;
- Finite elements in spatial discretizations that ensure local mass conservation in incompressible fluid flow and porous media flow.

5.2 The crack irreversibility constraint (entropy condition)

Basically five procedures exist:
1. Fixing crack nodes by Dirichlet values [72, 264];
2. Using a strain history function [313];
3. Using simple and augmented Lagrangian penalizations [318, 430];
4. Using a primal-dual active set method [221];
5. Employing the complementarity system with Lagrange multipliers [301].

Some comparisons were recently performed in [193] and we provide more in this book in Chapter 8.

https://doi.org/10.1515/9783110497397-005

5.2.1 Strain history function

Imposing the irreversibility constraint via a strain history field was discussed in Section 4.5.3.

5.2.2 Simple penalization and augmented Lagrangian penalization

The most simple procedure in the Propositions 19 and 20 is to regularize the inequality constraint $\partial_t \varphi \leq 0$ via a backward difference quotient:

$$\partial_t \varphi \approx \frac{\varphi - \varphi^{n-1}}{k} \leq 0 \quad \Rightarrow \quad \varphi \leq \varphi^{n-1} \quad \Rightarrow \quad \varphi - \varphi^{n-1} \leq 0.$$

This means that we need to penalize when $\varphi \leq \varphi^{n-1}$ is violated, namely $\varphi > \varphi^{n-1}$. Using penalization, we extend the solution space from the convex set $K := K(\varphi^{n-1})$ to the linear space V_φ. For the obstacle problem as prototype example, we refer to [260, Chapter 3, specifically Sections 3.4 and 3.5].

A simple penalization to act against this violation is

$$\gamma[\varphi - \varphi^{n-1}]_+,$$

with a penalization parameter $\gamma > 0$, $[x]_+ = [x]^+ := \max(0, x)$, and the time step size $k := t_n - t_{n-1}$ and $\varphi := \varphi^n := \varphi(t_n)$. Of course, if γ is too small, the penalization has no effect (Figure 6.4, Chapter 6), whereas if γ is too big will lead to ill-conditioned Jacobians and higher nonlinearities from which the nonlinear solver will suffer (see [340, p. 505]) and we also refer to computations and available code for the obstacle problem in our educational lecture notes [300, Section 8.19]. A mathematical analysis for phase-field fracture with supporting numerical simulations for $\gamma \to \infty$ in conjunction with an optimization setting was studied [337].

Proposition 26 (Simple penalization of the phase-field VI). *Let as before $U = (u, \varphi)$. Given $\gamma > 0$ and using simple penalization, the phase-field VI in Proposition* 19 *or* 20 *becomes*

$$A_2(U)(\Psi) + (\gamma[\varphi - \varphi^{n-1}]_+, \psi) = 0. \tag{5.1}$$

Specifically, the solution φ is not anymore sought in the convex set $K := K(\varphi^{n-1})$, but in the linear space $V_\varphi := H^1(B)$.

Thus, we have relaxed the inequality constraint by penalization, yielding now a phase-field equation.

Proposition 27 (Proposition 19 as a penalized version). *For the loading steps $n = 1, 2, 3, \ldots, N$, given φ^{n-1}, find $U := U^n = (u^n, \varphi^n) \in X$ with $\varphi(0) = \varphi_0$, such that*

$$A(U)(\Psi) = 0 \quad \forall \Psi = (w, \psi) \in X^0,$$

where the semilinear form is composed as

$$A(U)(\Psi) = A_1(U)(\Psi) + A_2(U)(\Psi),$$

with

$$\begin{aligned} A_1(U)(\Psi) &= (g(\varphi)\sigma(u), \nabla w), \\ A_2(U)(\Psi) &= ((1-\kappa)\varphi\sigma(u) : e(u), \psi) + \left(-\frac{G_C}{\varepsilon}(1-\varphi), \psi\right) + (G_C\varepsilon\nabla\varphi, \nabla\psi) \\ &\quad + (\gamma[\varphi - \varphi^{n-1}]_+, \psi). \end{aligned}$$

Proposition 28 (Proposition 20 as a penalized version). *All as before, the only difference is in the semilinear forms:*

$$\begin{aligned} A_1(U)(\Psi) &= (g(\varphi)\sigma^+(u), \nabla w) + (\sigma^-(u), \nabla w), \\ A_2(U)(\Psi) &= ((1-\kappa)\varphi\sigma^+(u) : e(u), \psi) + \left(-\frac{G_C}{\varepsilon}(1-\varphi), \psi\right) + (G_C\varepsilon\nabla\varphi, \nabla\psi) \\ &\quad + (\gamma[\varphi - \varphi^{n-1}]_+, \psi). \end{aligned}$$

Remark 29. On the energy level $(\gamma[\varphi - \varphi^n]_+, \psi)$ reads

$$\frac{\gamma}{2}\|[\varphi - \varphi^{n-1}]_+\|_{L^2}^2.$$

From a mathematical point of view, a fourth-order regularization (see, e. g., [336]) would yield a better regularization within Newton's method because the problem has better regularity to differentiate up to second-order derivatives:

$$\begin{aligned} &\frac{\gamma}{4}\|[\varphi - \varphi^{n-1}]_+\|_{L^2}^4 \\ \Rightarrow\quad &\text{PDE-level: } J'(\varphi)(\psi) = \gamma([\varphi - \varphi^{n-1}]_+^3, \psi)_{L^2} \\ \Rightarrow\quad &\text{Jacobian in Newton's method: } J''(\varphi)(\xi, \psi) = 3\gamma([\varphi - \varphi^{n-1}]_+^2 \xi, \psi)_{L^2}. \end{aligned}$$

For more details on the Newton's method, we refer the reader to the respective sections in Chapter 7; specifically to Section 7.9.1.1.

5.2.2.1 Augmented Lagrangian penalization

The augmented Lagrangian (AL) method is well known in numerical optimization dating back to [227] and [355] and proposed by [171] and [206] for use in discretized differential equations. In particular, we consider the augmented Lagrangian in a function space setting similar to [245, Chapter 4] and [340, Chapter 17]. Using the augmented Lagrangian approach in phase-field fracture was proposed in [430] and further investigated in [444].

Basically, we introduce a Lagrange multiplier function $\Xi \in L^2(B)$ such that we have as augmented penalization:

$$[\Xi + \gamma(\varphi - \varphi^{n-1})]_+. \tag{5.2}$$

Then we obtain the following.

Proposition 29 (AL penalization for the phase-field variational inequality). *Let us denote the solution by $U = (u, \varphi)$. Given $\gamma > 0$ and a Lagrange multiplier function $\Xi \in L^2(B)$. Then the augmented Lagrangian phase-field variational inequality in Proposition* 19 *or* 20 *reads*

$$A_2(U)(\Psi) + ([\Xi + \gamma(\varphi - \varphi^{n-1})]_+, \psi) = 0. \tag{5.3}$$

As before, the solution φ is not anymore sought in the convex set $K := K(\varphi^{n-1})$, but in the linear space $V_\varphi := H^1(B)$. The solution algorithm to calculate Ξ in practice is outlined in Section 7.11.

We remark that the augmented Lagrangian penalization is derived from the energy level and actually results into two contributions [430, p. 73]:

$$([\Xi + \gamma(\varphi - \varphi^{n-1})]_+, \psi) + (\min(0, \Xi + \gamma\varphi), \psi). \tag{5.4}$$

In our computational experiments, we could not detect a major influence of this second term and neglect this contribution in the remainder of this monograph. We specifically notice here as well that $\gamma > 0$ and $0 \leq \varphi \leq 1$. Starting with $\Xi \geq 0$ yields $\min(0, \Xi + \gamma\varphi) = 0$ on the space-continuous level since $\Xi + \gamma\varphi \geq 0$.

5.2.3 Primal-dual active set method

The augmented Lagrangian idea from the previous section can be further related to active set methods [244, Section 6]. A specific class are so-called primal-dual active set (PDAS) methods, which can be interpreted under certain assumptions as semismooth Newton methods [231, 232]. As previously illustrated for the obstacle problem, the domain is split into active (constraints) and inactive sets (solving the PDE) from which the name active set method originates. Applying these techniques to phase-field fracture was subject in [221] and details are given as well in Section 7.9.5.

5.2.3.1 Finite dimensional case

In the current section, we briefly recall the basic idea, following conceptionally closely Section 2 of [231]. We consider the following complementarity problem with dimension $d \geq 1$.

Formulation 14. Let $\varphi^{n-1} \in \mathbb{R}^d$ be a given constraint. Find a solution $\varphi : B \to \mathbb{R}^d$ and a Lagrange multiplier $\Xi : B \to \mathbb{R}^d$ such that

$$\begin{aligned} A\varphi + \Xi &= f, \\ \varphi &\le \varphi^{n-1}, \\ \Xi &\ge 0, \\ (\Xi, \varphi - \varphi^{n-1}) &= 0. \end{aligned}$$

Here, $A \in \mathbb{R}^{d\times d}$ is a well-posed matrix (for details see [231]), and let $f \in \mathbb{R}^d$.

If A is symmetric positive definite, the above system is the optimality system to

$$\begin{aligned} &\min_{\varphi} E(\varphi) = \frac{1}{2}(\varphi, A\varphi) - (f, \varphi), \\ &\text{subject to} \quad \varphi \le \varphi^{n-1}. \end{aligned}$$

The trick is now to reformulate the previous complementarity conditions in an equivalent fashion:

$$\begin{aligned} &\varphi \le \varphi^{n-1}, \quad \Xi \ge 0, \quad (\Xi, \varphi - \varphi^{n-1}) = 0, \\ \Leftrightarrow \quad &C(\varphi, \Xi) = 0, \end{aligned}$$

where

$$C(\varphi, \Xi) = \Xi - \left[\Xi + c(\varphi - \varphi^{n-1})\right]^{+}. \tag{5.5}$$

Here, $c > 0$ is a parameter and the max-operation has to be understood componentwise.

Remark 30. In (5.5), we observe a close relationship to (5.2), respectively (5.4).

With (5.5), we obtain the equivalent result:

Formulation 15. Find a solution $\varphi : B \to \mathbb{R}^d$ and a Lagrange multiplier $\Xi : B \to \mathbb{R}^d$ such that

$$\begin{aligned} A\varphi + \Xi &= f, \\ C(\varphi, \Xi) &= 0. \end{aligned}$$

Now, we are already in the position to formulate a primal-dual active set strategy. For a given pair (φ, Ξ), the constraint formulation $C(\varphi, \Xi) = 0$ is used as a predictor step to determine the next active and inactive sets. These two sets are given by

$$\begin{aligned} \mathcal{N} &= \{i : \Xi_i + c(\varphi - \varphi^{n-1})_i \le 0\}, \quad \text{(PDE is solved)} \\ \mathcal{A} &= \{i : \Xi_i + c(\varphi - \varphi^{n-1})_i > 0\}, \quad \text{(Constraint is active).} \end{aligned}$$

Here, i denotes the index of the solutions (for instance located at the DoFs when using Q_1 finite elements).

Algorithm 2 (Primal-dual active set algorithm). The basic primal-dual active set algorithm has the following steps:

1. Initialize φ^0 and Ξ^0. Set $j = 0$.
2. Determine $\mathcal{N}^j$ and $\mathcal{A}^j$.
3. Solve

$$
\begin{aligned}
A\varphi^{j+1} + \Xi^{j+1} &= f,\\
\varphi^{j+1} &= \varphi^{n-1} \text{ on } \mathcal{A}^j,\\
\Xi^{j+1} &= 0 \text{ on } \mathcal{N}^j.
\end{aligned}
$$

4. Stop, when an appropriate stopping criterion is fulfilled. If not, set $j \to j+1$ and go to Step 2.

Here, $\varphi^{j+1} = \varphi^{n-1}$ has to be understood componentwise for each index i: $\varphi_i^{j+1} = \varphi_i^{n-1}$ for $i \in \mathcal{A}^j$.

Remark 31. In [231], this algorithm is interpreted as a semismooth Newton method with corresponding convergence results.

5.2.3.2 Complementarity systems and Lagrange multipliers for phase-field fracture

In the following, we further elaborate on the previous concepts and formulate Proposition 19 or 20 by directly addressing the complementarity system. To this end the Lagrange multiplier becomes an unknown and is not anymore computed in a post-processing step as in the augmented Lagrangian iteration. Conceptionally, we follow [260, Section 3.4]. First, we need to introduce a function space for the Lagrange multiplier. Let Q^* be the dual space of a Banach space Q. We now define

$$V_\Xi = \{\chi \in Q^* \mid \chi \geq 0\}.$$

For the ordering relation $\leq$, we refer the reader to [260, Section 3]. To this end, we obtain

Proposition 30 (Lagrange multiplier system of Proposition 19). *Find* $(u, \varphi, \Xi) \in V_u^D \times V_\varphi \times V_\Xi$ *such that*

$$
\begin{aligned}
A_1(U)(w) &= 0 \quad \forall w \in V_u^0,\\
A_2(U)(\psi) + \langle \Xi, \psi \rangle &= 0 \quad \forall \psi \in V_\varphi,\\
\langle \Xi - \chi, \varphi - \varphi^{n-1} \rangle &\geq 0 \quad \forall \chi \in V_\Xi,
\end{aligned}
$$

where $\langle \cdot, \cdot \rangle := \langle \cdot, \cdot \rangle_{Q^*,Q}$ *denotes the duality pairing on* $Q^* \times Q$ *as introduced in Section* 3.2.

Remark 32. Such a complementarity system for phase-field fracture was recently introduced in [301] and further used in [299, 42].

If we write the last inequality in the previous proposition explicitly as complementarity system, then we obtain a link to the previous section with the primal-dual active set approach. In detail, we have

$$\begin{aligned} \varphi \leq \varphi^{n-1} &\quad \text{in } B \times I, \\ \Xi \geq 0 &\quad \text{in } B \times I, \\ \langle \Xi, \varphi - \varphi^{n-1} \rangle = 0 &\quad \text{in } B \times I. \end{aligned}$$

Then we obtain immediately from (5.5):

$$C(\varphi, \Xi) = 0,$$

where

$$C(\varphi, \Xi) = \Xi - [\Xi + c(\varphi - \varphi^{n-1})]^{+}. \tag{5.6}$$

5.3 Temporal discretization

So far, we mostly considered models for quasi-static phase-field fracture. Therein, temporal discretization results in a sequence of incremental steps, indexed by $n = 1, \ldots, N$, corresponding to a backward Euler discretization. An exception was the dynamic prototype model presented in Section 2.5. In view of this example and further nonstationary multiphysics PDEs/CVISs considered in Chapter 11, we formulate time-stepping schemes in a proper way in this section.

5.3.1 One-Step-θ and Fractional-Step-θ schemes

We start with

Definition 50. Find $U \in X$, e. g., $X = V_u^D \times V_\varphi$, for almost all times $t \in I$, such that

$$\int_0^T A(U)(\Psi)\, dt = \int_0^T F(\Psi)\, dt \quad \forall \Psi \in X^0. \tag{5.7}$$

The semilinear form $A(U)(\Psi)$ is the compact form that may consist of time-derivatives, explicit and implicit terms. To this end, we assume the decomposition:

$$A(U)(\Psi) := A_T(U)(\Psi) + A_E(U)(\Psi) + A_I(U)(\Psi).$$

Remark 33. In the previous definition, we assumed a fully time-dependent problem. For instance, dynamic phase-field fracture. In incremental quasi-static phase-field fracture, we could rather write: Given U^{n-1}, find $U^n \in X$ for $n = 1, 2, 3, \ldots, N$ such that

$$A(U^n)(\Psi) = F^n(\Psi) \quad \forall \Psi \in X^0,$$

where $F^n(\Psi) := (f(t_n), \Psi)$ are given right-hand side data. Further dependencies on previous time point data are neglected for the convenience of presentation.

Remark 34. We remark that the previous groupings into A_T, A_E, A_I are still time-continuous, space-continuous, and possibly nonlinear.

Example 9. In Proposition 27, we are concerned with

$$A(U)(\Psi) = A_1(U)(\Psi) + A_2(U)(\Psi).$$

Therein, we have only explicit terms, thus

$$A(U)(\Psi) := A_E(U)(\Psi) = A_1(U)(\Psi) + A_2(U)(\Psi).$$

When several PDEs/CVISs are coupled, it is worth to group them as proposed in [434].

Definition 51 (Arranging the semilinear form into groups). We formally define the following semilinear forms and group them into three categories:
1. time equation terms with time derivatives;
2. explicit terms: taken at current and old times t_n and t_{n-1}
→ stress terms, convection, reactions
3. implicit terms: taken only at the current time point t_n
→ e. g., incompressibility conditions or coupling to steady-state problems.

Arranging all terms into groups is explained in detail in [434, 443] in which also the pressure terms $A_P(U)(\Psi)$ from Navier–Stokes fluid flow are put into an own category; see also Section 11.8.1.6.

Definition 52 (One-Step-θ time-discretization). Given U^{n-1}, F^n, and F^{n-1}, and $\theta \in [0,1]$, we seek $U := U^n \in X$ by employing One-Step-θ splitting:

$$\begin{aligned} &A_T(U^{n,k})(\Psi) + \theta A_E(U^n)(\Psi) + A_I(U^n)(\Psi) \\ &= -(1-\theta)A_E(U^{n-1})(\Psi) + \theta F^n(\Psi) + (1-\theta)F^{n-1}(\Psi). \end{aligned} \tag{5.8}$$

The concrete schemes depend on θ:
- $\theta = 0$: first-order explicit forward Euler
- $\theta = 1$: first-order implicit, strongly A-stable, backward Euler
- $\theta = 0.5$: second-order implicit, A-stable, trapezoidal rule
- $\theta = 0.5 + k_{\text{char}}$: second-order implicit, strictly A-stable, shifted Crank–Nicolson, where k_{char} is the characteristic time stepsize with $k_{\text{char}} \ll 0.5$.

The last scheme was proposed and analyzed in [366].

A famous alternative is the Fractional-Step-θ scheme [86, 205]:

Definition 53 (Fractional-Step-θ scheme). Let us choose $\theta = 1 - \frac{\sqrt{2}}{2}$, $\theta' = 1 - 2\theta$, and $\alpha = \frac{1-2\theta}{1-\theta}$, $\beta = 1 - \alpha$. The time step is split into three consecutive subtime steps. Let U^{n-1} be given and the current time stepsize $k := k_n = t_n - t_{n-1}$. Find the new solution $U := U^n$ such that

$$\begin{aligned} A_T(U^{n-1+\theta,k})(\Psi) &+ \alpha\theta A_E(U^{n-1+\theta})(\Psi) \\ &+ A_I(U^{n-1+\theta})(\Psi) = -\beta\theta A_E(U^{n-1})(\Psi) + \theta F^{n-1}(\Psi), \\ A_T(U^{n-\theta,k})(\Psi) &+ \beta\theta' A_E(U^{n-\theta})(\Psi) \\ &+ A_I(U^{n-\theta})(\Psi) = -\alpha\theta' A_E(U^{n-1+\theta})(\Psi) + \theta' F^{n-\theta}(\Psi), \\ A_T(U^{n,k})(\Psi) &+ \alpha\theta A_E(U^n)(\Psi) \\ &+ A_I(U^n)(\Psi) = -\beta\theta A_E(U^{n-\theta})(\Psi) + \theta F^{n-\theta}(\Psi). \end{aligned} \tag{5.9}$$

Remark 35. This version of the Fractional-Step-θ scheme differs slightly from versions published for fluid problems involving pressure terms. For details, see [86, 205], [364, Section 6.4.1] and [420, Section 3.2.1] and for fluid-structure interaction [434, p. 59] as this becomes relevant in Chapter 11. Therein, explicit (convection, stress terms) and implicit terms (pressure) appear in the same governing equation (conservation of momentum). For this reason, these implicit pressure terms are weighted as well with θ (step 1 and 3) and θ' (step 2). We finally notice that some new interpretations can be found in [308, 309] and various older [421] and recent comparisons by Glowinski and coworkers were done in [202].

5.3.2 Example

We briefly demonstrate the procedure for a parabolic problem (the heat equation):

$$\partial_t u - \Delta u = f.$$

Time discretization using finite differences yields: Find $u := u^n$:

$$\frac{u - u^{n-1}}{k} - \theta\Delta u - (1-\theta)\Delta u^{n-1} = \theta f + (1-\theta) f^{n-1}.$$

Rearranging terms yields

$$\frac{u - u^{n-1}}{k} - \theta\Delta u = (1-\theta)\Delta u^{n-1} + \theta f + (1-\theta) f^{n-1}.$$

Classifying into time derivative terms A_T and stationary operator terms A_E, the problem reads in abstract form: Find $U \in X$ for $t \in (0, T)$ such that

$$A(U)(\Psi) = F(\Psi) \quad \Psi \in X$$

with

$$A(U)(\Psi) = A_T(U)(\Psi) + A_E(U)(\Psi),$$

where (for the above heat equation):

$$\begin{aligned} A_T(U)(\Psi) &= (\partial_t u, \Psi), \\ A_E(U)(\Psi) &= (\nabla u, \nabla \Psi), \\ F(\Psi) &= (f, \Psi). \end{aligned}$$

The backward difference quotient formulations for $A_T(U)(\Psi)$ read

$$\begin{aligned} A_T(U)(\Psi) \approx A_T(U^{n,k})(\Psi) &= \left(\frac{u^n - u^{n-1}}{k}, \Psi\right), \\ A_E(U^n)(\Psi) &= (\nabla u^n, \nabla \Psi), \\ F^n(\Psi) &= (f(t_n), \Psi). \end{aligned}$$

5.3.3 Nonlinear time derivatives

When the time derivative is nonlinear, we first apply the product rule and then apply the difference quotient with θ-weighting of the nonlinear coefficient.

Let $u \in V_u$ and $v \in V_v$ be solution variables of a nonlinear system. As an example, define $J := J(u)$ and let us consider the strong form

$$\partial_t(Jv) = J\partial_t v + v\partial_t J$$

respectively the weak form:

$$A_T(U)(\psi) = (\partial_t(Jv), \psi) = (J\partial_t v + v\partial_t J, \psi).$$

Applying the difference quotient and θ-approximation for the term in front of the time dervative yields

$$J^\theta\left(\frac{v - v^{n-1}}{k}\right) + v^\theta\left(\frac{J - J^{n-1}}{k}\right),$$

with

$$\begin{aligned} J^\theta &= \theta J^n + (1-\theta)J^{n-1}, \\ v^\theta &= \theta v^n + (1-\theta)v^{n-1}. \end{aligned}$$

Then

$$A_T(U)(\psi) \approx A_T(U^{n,k})(\psi) = \left(J^\theta\left(\frac{v - v^{n-1}}{k}\right), \psi\right) + \left(v^\theta\left(\frac{J - J^{n-1}}{k}\right), \psi\right).$$

Example 10. We consider as a full example:

$$\partial_t(Ju) - \Delta u = f,$$

with $J := J(u)$, thus J depends on u! For the One-Step-θ and Fractional-Step-θ schemes, we now time-discretize as just shown before:

$$A_T(u^{n,k})(\psi) = \left(J^\theta\left(\frac{u - u^{n-1}}{k}\right), \psi\right) + \left(u^\theta\left(\frac{J - J^{n-1}}{k}\right), \psi\right),$$
$$A_E(u^n)(\psi) = (\nabla u, \nabla \psi),$$

where as before we have used the convention $u := u^n = u(t_n)$.

5.3.4 Brief relation to quasi-static phase-field fracture

In deriving the incremental problems for quasi-static phase-field fracture, we have implicitly used the previous notation with $\theta = 1$ (backward Euler). The inequality constraint $\partial_t \varphi \leq 0$ corresponds to

$$A_T(\varphi)(\psi) = (\partial_t \varphi, \psi) \approx A_T(\varphi^{n,k})(\psi) = \left(\frac{\varphi - \varphi^{n-1}}{k}, \psi\right) \leq 0.$$

5.3.5 Importance of careful choices of time-stepping schemes in nonstationary multiphysics simulations

The choice of a good time-stepping scheme is particularly important for nonstationary multiphysics problems involving elasto-dynamics (i. e., wave type equations). Here, less dissipative schemes must be preferred in order to guarantee conservation of energy on the time-discrete level. For the above schemes, this holds true for $\theta = 0.5$ and $\theta = 0.5 + k_n$ and the Fractional-Step-θ scheme. But also for parabolic problems, with irregular initial conditions, care must be taken [359, 292].

For multiphysics fluid-structure interaction in terms of FSI-benchmark tests [240], we refer the reader for comparisons of various time-stepping schemes to [434, 379,

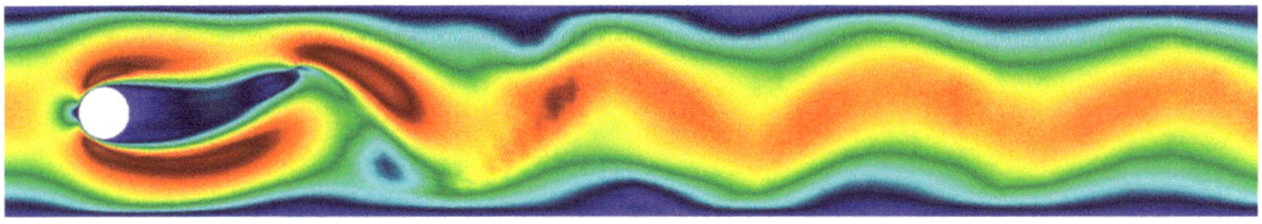

Figure 5.1: Numerical simulation of the FSI 2 benchmark problem: magnitude of the velocity field (displaying the van Karman vorticity street).

159]. The solution should result in periodic oscillations of an elastic beam and flow in the van Karman vorticity street as displayed in Figure 5.1. As expected, the backward Euler scheme with $\theta = 1$ is too dissipative as shown in Figure 5.2 for the FSI-2 benchmark. The shifted Crank–Nicolson scheme (see again Figure 5.2) produces solutions as shown in [240]. Also the Fractional-Step-θ scheme works well as shown in [434, p. 75]. However, the classical Crank–Nicolson scheme with $\theta = 0.5$ will blow-up as again shown in [434].

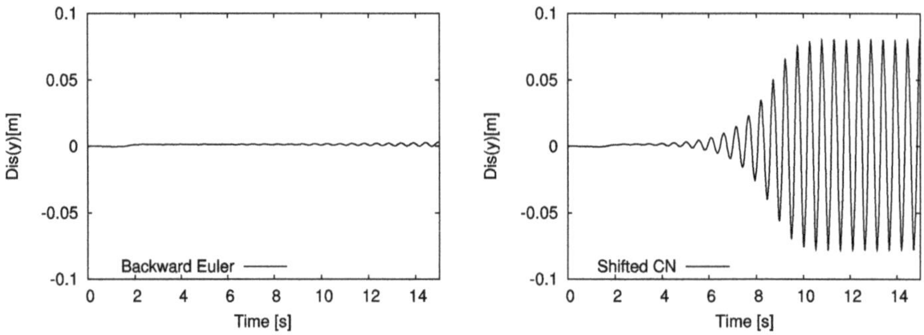

Figure 5.2: FSI-2 benchmark proposed in [240]: Comparison of the u_y displacement of the elastic beam tip using the backward Euler scheme (left) and the shifted Crank–Nicolson scheme (right). These findings show that it is important to choose the "correct" time-stepping scheme. For nonstationary flow, the backward Euler scheme is too dissipative and is not able to compute oscillatory flow as the shifted CN scheme does. The classical Crank–Nicolson scheme with $\theta = 0.5$ will blow-up as shown in [434, p. 75]. The results shown here are computed with [438].

For the FSI-3 benchmark the picture changes a bit in the sense that also the shifted Crank–Nicolson scheme exhibits some instabilities and in fact the time step size shift k_n must be very carefully chosen. Here, only the Rannacher time-stepping scheme (smoothing with Euler steps at every jth step, e. g., [228]; the original idea was published earlier at the beginning of the 1980s) and the Fractional-Step-θ scheme produce satisfactory solutions [159] shown in Figure 5.3.

In all of these discussions, it should however be noticed that smaller time steps, similar to explicit time-stepping, will increase numerical stability. Here, the overall time step sizes are however much larger than in explicit methods, making implicit methods still competitive. A supporting example is again given in [434, p. 76].

Remark 36. As previously mentioned, the considerations of this section are not relevant for quasi-static phase-field fracture. However, in dynamic fracture and multiphysics extensions the choice of good time-stepping schemes becomes important. In [441, Section 5], we observed some problems concerning time-stepping schemes and further studies are necessary.

5.4 Spatial discretization

In this section, we briefly introduce Galerkin finite elements for the spatial discretization. For details, we refer to classical books [101, 253, 115, 390, 79, 82, 241, 363]. Lecture notes with 'our' notation are available online as well: [447]. Isogeometric analysis [242, 118, 53, 66], virtual finite elements, meshless methods and related developments are alternatives to name a few.

The time-discretized equations are the starting point for a finite element Galerkin discretization method in space. So far, the equations are still posed in continuous spatial spaces V_u^D and V_φ. In the following, we discuss the spatial discretization of the semi-discrete problems. To this end, we construct finite dimensional subspaces $V_{u,h} \subset V_u^D$ and $V_{\varphi,h} \subset V_\varphi$ of piecewise polynomial functions up to order l.

5.4.1 Finite element spaces

The computational domain B is partitioned into open elements K that depend on the spatial dimension d. A mesh consists of quadrilateral or hexahedral elements K. They perform a nonoverlapping cover of the computation domain $B \subset \mathbb{R}^d$, $d = 2, 3$. The mesh $\mathcal{T}_h = \{K\}$ of B is formed by taking all elements. The element parameter h (spatial discretization parameter) is given as an elementwise constant function $h_K := \mathrm{diam}(K)$ (where $\mathrm{diam}(K)$ denotes the diameter h_K of an element K). The maximum diameter is denoted by $h := \max_{K \in \mathcal{T}_h} h_K$.

We follow the standard literature [79, 82, 115] to formulate the following statements.

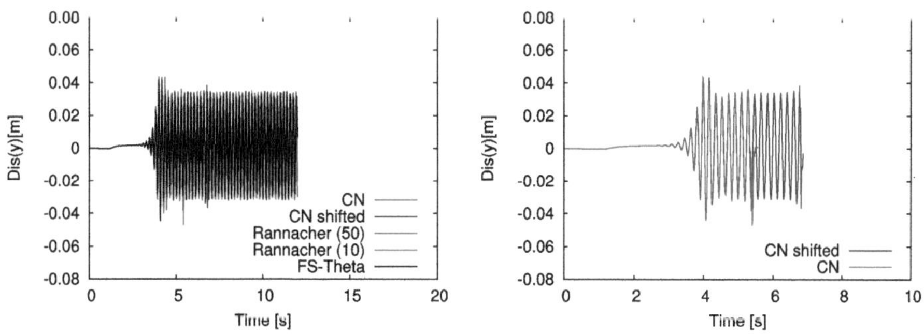

Figure 5.3: FSI-3 benchmark proposed in [240]: Comparison of the u_y displacement of the elastic beam tip using various time-stepping schemes. In comparison to the FSI-2 benchmark, also the shifted Crank–Nicolson scheme exhibits numerical instabilities and only the Rannacher time-stepping scheme and the Fractional-Step-θ scheme work satisfactory well. The results shown here are computed with [438] by augmenting the code with the Rannacher scheme and the Fractional-Step-θ scheme.

Definition 54 (Regularity). A mesh $\mathcal{T}_h = \{K\}$ is called regular if the following conditions are fulfilled:
(1) $\overline{B} = \bigcup_{K \in \mathcal{T}_h} \overline{K}$.
(2) $K_1 \cap K_2 = \emptyset$ for all cells $K_1, K_2 \in \mathcal{T}_h$ with $K_1 \neq K_2$.
(3) Any face of any cell $K_1 \in \mathcal{T}_h$ is either a subset of the boundary ∂B or a face of another cell $K_2 \in \mathcal{T}_h$.

The last condition is too restrictive for our purposes and is weakened for the following reason. To facilitate adaptive mesh refinement and to avoid connecting elements, we use the concept of hanging nodes. Cells are allowed to have nodes that lie on the midpoints of the faces or edges of neighboring cells. At most, one hanging node is allowed on each face or edge. In three dimensions, this concept is generalized to subplanes and faces because we must deal with two types of lower manifolds.

We define continuous H^1-conforming finite element spaces V_h^l by (see [82, 115, 253]):

$$V_h^l := \{v_h \in C(\overline{B}) \mid v_h|_K \in \mathcal{Q}(K)\ \forall K \in \mathcal{T}_h\} \subseteq H^1(B).$$

Here, $\mathcal{Q}(K)$ denotes the space of polynomial-like functions on $K \in \mathcal{T}_h$. In the following, we introduce the space $\mathcal{Q}_l(K)$ of tensor product polynomials up to degree l. On the reference element $K_{\text{unit}} = (0,1)^d$ they are defined as

$$\mathcal{Q}_l(K_{\text{unit}}) := \text{span}\left\{\prod_{i=1}^{d} x_i^{\alpha_i} \mid \alpha_i \in \{0,1,\ldots,l\}\right\}.$$

We consider for each $K \in \mathcal{T}_h$ the bilinear transformation $\sigma_K : K_{\text{unit}} \to K$. Then the Q_1^c element is defined as

$$Q_1^c(K) = \{q \circ \hat{\sigma}_K^{-1} : q \in \text{span}\langle 1, x, y, xy\rangle\} \qquad (d = 2),$$
$$Q_1^c(K) = \{q \circ \sigma_K^{-1} : q \in \text{span}\langle 1, x, y, z, xy, xz, yz, xyz\rangle\} \qquad (d = 3),$$

with dim $Q_1^c = 4$ (in 2D) and $Q_1^c = 8$ (in 3D) in which the dimension denotes the local degrees of freedom on a single cell. The Q_2^c element (in two dimensions) is defined as

$$Q_2^c(K) = \{q \circ \sigma_K^{-1} : q \in \text{span}\langle 1, x, y, xy, x^2, y^2, x^2y, y^2x, x^2y^2\rangle\},$$

with dim $Q_2^c = 9$. The P_1^{dc} element consists of linear functions defined by

$$P_1^{dc}(\hat{K}) = \{q \circ \hat{\sigma}_K^{-1} : q \in \text{span}\langle 1, x, y\rangle\},$$

with dim $P_1^{dc}(\hat{K}) = 3$.

Remark 37. If the transformation σ_K itself is an element of $\mathcal{Q}_l(K)^d$, the corresponding finite element space is called isoparametric.

Extending these concepts to finite element spaces in the case of hanging nodes requires some remarks. To enforce global continuity (i. e., global conformity), the degrees of freedom located on the interface between different refinement levels have to satisfy additional constraints. They are determined by interpolation of neighboring degrees of freedom. Therefore, hanging nodes do not carry any degrees of freedom. For more details on this, we refer to [101].

To ensure the approximation properties of the finite element spaces, additional conditions on the geometry of the cells are required. The two classical assumptions from the literature [82, 79] are the so-called *uniformity* and the weaker *quasi-uniformity*.

Definition 55 (Quasi-uniformity). A family of meshes $\{\mathcal{T}_h \mid h \searrow 0\}$ is called quasi-uniform if there is a constant κ such that the following two conditions are fulfilled:

(1) For each transformation $\sigma_K : K_{\text{unit}} \to K$, it holds

$$\frac{\sup\{\|\nabla\sigma_K(x)x\| \mid x \in K,\ \|x\| = 1\}}{\inf\{\|\nabla\widehat{\sigma}_K(x)x\| \mid x \in K,\ \|x\| = 1\}} \le \kappa, \qquad K \in \bigcup_h \mathcal{T}_h. \tag{5.10}$$

(2) It holds

$$\frac{h_K}{\rho_K} \le \kappa \quad \forall K \in \bigcup_h \mathcal{T}_h,$$

where ρ_K is the radius of the inner circle fitting into K.

5.4.2 Incremental, spatially discretized weak formulations of quasi-static brittle phase-field fracture

After spatial discretization, we arrive formally at the following incremental version.

Proposition 31 (Proposition 27, spatially discretized with FEM). *Let $V_{u,h}^D \subset V_u^D$ and $V_{\varphi,h} \subset V_\varphi$ be conforming finite element spaces. For the loading steps $n = 1,\dots,N$, given φ_h^{n-1}, find $U_h^n = (u_h, \varphi_h) \in V_{u,h}^D \times V_{\varphi,h}$ with $\varphi_h(0) = \varphi_{0,h}$, such that*

$$A(U_h^n)(\Psi_h) = 0 \quad \forall \Psi_h = (w_h, \psi_h) \in V_{u,h}^0 \times V_{\varphi,h},$$

where

$$A(U_h^n)(\Psi_h) = A_1(U_h^n)(w_h) + A_2(U_h^n)(\psi_h),$$

with

$$\begin{aligned}
A_1(U_h^n)(w_h) &= (g(\varphi_h^n)\sigma(u_h^n), \nabla w_h),\\
A_2(U_h^n)(\psi_h) &= ((1-\kappa)\varphi_h^n\sigma(u_h^n) : e(u_h^n), \psi_h)\\
&\quad + \left(-\frac{G_c}{\varepsilon}(1-\varphi_h^n), \psi_h\right) + (G_c\varepsilon\nabla\varphi_h^n, \nabla\psi_h)\\
&\quad + (\gamma[\varphi_h^n - \varphi_h^{n-1}]_+, \psi_h).
\end{aligned}$$

Proposition 32 (Proposition 28, spatially discretized with FEM). *All notation as just before, we obtain for the two semilinear forms:*

$$\begin{aligned} A_1(U_h^n)(w_h) &= (g(\varphi_h^n)\sigma^+(u_h^n), \nabla w_h) + (\sigma^-(u_h^n), \nabla w_h), \\ A_2(U_h^n)(\psi_h) &= ((1-\kappa)\varphi_h^n \sigma^+(u_h^n) : e(u_h^n), \psi_h) \\ &\quad + \left(-\frac{G_c}{\varepsilon}(1-\varphi_h^n), \psi_h\right) + (G_c \varepsilon \nabla \varphi_h^n, \nabla \psi_h) \\ &\quad + (\gamma[\varphi_h^n - \varphi_h^{n-1}]_+, \psi_h). \end{aligned}$$

Remark 38. For the augmented Lagrangian penalization, in both previous formulations, only the function Ξ_h must be added. Updating Ξ_h shall be discussed in Section 7.11.

Lastly, we also address the fully discrete version of Proposition 25.

Proposition 33. *For the loading steps $n = 1, 2, 3, \ldots, N$, given $\mathcal{H}_h^{n-1} := \mathcal{H}(u_h^{n-1})$, find $U_h^n = (u_h^n, \varphi_h^n) \in V_{u,h}^D \times V_{\varphi,h}$ with $\varphi_h(0) = \varphi_{0,h}$, such that*

$$A(U_h^n)(\Psi_h) = 0 \quad \forall \Psi_h = (w_h, \psi_h) \in V_{u,h}^0 \times V_{\varphi,h},$$

where

$$A(U_h^n)(\Psi_h) = A_1(U_h^n)(w_h) + A_2(U_h^n)(\psi_h),$$

and

$$\begin{aligned} A_1(U_h^n)(w_h) &= (g(\varphi_h^n)\sigma^+(u_h^n), \nabla w_h) + (\sigma^-(u_h^n), \nabla w_h), \\ A_2(U_h^n)(\psi_h) &= ((1-\kappa)\varphi_h^n \mathcal{H}_h^{n-1}, \psi_h) + \left(-\frac{G_c}{\varepsilon}(1-\varphi_h^n), \psi_h\right) + (G_c \varepsilon \nabla \varphi_h^n, \nabla \psi_h). \end{aligned}$$

Updating $\mathcal{H}_h^{n-1}$ is part of the nonlinear solution procedure and explained in Section 7.13.

5.5 Thoughts on the interaction of κ, ε and h

A key aspect in any phase-field method (fracture, fluids, materials sciences, and so forth) is the choice of the regularization parameter ε and its relation to the spatial discretization parameter h. A rigorous numerical analysis of Proposition 31, Proposition 32 or Proposition 33 with respect to ε and h is not known to the best of the author's knowledge. As we find out during this section, questions concerning "convergence" are subtle and finally depend on the field and research objectives. Often, choices like $\varepsilon = 2h$ or $\varepsilon = 8h$ are made in the literature and by ourselves. In the following subsection, we provide first some heuristic arguments, why such choices are reasonable and what else should be thought of. In the subsection that follows afterward, we restrict our attention to a (very) simplified prototype model problem. A summary and suggestions are finally provided in Section 5.5.3.

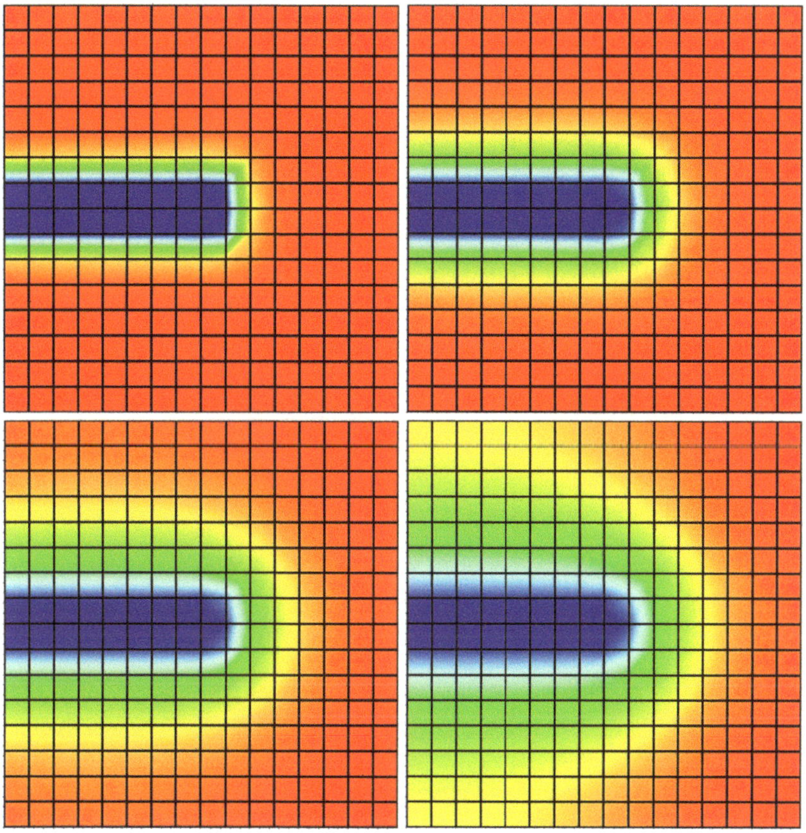

Figure 5.4: Approximation of a fracture with phase-field. Going from top left to bottom right: $\varepsilon = 0.5h, h, 2h, 4h$. We clearly see from the top left figure that for $\varepsilon < h$, we lose the approximation property. The limit case is $h \lesssim \varepsilon$, namely the top right figure.

5.5.1 Heuristic arguments for the relationship between ε and h

In this section, we provide some heuristic arguments how ε should be chosen. From Figure 5.4, we would heuristically think that the following should hold true.

Proposition 34 (Heuristic argument for size of ε). *From geometric and basic numerical considerations, using linear or bilinear finite elements, it must hold*

$$h \lesssim \varepsilon.$$

This means that we must capture the transition zone of size ε (i. e., the fracture boundary $\partial\mathcal{C}$) within the resolution of the finite element mesh.

Proof. The argument is as follows: finite element functions, such as $u_h(x)$ and $\varphi_h(x)$ are represented by their function values in the DoFs. For a given element $K \subset \mathcal{T}_h$, let h be the discretization parameter; respectively, a constant of square root of two, depending whether element edges or the diameter are denoted by h.

Let us exemplarily consider one spatial dimension $d = 1$. For linear finite elements, let x_{m-1} and x_m be two neighboring nodal points of the mesh. Then the mesh size is $h = x_m - x_{m-1}$ for the element $K_m = [x_{m-1}, x_m]$. For linear elements, the nodal points and DoFs coincide. Here, the phase-field function is well-defined with $\varphi_h(x_m)$ and $\varphi_h(x_{m-1})$. Furthermore, the transition zone ε basically determines the region in which φ_h changes from 0 to 1. For $\varepsilon > h$ (e. g., $\varepsilon = 2h$) and $\varepsilon = h$ there exist DoFs with $\varphi_h(x) < 1$ even when the fracture boundary ∂C is located within K_m. Specifically, the larger $\varepsilon \gg h$, the more DoFs cover the fracture area Ω_F.

However, if the transition zone is much smaller than h, i. e., $\varepsilon \ll h$, then φ_h changes from 0 to 1 in a small domain $\omega_\varepsilon \subset K_m$. Specifically, $\omega_\varepsilon := (-\varepsilon, \varepsilon)$. If ω_ε is a proper subset with $x_{m-1} \notin \omega_\varepsilon$ and $x_m \notin \omega_\varepsilon$, then $\varphi_h(x_{m-1}) = \varphi_h(x_m) = 1$.

Consequently, the fracture, marked by $\varphi_h < 1$, cannot be resolved because the finite element discretization is too coarse. □

Remark 39. As we shall see in the following section for a prototype, simplified situation, we must be able to represent the gradient $\nabla\varphi$ with the help of the spatial discretization. For this reason, this heuristic argument will be justified in a rigorous way.

Remark 40. Some other discussions were already provided in the first work [72]. Discussions for related problems can be found in [80] and [69]. Furthermore, we refer to a detailed numerical analysis in [111] and computational studies in [463] for the Allen–Cahn equation in which relationships between ε and h are investigated. Moreover, [164, 163] provide further numerical analysis for Allen–Cahn and phase-field models.

Proposition 35 (Higher-order finite elements). *The previous heuristic proposition may yield the hypothesis that*

$$\frac{h}{2^{l-1}} \lesssim \varepsilon$$

where l is the polynomial degree of the finite element. For $l = 1$, we obtain the previous result. For quadratic or biquadratic finite elements (see Figure 5.5), we have $l = 2$ and, therefore, $\frac{h}{2} \lesssim \varepsilon$.

Proof. Our arguments are based on the proof of Proposition 34. We work again in one dimension and consider $l = 2$ (quadratic finite elements) as an example. We now have as additional degree of freedom the midpoint $x_{m-1/2}$ in K_m. The distance to the neighboring DoFs is $h/2$. Consequently, following the arguments from Proposition 34, we can now choose as limiting size $\varepsilon = h/2$. These arguments can be extended to higher dimensions and higher polynomial degrees. □

Corollary 2. *When small $0 < \varepsilon \ll 1$ are of interest, then h must be very small requiring high resolution meshes in the crack region, where φ has steep gradients. This is partic-*

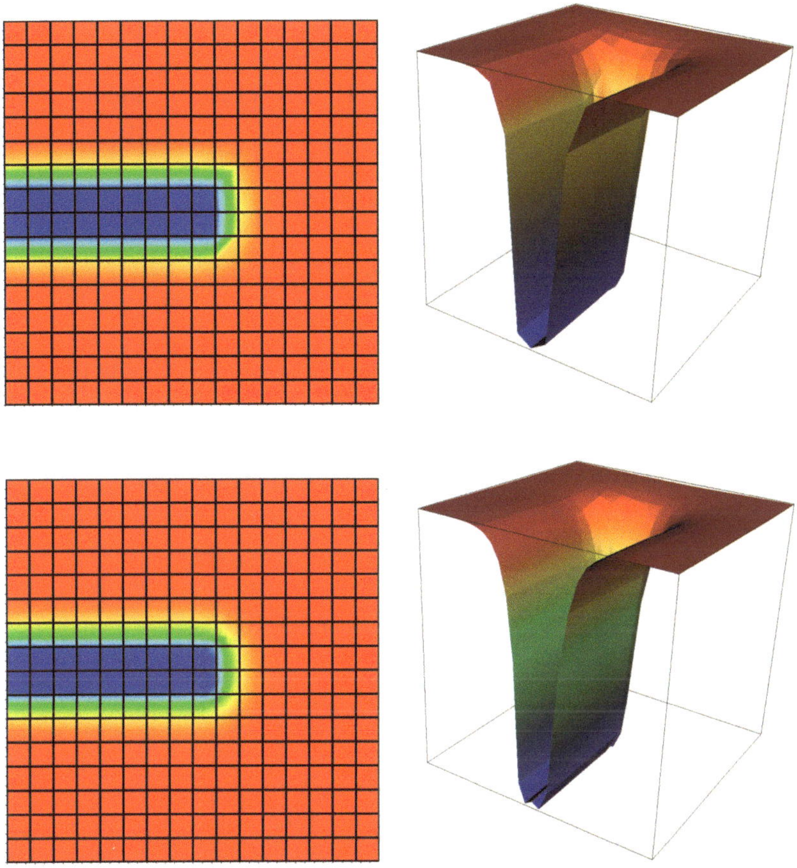

Figure 5.5: Comparison of the phase-field profile of Q_1 (top row) and Q_2 (bottom row) approximation for the case $\varepsilon = 0.5h$. In the right column, 3D surface plots better highlight the profile of the phase-field function and particularly its steep gradients $\nabla\varphi$.

ularly challenging in three dimensions and may result in a high computational cost. In our view, local mesh adaptivity is here an indispensable tool (Chapter 9).

Remark 41. Again we refer to [463, Section 2; specifically pp. 3044 and 3048 for comments on low-/higher-order discretizations] and we also notice that small-scale oscillations of size ε in relation to finite element mesh sizes are also important in numerical multiscale modeling [1].

5.5.2 Numerical analysis of a simplified, prototype problem

In this section, we derive some properties and error estimates for a simplified, prototype model problem. The objective is to obtain a better understanding of the interaction of the phase-field regularization parameters κ and ε with the spatial mesh size h.

5.5.2.1 Main objectives

The main objectives are:

1. Boundedness of $0 \le \varphi \le 1$ and behavior of the H^1 norm of φ (proofs are simplifications of results derived in [336])
2. Relation of $|\nabla u|$ to ε
3. Approximation results and finite element error estimates for the relation between κ and h (concerns the displacement PDE)
4. Approximation results and finite element error estimates for the relation between ε and h (concerns the phase-field variational inequality)

Some related known results, to the best of our knowledge, can be found in [69, 80]. Therein, qualitative rigorous results, in the form of $h \ll \kappa \ll \varepsilon$ were derived for similar-type equations using Γ convergence techniques. However, we neither send $\kappa \to 0$ nor $\varepsilon \to 0$, but rather work with classical finite element interpolation results for small, but fixed $\kappa > 0$ and $\varepsilon > 0$.

5.5.2.2 Problem statements and first results

In this first subsection, we outline our assumptions and state the governing equations.

Hypothesis 1. *We make the following assumptions:*

1. *Sufficient regularity of the domain B and its boundary ∂B;*
2. *Laplace terms; no elasticity;*
3. *No inequality constraint; thus "only" linear equations;*
4. *We assume as much regularity as needed; in particular for the term $\varphi|\nabla u|^2$;*
5. *Equations are decoupled and fully steady-state, thus both of classical elliptic type;*
6. *No proof that the coupled system is well-posed;*
7. *We assume from* [80, 21] *and* [173], *that*

$$0 < \kappa \ll \varepsilon$$

   and both are small, but fixed;
8. *For the finite element approximations and interpolation results, we make classical assumptions and specifically follow Braess* [79].

In the following, we state both prototype problems that we have in mind throughout this section.

Formulation 16 (Prototype steady-state phase-field displacements). Let the previous assumptions hold true. Let $\varphi : B \to \mathbb{R}$ be given and positive such that the governing differential operator for the solid is elliptic. Moreover, let the previously defined degradation (Definition 39) function be given

$$g(\varphi) = (1 - \kappa)\varphi^2 + \kappa.$$

Find $u : B \to \mathbb{R}$ such that

$$-\nabla \cdot (g(\varphi)\nabla u) = f \quad \text{in } B$$
$$u = 0 \quad \text{on } \partial B,$$

and the right-hand side data $f : B \to \mathbb{R}$.

Proposition 36 (Choice of κ). *For the degradation function, it holds per definition:*

$$\varphi = 1 \quad \Rightarrow \quad g(\varphi) = 1,$$
$$\varphi = 0 \quad \Rightarrow \quad g(\varphi) = \kappa.$$

We immediately obtain the following results. For $\varphi = 1$ it always holds $g(\varphi) = 1$. For $\varphi \to 0$ (fractured material), the regularization parameter should be chosen such that

$$\kappa \ll 1$$

because otherwise κ dominates $g(\varphi)$ (see Table 5.1) for $\varphi \ll 1$ and artificial stiffness is added in the fracture for $\varphi = 0$ as well as in the transition zone for $0 < \varphi < 1$. For this reason, choices like $\kappa = 10^{-10}$ are justified.

Example 11 (Quantification of the model error due to κ). We provide some hints for the model error caused by κ. Let us consider three values of $\varphi = 1.0, 0.1, 0.0$. These values are chosen on purpose:
- $\varphi = 1$: intact material;
- $\varphi = 0.1$: example as level-set isoline to determine fracture boundary (Section 11.2);
- $\varphi = 0$: (full) fracture.

As indicated, values $\varphi = 0.1, \dots, 0.3$ are interesting from the practical side because they are used to determine ∂C. Let us also consider three values of $\kappa = 10^0, 10^{-1}, 10^{-12}$. Our results are displayed in Table 5.1. The outcome is trivial, but interesting in view

Table 5.1: Evaluation of $g(\varphi)$ for different φ and κ.

κ	φ	$g(\varphi)$	Desired
1	1	1	1
1	0.1	1	small
1	0	1	TOL
0.1	1	1	1
0.1	0.1	0.109	small
0.1	0	0.1	TOL
10^{-12}	1	1	1
10^{-12}	0.1	0.01	small
10^{-12}	0	10^{-12}	TOL

of Theorem 6. For $\varphi = 1$, we obtain the desired values $g(1) = 1$ independently of κ. For $\varphi = 0.1$ (rows 2,5,8), let us compare the last two ones. Here, the absolute error introduced by κ is

$$|g_{\kappa=0.1}(0.1) - g_{\kappa=10^{-12}}(0.1)| = 0.099,$$

which is a model error of about 10 %. The relative error is 990 %.

Corollary 3 (Domain of influence of κ). *Due to the previous result, κ does only influence the solution u for $\kappa \approx \varphi^2$. This results in a small strip defined as*

$$\Omega_\kappa = \{x \in B \mid 0 \le \varphi^2 < \kappa \ll 1\}.$$

In practice, κ only serves the purpose to avoid singular system matrices in the discretization. Indeed, for $\varphi = 0$, we have

$$-\nabla \cdot (\kappa \nabla u) = f.$$

In theory, one can even work with $\kappa = 0$; see Remark 26 (Section 4.4.3.2).

Formulation 17 (Prototype steady-state phase-field equation). Let the previous assumptions hold true. Let $u : B \to \mathbb{R}$ be given and $|\nabla u|^2 > 0$. Find $\varphi : B \to \mathbb{R}$ such that

$$-\varepsilon \Delta \varphi - \frac{1}{\varepsilon}(1 - \varphi) + |\nabla u|^2 \varphi = 0 \quad \text{in } B,$$
$$\varepsilon \partial_n \varphi = 0 \quad \text{on } \partial B.$$

In the following, we derive weak forms with well-known techniques.

Proposition 37 (Weak form displacements). *Let $f \in L^2(B)$ and $g(\varphi) \in L^\infty(B)$. Find $u \in V := H_0^1(B)$ such that*

$$a(u, w) = l(w) \quad \forall w \in V$$

with

$$a(u, w) = (g(\varphi)\nabla u, \nabla w),$$
$$l(w) = (f, w).$$

Proof. The weak form follows immediately from standard arguments. The equivalence to the strong form can be shown with standard arguments using the fundamental lemma of calculus of variations.[1] □

For the phase-field variational inequality, we slightly reformulate the problem and write the term $-\frac{1}{\varepsilon}1$ as right-hand side functional.

1 The fundamental lemma of calculus of variations can be found, for instance, in [116].

Proposition 38 (Weak form phase-field equation). *Let $\varepsilon > 0$ and $|\nabla u|^2 \in L^\infty(B)$ be given. Find $\varphi \in H^1(B)$ such that*

$$a(\varphi, \psi) = l(\psi) \quad \forall \psi \in H^1(B)$$

with

$$a(\varphi, \psi) = \varepsilon(\nabla\varphi, \nabla\psi) + \frac{1}{\varepsilon}(\varphi, \psi) + (|\nabla u|^2\varphi, \psi),$$

$$l(\psi) = \frac{1}{\varepsilon}(1, \psi).$$

Proof. As before, the weak form follows immediately from standard arguments. The equivalence to the strong form can be shown again with standard arguments using the fundamental lemma of calculus of variations. □

5.5.2.3 Well-posedness

Since both equations are elliptic, their well-posedness follows trivially from the Lax-Milgram lemma. As norm, we use the usual H^1 norm (Section 2.1.6):

$$\|v\|_V^2 := \|v\|_{H^1}^2 = \int_B v^2 + |\nabla v|^2 \, dx.$$

The three assumptions are the following.

Hypothesis 2. *We suppose:*

1. *$l(\cdot)$ is a bounded linear form:*

$$|l(u)| \leq C\|u\|_V \quad \forall u \in V.$$

2. *$a(\cdot, \cdot)$ is a bilinear form on $V \times V$ and continuous:*

$$|a(u, v)| \leq \gamma\|u\|_V\|v\|_V, \quad \gamma > 0, \quad \forall u, v \in V.$$

3. *$a(\cdot, \cdot)$ is coercive (or V-elliptic):*

$$a(u, u) \geq \alpha\|u\|_V^2, \quad \alpha > 0, \quad \forall u \in V.$$

Proposition 39. *The problem formulated in Proposition 37 has a unique weak solution $u \in V$ with $H_0^1(B) \subset V \subset H^1(B)$ and the stability estimate*

$$\|u\|_V \leq \frac{\|l\|_{V^*}}{\alpha(\kappa)} = \frac{\|f\|_{L^2}}{\alpha(\kappa)},$$

with the usual dual norm $\|l\|_{V^*} := \sup_{u \neq 0} \frac{|l(u)|}{\|u\|_V}$. The constant is given by

$$\alpha(\kappa) := \frac{\kappa}{2} \min\left(1, \frac{1}{c_{PF}}\right),$$

where $c_{PF} > 0$ is the Poincaré–Friedrichs constant, e.g., [79].

Proof. We check the three assumptions to determine C, γ and α. In particular, they yield constants in terms of κ, which are useful for the later error estimates.

1. Continuity of the bilinear form.
   We use the usual arguments, i. e., applying inter alia the Cauchy–Schwarz inequality:
$$|a(u,w)| = |(g(\varphi)\nabla u, \nabla w)| \le \|g(\varphi)\|_{L^\infty}\|u\|_V\|w\|_V \le 1\|u\|_V\|w\|_V.$$
   Thus $\gamma := 1$.
2. For the coercivity, we make use of the Poincaré–Friedrichs inequality. In detail,
$$\begin{aligned} a(u,u) = (g(\varphi)\nabla u, \nabla u) &= \left(\frac{1}{2}(g(\varphi)\nabla u, \nabla u) + \frac{1}{2}(g(\varphi)\nabla u, \nabla u)\right) \\ &\ge \kappa\left(\frac{1}{2}(\nabla u, \nabla u) + \frac{1}{2c_{PF}}(u,u)\right) \\ &\ge \underbrace{\frac{\kappa}{2}\min\left(1, \frac{1}{c_{PF}}\right)}_{=:\alpha(\kappa)}\|u\|_V^2. \end{aligned}$$
   Thus, the coercivity depends mainly on κ.
3. The linearity of the right-hand side is obtained with standard arguments, yielding $|l(u)| \le \|f\|_{L^2}\|u\|_V$, i. e., $C := 1$. □

Proposition 40. *The problem formulated in Proposition 38 has a unique weak solution $\varphi \in H^1(B)$ with the stability estimate*
$$\|\varphi\|_{H^1} \le \frac{1}{\alpha\varepsilon}\|1\|_{L^2} \simeq \frac{1}{\varepsilon^2}\|1\|_{L^2}$$
where $\alpha := \alpha(\varepsilon) \sim \varepsilon$

Proof.

1. We show the stability estimate first. We use coercivity and the continuity of the right-hand side functional:
$$\alpha\|\varphi\|_V^2 \le a(\varphi,\varphi) = \frac{1}{\varepsilon}(1,\varphi) \le \frac{1}{\varepsilon}\|1\|_{L^2}\|\varphi\|_{L^2} \lesssim \frac{1}{\varepsilon}\|1\|_{L^2}\|\varphi\|_V.$$
   Dividing by α and $\|\varphi\|$ shows the assertion.
2. Now, we check the three assumptions of the Lax–Milgram lemma. Along with these proofs, we obtain specific constants in terms of ε that are useful for the later error estimates.
   (a) Continuity of the bilinear form.
$$|a(\varphi,\psi)| \le \left|\varepsilon(\nabla\varphi, \nabla\psi) + \frac{1}{\varepsilon}(\varphi,\psi) + (|\nabla u|^2\varphi, \psi)\right|$$

$$\le \underbrace{\max\left(\varepsilon, \frac{1}{\varepsilon}, |\nabla u|^2\right)}_{=:\gamma(\varepsilon)} \|\varphi\|_V \|\psi\|_V$$

We notice that we will see below, it is reasonable to assume $|\nabla u|^2 \sim \frac{1}{\varepsilon^2}$.

(b) Coercivity:

$$a(\varphi,\varphi) = \varepsilon\|\nabla\varphi\|_{L^2}^2 + \frac{1}{\varepsilon}\|\varphi\|_{L^2}^2 + (|\nabla u|^2\varphi,\varphi)$$
$$\ge \underbrace{\min\left(\varepsilon, \frac{1}{\varepsilon}, |\nabla u|^2\right)}_{=:\alpha(\varepsilon)} \|\varphi\|_{H^1}^2$$

Here, the coercivity mainly depends on the regularization parameter ε:

$$\alpha \approx \varepsilon.$$

(c) Continuity of the right-hand side:

$$|l(\psi)| = \left|\frac{1}{\varepsilon}(1,\psi)\right| \le \frac{1}{\varepsilon}\|1\|_{L^2}\|\psi\|_{H^1}.$$ □

5.5.2.4 First estimates

We arrive at the first set of main results.

Theorem 5. *With the previous problem statements and assumptions, it holds:*

1. *The phase-field function is bounded with*

$$0 \le \varphi \le 1.$$

2. *We have*

$$\|\nabla\varphi\|_{L^2}^2 + \frac{1}{2\varepsilon^2}\|\varphi\|_{L^2}^2 \le \frac{1}{2\varepsilon^2}\|1\|_{L^2}^2.$$

3. *The gradient of φ is bounded with*

$$\|\nabla\varphi\|_{L^2} \simeq \frac{1}{\varepsilon}.$$

4. *The gradient of the displacements can be approximated with*

$$\|\nabla u\|_{L^\infty} \simeq \frac{1}{\varepsilon}.$$

Proof.

1. The first result is a simplification of [336]. Since the proof is characteristic for such problems, we provide all details here. We start with the weak form in Proposition 38:

$$\varepsilon(\nabla\varphi, \nabla\psi) + \frac{1}{\varepsilon}(\varphi, \psi) + (|\nabla u|^2\varphi, \psi) = \frac{1}{\varepsilon}(1, \psi).$$

In the following, we apply cut-off arguments.

(a) First, we show that $\varphi \geq 0$. To this end, we consider $\varphi < 0$ and test with $\psi = \min(0, \varphi)$. This yields

$$\varepsilon\|\nabla\varphi\|^2_{\varphi\leq 0} + \frac{1}{\varepsilon}\|\varphi\|^2_{\varphi\leq 0} + |\nabla u|^2\|\varphi\|^2_{\varphi\leq 0} - \frac{1}{\varepsilon}(1, \varphi) = 0.$$

The first three terms are nonnegative and even positive for $\varphi < 0$. The last term is positive because of the assumption $\varphi < 0$. Consequently, the statement only holds true for $\varphi = 0$. Therefore, $\varphi \geq 0$ a. e. on B.

(b) We now show $\varphi \leq 1$. We consider $\varphi \geq 1$ and test with $\psi = \max(0, \varphi - 1)$. Furthermore, we combine the right-hand side value with the second term. This yields

$$\varepsilon(\nabla\varphi, \nabla(\varphi - 1)) + \frac{1}{\varepsilon}(\varphi - 1, \varphi - 1) + (|\nabla u|^2\varphi, \varphi - 1) = 0.$$

All terms are nonnegative and even positive for $\varphi > 1$. Consequently, the statement only holds true for $\varphi = 1$ and it can be inferred that $\varphi \leq 1$.

(c) In total, we have shown that $0 \leq \varphi \leq 1$.

2. The boundedness result follows from the following arguments. We start with Proposition 38 and take as test function $\psi := \varphi$. Then

$$\varepsilon\|\nabla\varphi\|^2_{L^2} + \frac{1}{\varepsilon}\|\varphi\|^2_{L^2} + |\nabla u|^2\|\varphi\|^2_{L^2} = \frac{1}{\varepsilon}(1, \varphi).$$

Using Youngs' inequality on the right-hand side yields

$$\varepsilon\|\nabla\varphi\|^2_{L^2} + \frac{1}{\varepsilon}\|\varphi\|^2_{L^2} + |\nabla u|^2\|\varphi\|^2_{L^2} \leq \frac{1}{2\varepsilon}\|1\|^2_{L^2} + \frac{1}{2\varepsilon}\|\varphi\|^2_{L^2}.$$

Therefore,

$$\varepsilon\|\nabla\varphi\|^2_{L^2} + \frac{1}{2\varepsilon}\|\varphi\|^2_{L^2} + |\nabla u|^2\|\varphi\|^2_{L^2} \leq \frac{1}{2\varepsilon}\|1\|^2_{L^2}.$$

Since $|\nabla u|^2\|\varphi\|^2_{L^2}$ is positive, we can neglect this term on the left-hand side. Dividing by ε yields

$$\|\nabla\varphi\|^2_{L^2} + \frac{1}{2\varepsilon^2}\|\varphi\|^2_{L^2} \leq \frac{1}{2\varepsilon^2}\|1\|^2_{L^2}.$$

3. The approximation of the gradient in terms of ε follows immediately from the previous result. We further neglect the positive term $\frac{1}{2\varepsilon^2}\|\varphi\|^2_{L^2}$. Then

$$\|\nabla\varphi\|^2_{L^2} \leq \frac{1}{2\varepsilon^2}\|1\|^2_{L^2}.$$

We take the square root and obtain

$$\|\nabla\varphi\|_{L^2} \leq \frac{C}{\varepsilon},$$

with $C := \frac{1}{\sqrt{2}}\|1\|_{L^2}$.

4. For the gradient behavior of u, we start from

$$\|\nabla\varphi\|_{L^2}^2 + \frac{1}{2\varepsilon^2}\|\varphi\|_{L^2}^2 + |\nabla u|^2\|\varphi\|_{L^2}^2 \leq \frac{1}{2\varepsilon^2}\|1\|_{L^2}^2,$$

which was previously derived in this proof. We now neglect the first two positive terms and obtain

$$|\nabla u|^2\|\varphi\|_{L^2}^2 \leq \frac{1}{2\varepsilon^2}\|1\|_{L^2}^2.$$

Dividing by $|\nabla u|^2 > 0$ brings us to

$$\|\varphi\|_{L^2}^2 \leq \frac{1}{2|\nabla u|^2\varepsilon^2}\|1\|_{L^2}^2.$$

Asking that the coefficient

$$\frac{1}{2|\nabla u|^2\varepsilon^2}$$

is bounded, $|\nabla u|^2\varepsilon^2 \sim C$, with some constant $C > 0$, i. e.,

$$0 < \frac{1}{2|\nabla u|^2\varepsilon^2} < \infty$$

we infer that the gradient of u should behave as

$$|\nabla u|^2 \simeq \frac{1}{\varepsilon^2}$$

thus

$$|\nabla u| \simeq \frac{1}{\varepsilon}.$$

Taking the supremum yields

$$\|\nabla u\|_{L^\infty} \simeq \frac{1}{\varepsilon}.$$

This last result is of course no proof in the asymptotic limit $\varepsilon \to 0$. But our result qualitatively confirms the findings from [173, Lemma 3.4]. □

Remark 42 (Motivation for a mesh refinement scheme). The bounds of the gradient $\|\nabla\varphi\|_{L^2} \leq \frac{C}{\varepsilon}$ is the motivation of a predictor-corrector mesh refinement scheme originally proposed in [221] and recapitulated in Section 9.5.1. Since the gradients of φ behave like $\frac{1}{\varepsilon}$ specifically in the transition zone, we need to resolve them with a sufficiently accurate discretization. This has led, as just written, to a very simple, but successful, mesh refinement scheme.

5.5.2.5 Approximation results: Céa's lemma

It holds

Proposition 41 (Céa lemma; displacement problem). *Let V_u be a Hilbert space and $V_h \subset V_u$ a finite dimensional subspace. Let the assumptions of the Lax–Milgram lemma hold true. Let $u \in V_u$ and $u_h \in V_h$ be the solutions of Proposition 37 and its discrete realization. Then*

$$\|u - u_h\|_{V_u} = C_{\mathrm{Cea}} \inf_{w_h \in V_h} \|u - w_h\|_{V_u}$$

with

$$C_{\mathrm{Cea}} = \frac{\gamma}{\alpha} = \frac{1}{\frac{\kappa}{2}\min(1, \frac{1}{c_{\mathrm{PF}}})} \simeq \frac{1}{\kappa}.$$

In more detail:

1. *For small $g(\varphi)$ and in particular $\varphi^2 \approx 0$, we expect errors much larger than the approximation error due to κ.*
2. *This effect is similar to locking, however these large errors only take place in a κ-strip Ω_κ around the crack.*
3. *Away from the crack region, we have $g(\varphi) = 1$ and there, we obtain the same results as for the classical Poisson problem.*

Proof. The result follows from standard arguments by employing the continuity and coercivity of the bilinear form and using Galerkin orthogonality. □

Proposition 42 (Céa lemma; phase-field problem). *Let V_φ be a Hilbert space and $V_h \subset V_\varphi$ a finite dimensional subspace. Let the assumptions of the Lax–Milgram lemma hold true. Let $\varphi \in V_\varphi$ and $\varphi_h \in V_h$ be the solutions of Proposition 38 and its discrete realization. Then*

$$\|\varphi - \varphi_h\|_{V_\varphi} = C_{\mathrm{Cea}} \inf_{\psi_h \in V_h} \|\varphi - \psi_h\|_{V_\varphi}$$

with

$$C_{\mathrm{Cea}} = \frac{\gamma}{\alpha} = \frac{\max(\varepsilon, \frac{1}{\varepsilon}, |\nabla u|^2)}{\min(\varepsilon, \frac{1}{\varepsilon}, |\nabla u|^2)} \simeq \frac{\max(\varepsilon, \frac{1}{\varepsilon}, \frac{1}{\varepsilon^2})}{\min(\varepsilon, \frac{1}{\varepsilon}, \frac{1}{\varepsilon^2})} \simeq \frac{1}{\varepsilon^3},$$

where we employed Theorem 5 with $|\nabla u|^2 \simeq \varepsilon^{-2}$.

Proof. As before, the result follows from standard arguments by employing the continuity and coercivity of the bilinear form and using Galerkin orthogonality. □

5.5.2.6 Finite element error estimates

We assume standard regularity[2] and employ classical interpolation estimates.[3]

Lemma 2 (Interpolation estimates). *It holds on a quasi-uniform triangulation:*

$$\|v - i_h v\|_{L^2} \le ch^2 |v|_{H^2} = \mathcal{O}(h^2), \tag{5.11}$$

$$\|v - i_h v\|_{H^1} \le ch |v|_{H^2} = \mathcal{O}(h). \tag{5.12}$$

Here, $i_h(\cdot) \in V_h$ is a uniquely determined interpolation.

We are now in a position to derive energy estimates. Again, we follow [79], now Chapter II, Paragraph 7.

Proposition 43 (A priori FEM estimate for displacements). *Let $\mathcal{T}_h$ be a quasi-uniform triangulation of the domain B and let as before $\kappa > 0$, small, but fixed. Let $u_h \in V_h$ be the discrete solution. Then it holds on triangular[4] elements:*

$$\|u - u_h\|_{V_u} \le cC_{\text{Cea}} h \|f\|_{L^2}.$$

In the κ-strip Ω_κ, we have specifically

$$\|u - u_h\|_{V_u} = O\left(\frac{h}{\kappa}\right).$$

Proof. If we assume H^2 regularity of the problem in Proposition 37 and insert the interpolation estimate into Céa's lemma, the result follows. □

Proposition 44 (A priori FEM estimate for the phase-field function). *Let $\mathcal{T}_h$ a quasi-uniform triangulation of the domain B and let as before $\varepsilon > 0$, small, but fixed. Let $\varphi_h \in V_h \subset H^1(B)$ be the discrete solution. Then it holds on triangular (quadrilateral) elements:*

$$\|\varphi - \varphi_h\|_{V_\varphi} \le cC_{\text{Cea}} h \frac{1}{\varepsilon} \|1\|_{L^2} = O\left(\frac{h}{\varepsilon^4}\right).$$

Proof. Assuming H^2 regularity of the problem in Proposition 38, we obtain first

$$\|\varphi\|_{H^2} \le c \frac{1}{\varepsilon} \|1\|_{L^2}.$$

Then we insert as usually the interpolation estimates into Céa's lemma from which we obtain the result. □

---

2 Braess [79, Chapter II, Paragraph 7].

3 Braess [79, Chapter II, Paragraph 6, Satz 6.4].

4 Similar results hold true for quadrilateral elements.

5.5.3 Summary and suggestions for κ, ε and h

Based on the literature and on what we have found in the error analysis of the simplified prototype problems, the following summary and suggestions can be made for the relationship between κ, ε, and h.

5.5.3.1 Decisions to be made

The most important aspect is to pose the right question of what is of interest as specific research goal. In view of the previous discussions, three major directions can be distinguished:

1. Studying limit cases with $\kappa \to 0$ and $\varepsilon \to 0$ and $h \to 0$, i. e., Γ convergence (calculus of variations)
2. Studying the numerical discretization, fixing $\varepsilon > 0$ and $\kappa > 0$, classical mesh refinement studies in h and time step k
3. (Rough) predictions and proof of concepts of damaged regions, fracture paths, and approximate displacements and stresses (engineering-related)

5.5.3.2 Theory: calculus of variations

Based on the studies and results presented in [80], [21], and [69] for related problems, it is reasonable to suggest the following.

Theorem 6. *For the classical phase-field fracture model derived from* [72] *(see Proposition* 19*) with Ambrosio–Tortorelli approximations, it holds asymptotically*

$$h \ll \kappa \ll \varepsilon,$$

which means in other words

$$h = o(\varepsilon), \; h = o(\kappa), \; \kappa = o(\varepsilon).$$

These results follow from Γ *convergence considerations.*

5.5.3.3 Numerical analysis of the simplified, prototype problems

In our previous numerical analysis, we have established the following results.

Theorem 7 (Choice of fixed κ in relation to h). *In the crack region* Ω_κ *with* $\varphi^2 \ll 1$ *and* $\kappa \approx \varphi^2$, *it should hold*

$$h \ll \kappa$$

in order to avoid dominating errors in κ *as a consequence of Céa's lemma. In* $B \setminus \Omega_\kappa$ *with* $\varphi \lesssim 1$, *a choice*

$$\kappa \ll 1$$

independent of h can be made. For instance, $\kappa = 10^{-12}$*; see also Proposition 36 with the accompanying example. However,* Ω_κ *is nearly irrelevant in practice and, moreover,* $h = o(\kappa)$ *is computationally infeasible when small* κ *are of interest in order to avoid model errors in the degradation function* $g(\varphi)$.

Theorem 8 (Choice of fixed ε in relation to h). *For the choice of the phase-field regularization parameter* ε*, the heuristic argument can be confirmed with the help of Theorem 5, namely*

$$h \lesssim \varepsilon$$

such that $\|\nabla\varphi\|_{L^2} \sim \frac{1}{\varepsilon}$ *is approximated with sufficient accuracy. As supporting argument, we also refer to Proposition 44 in which we found* $\|\varphi - \varphi_h\|_{V_\varphi} = O(\frac{h}{\varepsilon^4})$.

5.5.3.4 Practice and numerical experiments

Proposition 45. *Let h be the spatial discretization parameter. In practice, the following choices are frequently made:*

$$\kappa = 10^{-12}, \dots, 10^{-8},$$
$$\varepsilon = 2h, \dots, 8h.$$

These choices infer the following comments:
1. *Both choices violate Theorem 6 in the asymptotic limits.*
2. *In view of Theorem 7, the choice of* κ *is justified except in the* κ *strip* Ω_κ*, where* $\varphi^2 \ll 1$*. However, the material is almost fully cracked and it does not seem necessary to have highly accurate approximations in practice in* Ω_κ.
3. *Numerically, setting* $\varepsilon = ch$*, with* $c > 1$*, is justified because* $h \lesssim \varepsilon$ *holds true. In view of asymptotic considerations as in Theorem 6, choices* $\varepsilon = ch$ *violate* Γ *convergence. Therefore, options such as* $h = o(\varepsilon)$ *realized via*

   $$\varepsilon = ch^\beta, \quad 0 < \beta < 1,$$

   would rather be justified. Further ideas can be obtained from Proposition 44, where we emphasize however that this estimate only holds true as long as $\varepsilon > \varepsilon_0 \geq 0$.

Remark 43. For Allen–Cahn phase-field models, similar relationships were proven in [463, 163, 111].

5.5.3.5 Further decomposition into model and discretization errors

In case ε is varied, we have no proofs established in this monograph. However, the previous protopye estimates may lead to some ideas. Importantly, however, is the distinction of different error terms.

Lemma 3. *Let $U = (u, \varphi)$ or $U := u$ or $U := \varphi$ be the continuous solution for $\varepsilon = 0$ (the limit case) and U_ε the continuous solution with $\varepsilon > 0$. Finally, let $U_{\varepsilon,h}$ be the discrete solution with $\varepsilon > 0$. If ε is varied, then, we must consider*

$$\|U - U_{\varepsilon,h}\| \leq \underbrace{\|U - U_\varepsilon\|}_{\text{Model error}} + \underbrace{\|U_\varepsilon - U_{\varepsilon,h}\|}_{\text{Discretization error}}. \tag{5.13}$$

In general, the limit U does not belong to classical function spaces (see references in 4.4.3.2) we were considering so far. For this reason, investigating the model error becomes technical.

Remark 44 (Notation U_ε and U). Except in this section, we do not explicitly write U_ε when we work with $\varepsilon > 0$. We understand that in general when using U in the other parts of this monograph, we implicitly work with $\varepsilon > 0$.

Remark 45. The last part on the right-hand side of (5.13) can be obtained with the help of classical discretization error estimates as we had before. The first part is challenging in terms of convergence rates because Γ convergence does not provide such rates. In computations, we have observed rates according to β in the choice $\varepsilon = ch^\beta$. For instance in [442, Section 6.1], we obtained numerically indeed rates of β. In [222, Section 4.1, Figure 1] using parallel computing and adaptive mesh refinement for $\varepsilon = 2h$, i. e., $\beta = 1$, a clearly-visible linear rate in ε is observed. According to these simulations, this yields the hypothesis that

$$\|U - U_\varepsilon\| = O(h^\beta).$$

Specifically, it is clear, that $\beta = 0$ yields a fixed ε and consequently the limit $\varepsilon \to 0$ cannot be attained and we have $\|U - U_\varepsilon\| = O(1)$.

Remark 46. Rigorous proofs and computational analyses for the related Allen–Cahn phase-field model were established in [111, 163, 164, 463]. These confirm our previous attempts for splitting the error into a model and discretization part; see, for example, [463, Section 2]. Moreover, linear convergence rates in ε are shown in [111, Theorem 2.1].

5.5.3.6 Final recommendations

Let us summarize and close the loop to Section 5.5.3.1.

Proposition 46. *The previous results further yield the following suggestions:*

1. *For predictive engineering computations the choices $\kappa = 10^{-12}$ and $\varepsilon = ch$, with $c = 2$ for instance, are fine.*
2. *For numerical convergence studies, take $\kappa = 10^{-12}$ again, and compare the choices*
   (a) *Fixed $\varepsilon = 2h_{\text{coarse}}$ with h_{coarse} being the mesh size parameter of the coarsest mesh; send now $h \to 0$*
   (b) *Varying $\varepsilon = 2h$*

(c) *Varying $\varepsilon = ch^\beta$, $0 < \beta < 1$, that is covered by Γ convergence.*

3. *For refinement studies, compute on at least three meshes. If applicable: for time-dependent configurations, study at least three time step sizes k.*
4. *Reasonable choices of global norms or technical quantities of interest (goal functionals) are important. As we have seen, for very small ε, the bounds for the gradients become extremely large. See also Figure 5.5, which highlights the steep gradients. Therefore, it does not make sense to measure a global H^1 norm for small ε. Rather, an ε-strip with, e. g., $\varphi < 0.9$ must be cut out or other goal functionals should be adopted.*
5. *A computational analysis that confirms an asymptotic analysis is currently not feasible. The requirement of $h \ll \kappa \ll \varepsilon$ is computationally too demanding, even with local mesh adaptivity and parallel computing. Some indications of current feasibility using an academic test case were presented in [222]; see also Remark 45. Further older computations are provided in [221] and [442].*

Several aspects are numerically substantiated in Chapter 6 and Chapter 8.

The following three considerations are also of interest:

- The interpretation of ε being a physical length-scale parameter. Interpreting ε as a characteristic damage scale was already mentioned in the early work [72, p. 807]. As later indicated in [317], and also in other fields such as peridynamics [62], the parameter ε might have a physical meaning and is therefore determined by material properties. In this case, there is no reason to send ε to zero from an engineering perspective.
- On the other hand, mathematical modeling might be restricted to ε being bounded away from zero. Using upscaled models for instance, like in porous media, the mathematical model would change for $\varepsilon \to 0$. For detailed discussions, we refer to [323, 322].
- For other reasons as well, it might be sufficient to consider fixed $\varepsilon > 0$. One example are situations in which the fracture is considered as a thick object on purpose as for instance in fluid-filled fractures. Therein, the phase-field variable is interpreted as a level-set function that helps to determine the fluid-region inside the fracture [321, 286]; see as well Section 11.2.

In all these cases, numerical and computational analysis reduces to classical concepts of h (spatial) and k (time step) refinement studies. However, this does not mean that the problem becomes simpler. We are still dealing with nonstationary, nonlinear, multiphysics problems with inequality constraints for which rigorous results are challenging and numerous open questions remain.

5.6 A phase-field model for dynamic fracture

In this section, we introduce a dynamic fracture model and the corresponding variational form and temporal discretization.

Proposition 47. *Let $B \subset \mathbb{R}^d$ be open and let $I := (0,T)$ with $T > 0$ being the end time value. Find $u : B \times I \to \mathbb{R}^d$ and $\varphi : B \times I \to \mathbb{R}$ such that*

$$
\begin{aligned}
\rho_s \partial_t^2 u - \nabla \cdot (g(\varphi)\sigma(u)) = f \quad & \text{in } B \times I,\\
\partial_t \varphi + (1-\kappa)\varphi\sigma(u) : e(u) - G_c \varepsilon \Delta\varphi - \frac{G_c}{\varepsilon}(1-\varphi) \le 0 \quad & \text{in } B \times I,\\
\partial_t \varphi \le 0 \quad & \text{in } B \times I,\\
\left(\partial_t \varphi + (1-\kappa)\varphi\sigma(u) : e(u) - G_c \varepsilon \Delta\varphi - \frac{G_c}{\varepsilon}(1-\varphi)\right) \cdot \partial_t \varphi = 0 \quad & \text{in } B \times I,\\
u = 0 \quad & \text{on } \partial B \times I,\\
\varepsilon \partial_n \varphi = 0 \quad & \text{on } \partial B \times I,\\
u = u_0 \quad & \text{in } B \times \{0\},\\
\partial_t u = v_0 \quad & \text{in } B \times \{0\},\\
\varphi = \varphi_0 \quad & \text{in } B \times \{0\}.
\end{aligned}
$$

We redefine the convex set for the time-continuous case in the remainder of this section:

$$
K := \{\varphi \in V_\varphi \mid \partial_t \varphi \le 0 \text{ a. e. in } B \times I\}.
$$

A variational formulation is formally derived in the usual way.

Proposition 48. *Find $(u,\varphi) \in V_u^D \times K$ such that $u(0) = u_0$, $\partial_t u(0) = v_0$ and $\varphi(0) = \varphi_0$ are satisfied, and for almost all times $t \in I$, it holds*

$$
\begin{aligned}
(\rho_s \partial_t^2 u, w) + (g(\varphi)\sigma(u), \nabla w) = (f, w) \quad & \forall w \in V_u^0,\\
(\partial_t \varphi, \psi - \varphi) + (1-\kappa)(\varphi\sigma(u) : e(u), \psi - \varphi) &\\
+ \varepsilon(G_c \nabla\varphi, \nabla(\psi - \varphi)) - \frac{1}{\varepsilon}(G_c(1-\varphi), \psi - \varphi) \ge 0 \quad & \forall \psi \in K.
\end{aligned}
$$

The next step is a mixed formulation for the displacements in order to apply a One-Step-θ or Fractional-Step-θ scheme.

Proposition 49. *Find $(u,v,\varphi) \in V_u^D \times L^2(B) \times K$ such that $u(0) = u_0$, $v(0) = v_0$ and $\varphi(0) = \varphi_0$ are satisfied, and for almost all times $t \in I$, it holds*

$$
\begin{aligned}
(\partial_t u, w^u) - (v, w^u) = 0 \quad & \forall w^u \in L^2(B),\\
(\rho_s \partial_t v, w) + (g(\varphi)\sigma(u), \nabla w) = (f, w) \quad & \forall w \in V_u^0,
\end{aligned}
$$

$$(\partial_t\varphi, \psi - \varphi) + (1-\kappa)(\varphi\sigma(u) : e(u), \psi - \varphi)$$
$$+ \varepsilon(G_c\nabla\varphi, \nabla(\psi - \varphi)) - \frac{1}{\varepsilon}(G_c(1-\varphi), \psi - \varphi) \geq 0 \quad \forall \psi \in K.$$

As before, we sum-up all equations in order to obtain a common semilinear form:

$$A(U)(\Psi - \Phi) := A_1(U)(w^u) + A_2(U)(w) + A_3(U)(\psi - \varphi)$$

with $\Phi = (0, 0, \varphi)$ and where A_1, A_2, A_3 are the semilinear forms representing the previous weak forms.

Discretization in time is undertaken in combination with an augmented Lagrangian regularization of the variational inequality constraint. The latter one is done as before in Section 5.2.2. Specifically, we approximate

$$\gamma[\partial_t\varphi]^+ \quad \to \quad \gamma\left[\frac{\varphi - \varphi^{n-1}}{k}\right]^+.$$

Then, semidiscretization in time, using the notation and groupings from Section 5.3, yields the following.

Proposition 50 (Semidiscrete dynamic phase-field fracture). *Let $u^{n-1}, v^{n-1}, \varphi^{n-1}$ be the previous time step solutions. For $n = 1, 2, 3, \ldots, N$, find $U^n \in X^D$ such that*

$$A(U^n)(\Psi) = F^{n,n-1}(\Psi) \quad \forall \Psi \in X^0$$

with

$$\begin{aligned}
A(U^n)(\Psi) &:= A_T(U^{n,k})(\Psi) + A_E(U^n)(\Psi)\\
&= \left(\frac{u^n - u^{n-1}}{k}, w^u\right) - \theta(v^n, w^u) - (1-\theta)(v^{n-1}, w^u)\\
&\quad + \left(\rho_s\frac{v^n - v^{n-1}}{k}, w\right)\\
&\quad + \theta(g(\varphi^n)\sigma(u^n), \nabla w) + (1-\theta)(g(\varphi^{n-1})\sigma(u^{n-1}), \nabla w)\\
&\quad + \left(\frac{\varphi^n - \varphi^{n-1}}{k}, \psi\right)\\
&\quad + \theta(1-\kappa)(\varphi^n\sigma(u^n) : e(u^n), \psi)\\
&\quad + (1-\theta)(1-\kappa)(\varphi^{n-1}\sigma(u^{n-1}) : e(u^{n-1}), \psi)\\
&\quad + \theta\varepsilon(G_c\nabla\varphi^n, \nabla\psi) + (1-\theta)\varepsilon(G_c\nabla\varphi^{n-1}, \nabla\psi)\\
&\quad - \theta\frac{1}{\varepsilon}(G_c(1-\varphi^n), \psi) - (1-\theta)\frac{1}{\varepsilon}(G_c(1-\varphi^{n-1}), \psi)\\
&\quad + \left(\left[\Xi + \gamma\frac{\varphi^n - \varphi^{n-1}}{k}\right]^+, \psi\right),
\end{aligned}$$

and

$$F^{n,n-1}(\Psi) := \theta(f, w) + (1-\theta)(f^{n-1}, w).$$

This is now a sequence of quasi-stationary problems, which can be discretized in space with a Galerkin finite element scheme. Formally, the fully discrete problem reads the following.

Proposition 51 (Temporally and spatially discretized dynamic phase-field fracture). *Let u_h^{n-1}, v_h^{n-1}, φ_h^{n-1} be the previous spatially-discretized time step solutions. For $n = 1, 2, 3, \ldots, N$, find $U_h^n \in X_h^D$ such that*

$$A(U_h^n)(\Psi_h) = F^{n,n-1}(\Psi_h) \quad \forall \Psi_h \in X_h^0.$$

This is still a nonlinear discrete problem, which can finally be solved with the methods explained in Chapter 7.

6 Simulations I: a phase-field slit domain

In the previous chapter, we focused on regularizations and discretizations. In the current part, we study some properties in more detail with the help of a numerical example. In discussions with Ivo Babuska,[1] we identified a simple slit domain for which theoretical work exists [65]. Preparing the work in [442], the work [24] appeared. Therein, a manufactured solution for a crack slit is proposed and analyzed. Despite the fracture being fully steady-state, this example offers various features to better understand phase-field fracture.

6.1 Motivation and objectives

The calculations in [24, 65] yield a manufactured solution for a Poisson-type slit problem with nonhomogeneous Dirichlet conditions. This is, first of all, a simplification of the elastic phase-field fracture models, we previously derived. However, as it is well known, Poisson's problem and elasticity share some important features. The manufactured solution of [24, 65] corresponds to the case $\varepsilon = 0$ (the limit case), i. e., the continuous solution given by $\varphi = 1$ in B and $\varphi = 0$ in $\mathcal{C}$.

The following features can be studied and provide useful insight to the elastic phase-field fracture problem as well:

1. Prescription of the initial fracture with the help of the phase-field function $\varphi^0 = \varphi(t_0)$ in B. This seems too trivial, but the main question is how to prescribe φ^0 in relation to the mesh size h.
2. Influence of the penalization parameter γ.
3. Fixing κ and ε allows us to compute a numerical reference solution on a very fine mesh with sufficiently small h.
4. The previous task allows for classical mesh refinement studies with $h \to 0$ while keeping γ, κ and ε fixed.
5. Computational analysis with variations in κ and ε according to the suggestions made in Section 5.5.2.
6. Designing adaptive schemes and studying local mesh refinement and error control; see Chapter 10.
7. Testing various quantities of interest (goal functionals) such as global error norms, total crack volume, crack opening displacements, local point displacements, or stresses.
8. Performance studies of linear and nonlinear solvers. Some hints on the nonlinear solver with respect to γ are provided.

1 At this place, I want to express my appreciation to Ivo Babuska, UT Austin, Texas. He forced me in December 2013 to look closer into numerical convergence and gave several useful hints during blackboard discussions in his office.

https://doi.org/10.1515/9783110497397-006

6.2 Configuration and manufactured solution

The domain is $B := (-1,1)^2$ with a slit at $-1 \leq x \leq 0$ for $y = 0$. This slit causes a discontinuity of the boundary conditions at $(-1,0)$. The analytical solution on the slit domain $(-1,1) \setminus \{(x,0) \mid -1 \leq x \leq 0\}$ is given by [65]:

$$(\lambda_{G_c} r^{1/2} \sin \phi/2; \{(x,0) \mid -\infty \leq x \leq 0\}),$$

where polar coordinates with $r^2 = x^2 + y^2$ are used.

6.3 Boundary conditions

Employing the boundary function $u_D := g = \lambda_{G_c} \sin(\phi/2)$ on ∂B, we prescribe non-homogeneous Dirichlet conditions on all parts. Specifically, transforming g into Cartesian coordinates we have

$$x \leq 0 \text{ and } y \geq 0 : g(x,y) = \lambda_{G_c}/\sqrt(2) * \sqrt{\sqrt{x^2+y^2} - x},$$

$$x \leq 0 \text{ and } y \leq 0 : g(x,y) = -\lambda_{G_c}/\sqrt(2) * \sqrt{\sqrt{x^2+y^2} - x},$$

$$x \geq 0 \text{ and } y \geq 0 : g(x,y) = \lambda_{G_c}/\sqrt(2) * \sqrt{\sqrt{x^2+y^2} - x},$$

$$x \geq 0 \text{ and } y \leq 0 : g(x,y) = -\lambda_{G_c}/\sqrt(2) * \sqrt{\sqrt{x^2+y^2} - x}.$$

For the phase-field variable, traction-free conditions (i. e., homogeneous Neumann conditions) are applied on ∂B, i. e., $\varepsilon \partial_n \varphi = 0$ on ∂B. Plots of the numerical solutions are provided in the Figures 6.2 and 6.3.

6.4 Initial condition

For the phase-field function, we need an initial value φ^0 to prescribe the slit in the initial geometry. Here, we use

$$\varphi^0 := \{(x,y) \in B \mid \; -1 \leq x \leq 0; -h \leq y \leq h\},$$

where h is the mesh size parameter defined in the next subsection; see also Figure 6.1 for an illustration.

6.5 Model, material, discretization parameters

As model parameters, we have γ, κ, ε. As material parameters, we have G_c, λ_{G_c}, μ. Lastly, the discretization parameter is h. In detail, the model and material parameters

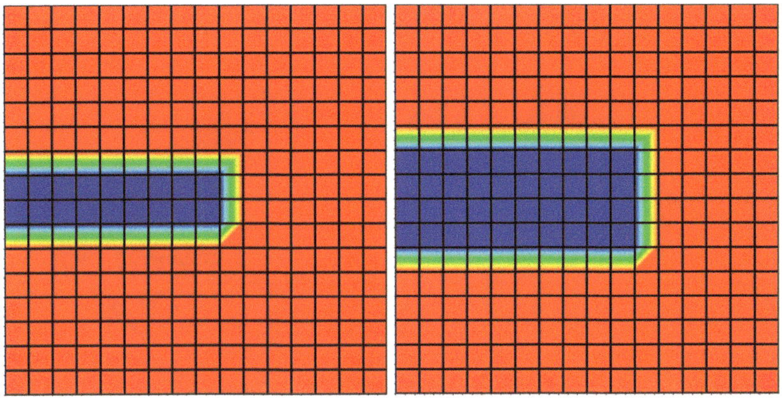

Figure 6.1: Initial conditions φ_0 in the domain $B = (-1,1)^2$. On the left, the initial fracture is initialized with the mesh size h. On the right, the initial fracture is initialized with the mesh size $2h$.

are given as $\kappa = 10^{-12}$, $G_c = \lambda_{G_c}^2 \times \sqrt{\pi/2}$, $\lambda_{G_c} = 1.0$, and $\mu = 1.0$. The penalization parameter is $\gamma = 10^3$. The phase-field regularization parameter ε will be adapted in the various examples and will be provided individually.

6.6 Quantities of interest

We consider the following norms:
1. L^2 norm of φ: $\|\varphi - \varphi_{\varepsilon,h}\|_{L^2(B)}$;
2. L^2 norm of u: $\|u - u_{\varepsilon,h}\|_{L^2(B)}$;
3. H^1 norm of u: $\|u - u_{\varepsilon,h}\|_{H^1(B)}$;
4. The global H^1 norm is disputable in the entire domain B as discussed in Section 5.5; see specifically Subsection 5.5.3.6. For this reason, we also investigate a cut-off on the domain $\tilde{B} := \{x \in B \mid x \geq 0.5 \cup x \leq 0.5\}$:

$$\|u - u_{\varepsilon,h}\|_{H^1(\tilde{B})}.$$

Furthermore, we have the following functionals in mind:
1. Point value: $J^P(u) = u(0.75, -0.5)$,
2. Half domain functionals: $J^D(u) = \int_{\{x\geq 0\}} u\,dx$,
3. Crack energy: $J^C(\varphi) = E_s(\varphi) = \frac{1}{2}\int_B \varepsilon|\nabla\varphi|^2\,dx + \frac{1}{2\varepsilon}\int_B (1-\varphi)^2\,dx$.

6.7 Discussion of our findings

We ran several computations and will analyze and interpret them in the following. Some results are basic, but very illustrative and perfect for a better understanding of the phase-field fracture approach.

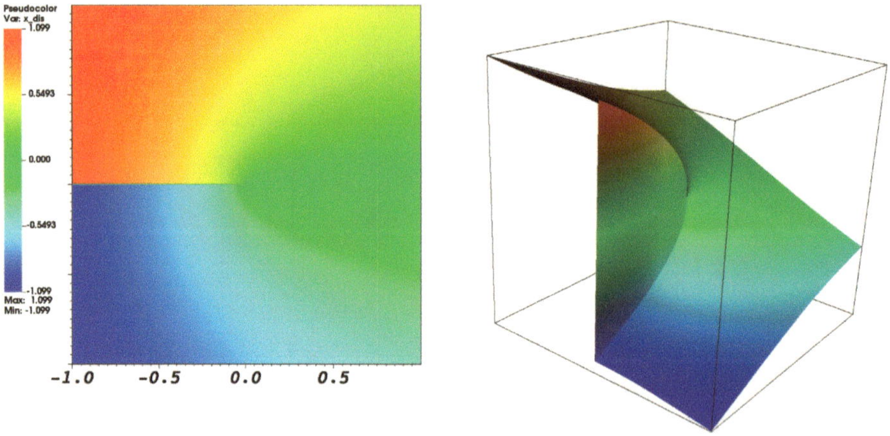

Figure 6.2: Plots of u on the finest mesh. Left: standard plot. Right: 3D surface plot in order to highlight the jump of u across the fracture with steep gradients ∇u.

Solution plots for $\varepsilon = 2h$

The physical displacements are plotted twice in Figure 6.2. First, in a standard 2D plot and additionally in a 3D surface plot that highlights the discontinuity in the displacement field. The phase-field solution is displayed in Figure 6.3 for six refinement steps. Therein, the effect of linking ε to h is clearly visible.

Global norms and point value for $\varepsilon = 2h$ with uniform refinement, Q_1^c discretization

In this first set of computations, we study an $\varepsilon - h$ relation, which is often used in the literature. As we observe in the Tables 6.1 and 6.2, for global L^2 norms and the point value, we obtain convergence rates between 0.5 and 1. As hypothesized in Section 5.5.3.6, we do not obtain convergence for the displacements in the global H^1 norm. However, when we cut out the fracture region, we obtain a rate for H^1 around 0.85.

Global norms and point value for $\varepsilon = 2h$ with uniform refinement, Q_2^c discretization

In this second set of numerical tests (Tables 6.3 and 6.4), we observe again convergence rates less than 1 in general and for the global H^1 norm again no convergence. Overall, there is no benefit using a higher-order discretization. However, these results give some indication about the validity of Remark 45. We still work with $\varepsilon = 2h$, which means a linear relationship between ε and h. Therefore, the entire error (model and discretization) might be bounded by $\beta = 1$, i. e., we have linear convergence in ε.

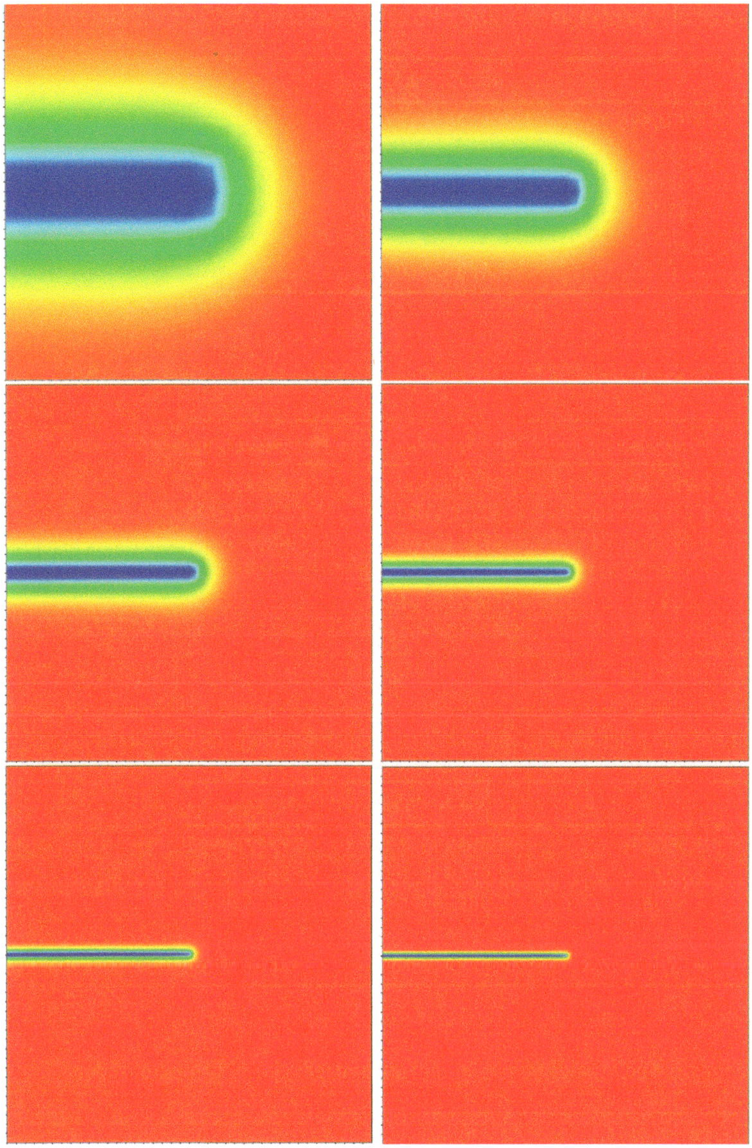

Figure 6.3: Plots of φ for $\varepsilon = 2h$ on uniformly refined meshes. On the finest mesh, we have 262 144 elements, 263 169 Dofs for u and 263 169 Dofs for φ with Q_1^c discretization. The mesh size parameter is $h = 5.52427 \cdot 10^{-3}$ and $\varepsilon = 1.10485 \cdot 10^{-2}$.

Global norms and point value for $\varepsilon = 2h_{\text{coarse}} = 0.353553$ with uniform refinement, Q_1^c discretization

In this numerical test, ε is fixed to the coarse mesh and only h is varied. Again, the total error from Lemma 3 is observed. As indicated in Remark 45, the total error should be

Table 6.1: Global norms and point value for $\varepsilon = 2h$ with uniform refinement, Q_1^c discretization.

h	ε	$u(0.75, -0.5)$	$\|\varphi - \varphi_{eps,h}\|_{L^2}$
1.76776e−01	3.53553e−01	9.92766e−03	6.65147e−01
8.83883e−02	1.76776e−01	5.33744e−03	4.07929e−01
4.41941e−02	8.83883e−02	2.87674e−03	2.64571e−01
2.20970e−02	4.41941e−02	1.56450e−03	1.77630e−01
1.10485e−02	2.20970e−02	8.54456e−04	1.22886e−01
5.52427e−03	1.10485e−02	4.66603e−04	8.66212e−02

Table 6.2: Global norms and point value for $\varepsilon = 2h$ with uniform refinement, Q_1^c discretization.

h	$\|u - u_{eps,h}\|_{L^2}$	$\|u - u_{eps,h}\|_{H^1}$	$\|u - u_{eps,h}\|_{H^1(\tilde{B})}$
1.76776e−01	2.25360e−01	3.28350e+00	1.28949e−01
8.83883e−02	1.57825e−01	4.39123e+00	7.17441e−02
4.41941e−02	1.09320e−01	5.97844e+00	3.98377e−02
2.20970e−02	7.57405e−02	8.25670e+00	2.20526e−02
1.10485e−02	5.26920e−02	1.15146e+01	1.21450e−02
5.52427e−03	3.68272e−02	1.61551e+01	6.64832e−03

Table 6.3: Global norms and point value for $\varepsilon = 2h$ with uniform refinement, Q_2^c discretization.

h	ε	$u(0.75, -0.5)$	$\|\varphi - \varphi_{eps,h}\|_{L^2}$
1.76776e−01	3.53553e−01	8.30593e−03	6.77870e−01
8.83883e−02	1.76776e−01	4.41495e−03	4.18879e−01
4.41941e−02	8.83883e−02	2.35965e−03	2.73134e−01
2.20970e−02	4.41941e−02	1.27261e−03	1.84253e−01
1.10485e−02	2.20970e−02	6.89267e−04	1.28390e−01

Table 6.4: Global norms and point value for $\varepsilon = 2h$ with uniform refinement, Q_2^c discretization.

h	$\|u - u_{eps,h}\|_{L^2}$	$\|u - u_{eps,h}\|_{H^1}$	$\|u - u_{eps,h}\|_{H^1(\tilde{B})}$
1.76776e−01	2.12162e−01	3.35415e+00	1.07176e−01
8.83883e−02	1.50870e−01	4.42787e+00	5.89949e−02
4.41941e−02	1.05871e−01	5.99364e+00	3.24509e−02
2.20970e−02	7.41070e−02	8.25971e+00	1.78001e−02
1.10485e−02	5.19459e−02	1.15113e+01	9.71886e−03

$O(1)$. Indeed, we observe no (significant) convergence in any goal functional as shown in the Tables 6.5 and 6.6.

Table 6.5: Global norms and point value for $\varepsilon = 2h_{\text{coarse}} = 0.353553$ with uniform refinement, Q_1^c discretization.

h	ε	$u(0.75, -0.5)$	$\|\varphi - \varphi_{\text{eps},h}\|_{L^2}$
1.76776e−01	3.53553e−01	9.92766e−03	6.65147e−01
8.83883e−02	3.53553e−01	1.66914e−02	7.77520e−01
4.41941e−02	3.53553e−01	1.75405e−02	8.30471e−01
2.20970e−02	3.53553e−01	1.75858e−02	8.53918e−01
1.10485e−02	3.53553e−01	1.75052e−02	8.63479e−01
5.52427e−03	3.53553e−01	1.73946e−02	8.66369e−01

Table 6.6: Global norms and point value for $\varepsilon = 2h_{\text{coarse}} = 0.353553$ with uniform refinement, Q_1^c discretization.

h	$\|u - u_{\text{eps},h}\|_{L^2}$	$\|u - u_{\text{eps},h}\|_{H^1}$	$\|u - u_{\text{eps},h}\|_{H^1(\tilde{B})}$
1.76776e−01	2.25360e−01	3.28350e+00	1.28949e−01
8.83883e−02	2.01387e−01	4.28395e+00	1.71584e−01
4.41941e−02	1.81495e−01	5.98922e+00	1.74764e−01
2.20970e−02	1.72314e−01	6.63652e+00	1.74176e−01
1.10485e−02	1.68434e−01	6.08944e+00	1.72781e−01
5.52427e−03	1.66089e−01	5.82924e+00	1.71000e−01

Global norms and point value for $\varepsilon = 2\sqrt{h}$ with uniform refinement, Q_1^c discretization

In this numerical test, we choose ε according to requirements for the discretized problem, namely that h should tend faster to zero than ε, i. e., $h = o(\varepsilon)$; see Section 5.5.3.6. The power is $\beta = 0.5$. For global L^2 norms in the entire domain, we observe a significantly reduced convergence order (see Tables 6.7 and 6.8). As in the other examples, no convergence is obtained in the global H^1 norm. For the point value and the cut H^1 norm, orders of about $\alpha = 0.85$ are observed. This indicates that away from the crack, the choice of ε plays a minor role and we obtain orders similar to the first numerical test.

Table 6.7: Global norms and point value for $\varepsilon = 2\sqrt{h}$ with uniform refinement, Q_1^c discretization.

h	ε	$u(0.75, -0.5)$	$\|\varphi - \varphi_{\text{eps},h}\|_{L^2}$
1.76776e−01	8.40896e−01	9.92766e−03	1.20001e+00
8.83883e−02	5.94603e−01	5.33744e−03	9.21985e−01
4.41941e−02	4.20448e−01	2.87674e−03	7.16836e−01
2.20970e−02	2.97301e−01	1.56450e−03	5.77596e−01
1.10485e−02	2.10224e−01	8.54456e−04	4.78872e−01
5.52427e−03	1.48650e−01	4.66603e−04	4.01274e−01
	Order α	0.87240	0.34738

Table 6.8: Global norms and point value for $\varepsilon = 2\sqrt{h}$ with uniform refinement, Q_1^c discretization.

h	$\|u - u_{\text{eps},h}\|_{L^2}$	$\|u - u_{\text{eps},h}\|_{H^1}$	$\|u - u_{\text{eps},h}\|_{H^1(\tilde{B})}$
1.76776e−01	2.25360e−01	3.28350e+00	1.28949e−01
8.83883e−02	1.57825e−01	4.39123e+00	7.17441e−02
4.41941e−02	1.09320e−01	5.97844e+00	3.98377e−02
2.20970e−02	7.57405e−02	8.25670e+00	2.20526e−02
1.10485e−02	5.26920e−02	1.15146e+01	1.21450e−02
5.52427e−03	3.68272e−02	1.61551e+01	6.64832e−03
α	0.53884	no conv.	0.84998

Point value, half-domain functional, crack energy: focusing on discretization error estimates

In this section, we fix $\varepsilon = 2h_{\text{coarse}} = 0.353553$ and use a Q_1^c discretization. In comparison to the previous computations, we now do not use anymore U with $\varepsilon = 0$ as reference value, but compute a reference solution numerically on a fine grid with small h, but fixed $\varepsilon = 2h_{\text{coarse}} = 0.353553$. This allows us to study classical convergences only in h.

We computed the following reference values on a nine times globally refined mesh:

$$J^P_{\text{ref}}(u) = 0.40491845333765075,$$
$$J^D_{\text{ref}}(u) = 1.1820392158614590,$$
$$J^C_{\text{ref}}(\varphi) = 1.8152453540665945.$$

Since we numerically computed the reference solution on a grid that is only one step finer than the finest shown below, the convergence orders are less reliable. Overall, we observe in the final steps linear convergence as seen in the Tables 6.9, 6.10 and 6.11.

Table 6.9: Fixed $\varepsilon = 0.353553$ and h refinement in $J^P_{\text{ref}}(u)$.

h	$J^P(u_h)$	$\|J^P(u_h) - J^P_{\text{ref}}(u_h)\|$
1.76776e−01	4.00505e−01	4.41306e−03
8.83883e−02	4.04598e−01	3.19599e−04
4.41941e−02	4.05037e−01	1.18667e−04
2.20970e−02	4.05041e−01	1.22623e−04
1.10485e−02	4.04987e−01	6.90576e−05

Table 6.10: Fixed $\varepsilon = 0.353553$ and h refinement in $J^D_{\text{ref}}(u)$.

h	$J^D(u_h)$	$\lvert J^D(u_h) - J^D_{\text{ref}}(u_h)\rvert$
1.76776e−01	1.13282e+00	4.92143e−02
8.83883e−02	1.17345e+00	8.57989e−03
4.41941e−02	1.18937e+00	7.33318e−03
2.20970e−02	1.19160e+00	9.56847e−03
1.10485e−02	1.18669e+00	4.66065e−03

Table 6.11: Fixed $\varepsilon = 0.353553$ and h refinement in $J^C_{\text{ref}}(\varphi)$.

h	$J^C(\varphi_h)$	$\lvert J^C(\varphi_h) - J^C_{\text{ref}}(\varphi_h)\rvert$
1.76776e−01	2.18138e+00	3.66142e−01
8.83883e−02	2.03920e+00	2.23964e−01
4.41941e−02	1.95764e+00	1.42396e−01
2.20970e−02	1.90215e+00	8.69079e−02
1.10485e−02	1.85612e+00	4.08773e−02

Nonlinear solver behavior

In this subsection, we study exemplarily some findings regarding the nonlinear solver behavior. Nonlinear solution algorithms themselves will be discussed in detail in Chapter 7. We specifically use the residual-based Newton solver from Section 7.9.1. We use $\varepsilon = 2h$ and $\gamma = 10^3$ and a Q_1^c discretization. The minimal element diameter in refinement level 0 is $h = 0.176777$ with 256 elements. The nonlinear iteration history is displayed in the following verbatim tables.

```
==================================================================
Refinement cycle 0:
Iter    Residual        Reduction     Mat.build  LS       CPU time
1       1.76271e+00     1.00000e+00   r          6        2.09250e-02
2       1.68046e+00     9.53344e-01   r          0        1.12610e-02
3       7.34867e-01     4.37300e-01   r          0        1.13100e-02
4       3.19290e-01     4.34487e-01   r          0        1.13130e-02
5       2.59225e-01     8.11878e-01   r          0        1.01670e-02
6       1.65960e-01     6.40218e-01   r          0        1.09050e-02
7       1.57344e-01     9.48083e-01   r          0        9.82000e-03
8       8.68526e-02     5.51991e-01   r          0        8.60100e-03
        2.91434e-16
==================================================================
```

On refinement level 5, we have $h = 5.5242717 \times 10^{-3}$ with 262 144 elements. Due to the interpolation of initial values to the next mesh, the nonlinear convergence improves during mesh refinement as illustrated in the final refinement cycle 5.

```
==========================================================================
Refinement cycle 5:
Iter     Residual          Reduction      Mat.build  LS        CPU time
1        1.82899e+00       1.00000e+00    r          0         8.09725e+01
2        1.05220e+00       5.75289e-01    r          0         9.11733e+01
3        4.29467e-03       4.08161e-03               0         9.14439e+01
4        3.82478e-04       8.90588e-02               0         9.11363e+01
5        2.66439e-07       6.96612e-04               0         9.10128e+01
         5.88071e-16
==========================================================================
```

We also briefly explain the notation with the help of Section 7.9.1:

1. `Iter` denotes the Newton iteration step j.
2. `Residual` is measured as $\|A(U^j)(\Psi)\|_{l^\infty}$.
3. `Reduction` is the rate

$$\theta_j = \frac{\|A(U^j)(\Psi)\|}{\|A(U^{j-1})(\Psi)\|}.$$

4. The next column is `Mat.build` indicates whether the Jacobian matrix was recomputed or the old one is still used; see Remark 51. As factor, we chose $\theta_{max} = 0.1$.
5. In the fifth column, the number l of line search (`LS`) steps is shown. In the very first evaluation, we observe $l_M = 6$. But in general line search is not necessary.
6. In the last column, the CPU time per Newton step is shown.

Illustrations of the influence of γ

In this final section in Figure 6.4, some illustrative insights to the influence of the penalization parameter γ are presented. Clearly, when γ is too small, the irreversibility condition $\partial_t \varphi \le 0$ will not be imposed sufficiently strongly. In practice this means, that the fracture will vanish over time, in case there are no other forces that cause opening of the fracture. If γ is sufficiently large, then the constraint is much better enforced. However, the nonlinear solver, shown in the previous subsection, will suffer. A standard procedure therefore is to start with sufficiently small γ and to iterate (e. g., see standard literature in numerical optimization as for instance [340]) either in simple penalization or the augmented Lagrangian fashion (Section 5.2.2).

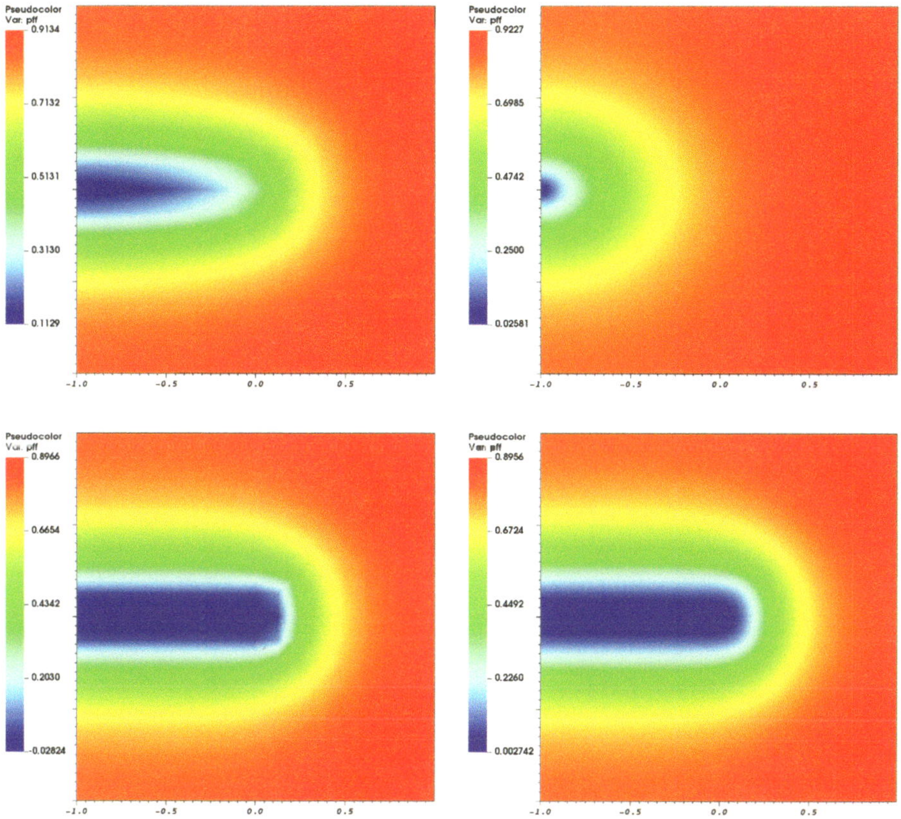

Figure 6.4: Illustration of the influence of $\gamma = 1$ and $\gamma = 1000$ on the phase-field function. The smaller γ, the less the irreversibility condition is enforced and the crack finally vanishes. In the left column, the first coarse grid solution is shown. In the right column, the solutions after three times uniform refinement are plotted. For $\gamma = 10^4$, the nonlinear solver did not converge anymore.

7 Numerical modeling part II: linearizations, nonlinear solvers, and linear solution

In this chapter, we address the nonlinear and linear numerical solution of discretized coupled problems. As examples, we may have the fully discrete versions of the Propositions 31, 32, or 33 in mind. Alternatively, we may consider Proposition 51 for dynamic fracture. We shall discuss three topics:

1. Solving coupled PDE systems or CVISs;
2. Solving nonlinear PDEs/CVISs (single and/or systems);
3. Solving linear PDE systems including the construction of preconditioners.

For the first topic, two basic approaches exist:

- Partitioned schemes: sequential solving of the equations, also known as staggered schemes or iterative coupling;
- Monolithic schemes: all-at-once solution also known as fully-coupled.

These two concepts are a priori independent whether the coupled problem is linear/nonlinear or nonstationary/steady-state. Most coupled problems are nonlinear as extensively discussed in Chapter 3. An example of a linear coupled problem are Biot's equations as presented in Section 3.4.

Concerning the second topic, we notice that if the entire coupled problem (in a monolithic fashion) or parts of it (subproblems in a partitioned scheme) are nonlinear, the following linearization methods are often used:

- Linearizations via decoupling, time-lagging or extrapolation;
- Fixed-point iterations;
- Functional iterations or operator splitting with iterative coupling;
- Newton's method.

These approaches can be found in both textbooks and research articles. Our preferred choices are [365, 131, 340, 381, 55].

Regarding the third topic, namely linear solvers, we rather present some characteristic results and refer for technical details to the literature. Iterative linear solvers require the development of preconditioners. Due to memory consumption and computational cost, direct solvers often can only be used for prototype academic problems. For two-dimensional configurations, approximately up to 100k DoFs direct solvers such as UMFPACK [125] are competitive. Some current numbers of the computational cost for general linear systems $Ax = b$ are listed in [381, pp. 68–69 in the Tables 3.1–3.2].

https://doi.org/10.1515/9783110497397-007

7.1 Partitioned versus monolithic coupling

A principle design decision is to solve the problem in an all-at-once fashion (monolithically) or in a partitioned way. Discussions were already provided in Section 3.3.

Monolithic approaches are implicit techniques and are classified as strongly-coupled methods. Partitioned techniques can be either loosely-coupled (no iterations or just a few subiterations) or as well strongly-coupled (performing several subiterations). A flow-chart is provided in Figure 7.1.

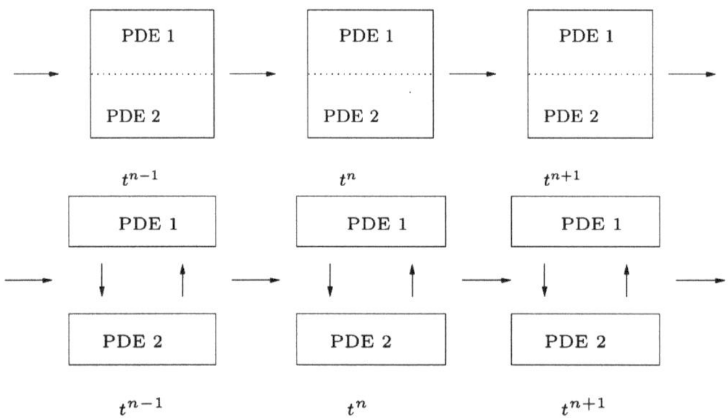

Figure 7.1: Monolithic and partitioned coupling schemes. In monolithic coupling, the coupled system is solved all-at-once, whereas in the partitioned scheme subiterations are often required.

7.2 Linearizations via decoupling, time-lagging, or extrapolation

The simplest methods consist of replacing nonlinear terms by known or guessed information. Such information can be previous time step solutions (known as time-lagging), interpolations or extrapolations, or analytical information.

Let us consider the following time-dependent PDE as an example: Find $u \in V$ for almost all $t \in I$ such that

$$a(u)(\phi) := (\partial_t u, \phi) + (u^2, \phi) = (f, \phi) =: l(\phi) \quad \forall \phi \in V.$$

A simple linearization of the middle term is

$$(u^2, \phi) \to (uu^{n-1}, \phi),$$

where $u := u^n := u(t_n)$ is the current unknown,[1] and $u^{n-1} := u(t_{n-1})$ is the previous time step solution. This is an explicit, loosely-coupled, scheme. We arrive at a functional

1 Please do not confuse the power u^2 with u^n indicating the iterate.

iteration, when additional iterations are carried out: Set $u^{n,0} := u^{n-1}$, iterate for $j = 1, 2, 3, \dots$ such that we hopefully obtain

$$\lim_{j\to\infty} uu^{n,j} = u^2.$$

Functional iterations and iterative coupling are discussed below in more detail.

7.3 Fixed-point iterations

The classical idea of fixed-point iterations[2] is well known.

Definition 56 (Fixed point in $\mathbb{R}$). Given a function $f : \mathbb{R} \to \mathbb{R}$. A fixed point is

$$x = f(x).$$

An iterative scheme is obtained by constructing a sequence $(x_j)_{j\in\mathbb{N}}$ with

$$x^j = f(x^{j-1}), \quad j = 1, 2, 3, \dots.$$

This idea can immediately be extended to solving linear equation systems.

Proposition 52 (Fixed-point iteration for $Ax = b$). *Given $Ax = b$ with $A \in \mathbb{R}^{m\times m}$, $x \in \mathbb{R}^m$, $b \in R^m$. Then*

$$x^j = x^{j-1} + Cd^{j-1}, \quad j = 1, 2, 3, \dots,$$

where the defect is defined by $d^{j-1} = b - Ax^{j-1}$ and $C \in \mathbb{R}^{m\times m}$ is an invertible preconditioner matrix.

Proof. Given $Ax = b$. Then

$$0 = b - Ax \quad \Leftrightarrow \quad x = x + b - Ax \quad \Leftrightarrow \quad x = x + d \quad \Rightarrow \quad x^j = x^{j-1} + Cd^{j-1}. \qquad \square$$

Corollary 4 (Fixed point as defect correction). *The fixed-point iteration can be written as defect-correction scheme: Given an initial guess x^0, iterate for $j = 1, 2, 3, \dots$ such that*

$$\delta x^{j-1} = Cd^{j-1}$$
$$x^j = x^{j-1} + \delta x^{j-1}$$

Remark 47. It is well known (see, e. g., [381]) that convergence can be shown with Banach's fixed-point theorem. Second, the speed of convergence depends on the design of the preconditioner matrix C. Third, in Newton's method, the matrix C is simply the Jacobian matrix.

2 In ODEs, fixed-point iterations are known as Picard iterations.

We now extend these concepts to semilinear forms.

Proposition 53. *Let the following problem be given: Find $U \in X$ such that*

$$A(U)(\Psi) = F(\Psi) \quad \forall \Psi \in X.$$

Then, a fixed-point iteration is defined as

$$(U^j, \Psi) = (U^{j-1}, \Psi) + C(F(\Psi) - A(U^{j-1})(\Psi)),$$

or in defect correction notation:

$$C^{-1}(\delta U^{j-1}, \Psi) = F(\Psi) - A(U^{j-1})(\Psi),$$
$$U^j = U^{j-1} + \delta U^{j-1},$$

where C, respectively, $C^{-1}(\delta U^j, \Psi)$ is a suitable bilinear form serving as preconditioner. The simplest choice is a Richardson type preconditioner with $C = \omega I$, where I is the identity matrix and $\omega > 0$ a relaxation parameter.

Remark 48 (Relation to Newton's method). In Newton's method, we have

$$C^{-1}(\delta U^{j-1}, \Psi) := A'(U^{j-1})(\delta U^{j-1}, \Psi),$$

where A' is the Jacobian matrix.

Proposition 54 (Stopping criteria of nonlinear iterative methods). *We have basically two stopping criteria. Residual-based criteria:*

$$\|A(U^j)(\Psi) - F(\Psi)\| < \text{TOL} \quad \text{or} \quad \|A(U^j)(\Psi) - F(\Psi)\| < \|F(\Psi)\|\,\text{TOL}$$

or error iteration criteria:

$$\|U^j - U^{j-1}\| < \text{TOL} \quad \text{or} \quad \|U^j - U^{j-1}\| < \|U^j\|\,\text{TOL}.$$

The tolerance is often chosen in the range of $\text{TOL} = 10^{-12}, \dots, 10^{-8}$.

7.4 Functional iteration

The basic idea is to introduce an iteration indexed by j and to linearize the nonlinear terms by taking them from the previous iteration $j - 1$. This introduces a sequence of iterates U^j, $j = 1, 2, 3, \dots$ and hopefully we obtain

$$\lim_{j\to\infty} U^j = U.$$

This procedure can be applied to both steady-state and time-dependent problems.

7.4.1 Abstract procedure

We are given the following abstract form:

Formulation 18. Let the following problem be given: Find $U \in X$ such that

$$A(U)(\Psi) = F(\Psi) \quad \forall \Psi \in X.$$

Given an initial guess $U^0 \in X$, for $j = 1, 2, 3, \ldots$, given U^{j-1}, find $U^j \in X$ such that

$$A(U^j, U^{j-1})(\Psi) - F(\Psi) = 0 \quad \forall \Psi \in X.$$

In the semilinear form $A(U^j, U^{j-1})(\Psi)$ some terms are taken at the current iteration j and some are known by taking them from iteration $j - 1$. We stop the iteration with a criterion chosen from Proposition 54.

7.4.2 Example

This is best illustrated in terms of an example. We consider

$$-\Delta u + u^2 = f.$$

The variational formulation reads: Find $u \in V$ such that

$$A(u)(\psi) = F(\psi) \quad \forall \psi \in V,$$

with

$$\begin{aligned} A(u)(\psi) &= (\nabla u, \nabla \psi) + (u^2, \psi), \\ F(\psi) &= (f, \psi). \end{aligned}$$

An iterative scheme is constructed as follows.

Algorithm 3 (Functional iteration). For $j = 1, 2, 3, \ldots$, given u^{j-1}, we seek $u^j \in V$ such that

$$A(u^j, u^{j-1})(\psi) = F(\psi) \quad \psi \in V,$$

with

$$\begin{aligned} A(u^j, u^{j-1})(\psi) &= (\nabla u^j, \nabla \psi) + (u^j u^{j-1}, \psi) \\ F(\psi) &= (f, \psi) \end{aligned}$$

until a stopping criterion chosen from Proposition 54 is fulfilled.

7.5 Iterative coupling for solving PDE systems

The previous concept of a functional iteration was introduced having a single PDE in mind. This concept can be extended to multiple PDEs.

Formulation 19. Let us explicitly assume $U = (u, \varphi) \in X$. Let the following problem be given: Find $U \in X$ such

$$A(U)(\Psi) = F(\Psi) \quad \forall \Psi \in X.$$

Given an initial guess $U^0 \in X$, find $U^j \in X$ for $j = 1, 2, 3, \ldots$, such that

$$A(U^j, U^{j-1})(\Psi) - F(\Psi) = 0 \quad \forall \Psi \in X.$$

Here, the operator $A(\cdot)(\cdot)$ can be split into the single operators $A_1(u^j, \varphi^{j-1})(\psi_1)$ and $A_2(u^j, \varphi^j)(\psi_2)$ such that

$$\begin{aligned} \text{given } \varphi^{j-1}, \text{ find } u^j: \quad & A_1(u^j, \varphi^{j-1})(\psi_1) = F_1(\psi_1), \\ \text{given } u^j, \text{ find } \varphi^j: \quad & A_2(u^j, \varphi^j)(\psi_2) = F_2(\psi_2). \end{aligned}$$

This procedure can be readily extended to multiple PDEs and CVISs.

7.6 Iterative coupling for phase-field fracture

We apply the previous algorithmic concepts to phase-field fracture. For quasi-brittle fracture, partitioned schemes, in the fracture community better known as alternating minimization, have been investigated in great detail. The early studies employed partitioned schemes [72, 73]. Convergence proofs were established in [70] and [94]. Numerical improvements were presented in [311]. We used in our work partitioned schemes in [430] including a mathematical analysis in [323]. Due to huge numbers of iterations, the schemes are robust, but not efficient, a stabilized version was proposed in [88] and accompanied with a rigorous numerical analysis. The scheme is sufficiently accurate at reasonable computational cost. An improvement in efficiency is achieved with an dynamic update of the stabilization parameters as shown in [154]. The latter is recapitulated below, however, a rigorous convergence proof has not been shown yet. To this end, we consider Proposition 19 and solve the CVIS

$$\begin{aligned} A_1((u, \varphi))(w) &= (g(\varphi)\sigma(u), \nabla w), \\ A_2((u, \varphi))(\psi - \varphi) &= ((1 - \kappa)\varphi\sigma(u) : e(u), \psi - \varphi) \\ &\quad + \left(-\frac{G_c}{\varepsilon}(1 - \varphi), \psi - \varphi\right) + (G_c \varepsilon \nabla \varphi, \nabla(\psi - \varphi)), \end{aligned}$$

with iterative coupling.

Algorithm 4. At a given loading (time) t_n, let the initial iteration values be given:

$$u^0 := u^{n,0} := u^{n-1}, \qquad \varphi^0 := \varphi^{n,0} := \varphi^{n-1}.$$

We iterate for $j = 1, 2, 3 \ldots$:

$$\begin{aligned} \text{Given } \varphi^{j-1} \in K, \quad \text{find } u^j \in V_u^D: \quad & A_1((u^j, \varphi^{j-1}))(w) = 0 \quad \forall w \in V_u^0, \\ \text{Given } u^j \in V_u^D, \quad \text{find } \varphi^j \in K: \quad & A_2((u^j, \varphi^j))(\psi - \varphi^j) \geq 0 \quad \forall \psi \in K. \end{aligned}$$

Remark 49. We notice that in this realization, both subequations may be still nonlinear and are required to be solved with another nonlinear method. For instance, $A_1(\cdot)(\cdot)$ can be nonlinear due to the nonlinear constitutive stress tensor (stress splitting). The second problem $A_2(\cdot)(\cdot)$ is nonlinear due to the nonlinear constitutive stress tensor and the variational inequality constraint.

7.6.1 A partitioned scheme with updating stabilizing parameters: *L*-scheme

In this section, we present the previously mentioned partitioned scheme with stabilizing parameters. The stabilized decoupled method is known as L-scheme and was originally proposed in [354] for solving Richards' equation, which is a nonlinear elliptic-parabolic equation. The performance against other nonlinear methods, such as for instance Newton's method, was investigated in [290]. These ideas were applied in [88] to phase-field fracture. Here, the PDEs were augmented with additional stabilization terms including two subsequent iterates at a time point t_n:

$$\begin{aligned} & L_u(u^j - u^{j-1}), \\ & L_\varphi(\varphi^j - \varphi^{j-1}). \end{aligned}$$

Here, the fixed $L_u, L_\varphi > 0$ are the 'L' stabilization parameters and j the iteration index. Clearly, in the case of convergence, we have

$$\begin{aligned} & \lim_{j\to\infty} L_u \|u^j - u^{j-1}\| = 0, \\ & \lim_{j\to\infty} L_\varphi \|\varphi^j - \varphi^{j-1}\| = 0. \end{aligned}$$

In [88], the stabilization parameters L_u and L_φ where chosen as constants, mainly because, we could establish a rigorous numerical analysis. Computationally, we however observed an increase in efficiency for variable L_u and L_φ, specifically when these were updated in a dynamic way during the iterations $j = 1, 2, 3, \ldots$. These results were recently summarized in [154]. In the following, we consider exemplarily Proposition 32 with an augmented Lagrangian penalization.

Proposition 55. *Let the positive stabilization parameters L_u and L_φ and the penalization parameter $\gamma > 0$ be given. The semilinear forms from Proposition 32 are denoted by $A_1(U_h^n)(w_h)$ and $A_2(U_h^n)(\psi_h)$. We solve the following problems iteratively at time t_n. Set the initial guesses $u_h^{n,0} := u_h^{n-1}$ and $\varphi_h^{n,0} := \varphi_h^{n-1}$. Given $(u_h^{n,j-1}, \varphi_h^{n,j-1}) \in V_{u,h}^D \times V_{\varphi,h}$, and given φ_h^{n-1}, find $u_h^j \in V_{u,h}^D$ such that*

$$A_{L_u}(u_h^{n,j}, \varphi_h^{n,j-1})(w_h) := L_u(u_h^{n,j} - u_h^{n,j-1}, w_h) + A_1(u_h^{n,j}, \varphi_h^{n,j-1})(w_h) = 0 \quad \forall w_h \in V_{u,h}^0. \tag{7.1}$$

Given $(\varphi_h^{n,j-1}, u_h^{n,j}, \varphi_h^{n-1}) \in V_{\varphi,h} \times V_{u,h}^D \times V_{\varphi,h}$, find $\varphi_h^j \in V_{\varphi,h}$ such that

$$A_{L_\varphi}(U_h^{n,j})(\psi_h) := L_\varphi(\varphi_h^{n,j} - \varphi_h^{n,j-1}, \psi_h) + A_2(U_h^{n,j})(\psi_h) = 0 \quad \forall \psi_h \in V_{\varphi,h}. \tag{7.2}$$

This result can be used to design the following algorithm. Since we have an iteration on j due to iterative coupling, this can be conveniently combined with the update of the augmented Lagrangian function $\Xi \in L^2(B)$.

Algorithm 5.

Choose $\gamma > 0$, $a > 1$, as well as Ξ_h^0 and $L^0 > 0$. We use specifically $L = L_u = L_\varphi$. Set $j = 0$.

repeat

Let $j = j + 1$;

Solve the two problems, namely

Solve the nonlinear elasticity problem in (7.1)

Solve the nonlinear phase-field problem in (7.2)

Update $L^j = aL^{j-1}$

Update $\Xi_h^j = [\Xi_h^{j-1} + \gamma(\varphi_h^{n,j} - \varphi_h^{n-1})]^+$

until

$\max\{\|A_{L_u}(u_h^{n,j}, w_h)\|, \|A_{L_\varphi}(\varphi_h^{n,j}, \psi_h)\|\} \leq \text{TOL}.$

Set $U_h^n := (u_h^n, \varphi_h^n) := (u_h^{n,j}, \varphi_h^{n,j})$.

Increment $t_n \to t_{n+1}$.

When working with the dynamic updates [154], we found an a priori counterintuitive behavior. The natural idea would be that $L^j \to 0$ for $j \to \infty$ (or at least $L^j \to$ const), such that

$$L_u^j(u^j - u^{j-1}) \to 0,$$
$$L_\varphi^j(\varphi^j - \varphi^{j-1}) \to 0,$$

in order to satisfy the primary PDE system, namely

$$A_1(U_h^n)(w_h) = 0,$$
$$A_2(U_h^n)(\psi_h) = 0.$$

This is somewhat related to numerical continuation methods, e. g., [13], which start with a regular perturbed system and during the iterations, regularization parameters are sent to zero. In [154], we found the opposite, namely $a > 1$, which means that the stabilization is increased during the iterations:

$$L^j \to \infty \quad \text{for } j \to \infty.$$

Our current interpretation is that the phase-field fracture CVIS is nonlinear and in general the solutions are nonunique. Starting with a small stabilization L^j, solves the physical system, and increasing L^j later steers the system into some unique solution. Computationally, we observed supporting findings, but a rigorous justification is missing to date and under ongoing discussions.

7.7 Monolithic schemes

As before in Formulation 19, we consider a common semilinear form:

$$A(U)(\Psi) = F(\Psi). \tag{7.3}$$

But now, we do not split into U^j and U^{j-1} anymore, but only consider U^j in the above operator A. To this end, all variables and thus all equations are solved all-at-once in an implicit fashion. Therefore, monolithic treatments require fixed-point iterations on the full operator $A(U)(\Psi)$ or Newton-type methods.

Definition 57 (Monolithic solution). A truly monolithic scheme solves (7.3) with a nonlinear method, with a monolithic linear solver, and a monolithic preconditioner. But monolithic schemes are commonly also denoted as those approaches, which have a monolithic formulation, and a common nonlinear solver, but which decouple in the linear solver or the preconditioner. However, once the nonlinear iteration is decoupled, then we arrive at the previously described schemes such as iterative coupling.

Remark 50 (Implicit coupling). In monolithic schemes, all terms are implicitly coupled, which means that we deal only with the current iterate U^j and no explicit linearizations.

7.7.1 Challenges in phase-field fracture

For phase-field fracture, the objective of having monolithic methods results into various challenges. In the energy form, Proposition 15, the bulk part E_b (Section 4.4.4) is a fourth-order term:

$$\frac{1}{2}\int_B g(\varphi)\sigma(u) : e(u)\,dx.$$

In the first variation, the first term in the displacement equation $A_1(U)(w)$ causes these difficulties, namely

$$\int_B ((1-\kappa)\varphi^2 + \kappa)\, \sigma(u) : e(w)\, dx,$$

when φ and u are simultaneously unknown; see Proposition 19. In the Propositions 20 or 25, it is

$$\int_B ((1-\kappa)\varphi^2 + \kappa)\, \sigma^+(u) : e(w)\, dx.$$

The related energy term is not strictly convex in both solution variables u and φ simultaneously. As previously described, decoupling results in a partitioned scheme. A prototype simplification is to consider:

$$((1-\kappa)\varphi^2 + \kappa)\, \sigma(u) : e(u) \sim \varphi^2 |\nabla u|^2.$$

Fixing φ is denoted by $\bar{\varphi}$ and yields

$$\bar{\varphi}^2 |\nabla u|^2,$$

which means classical Poisson on the PDE level with a given coefficient $\bar{\varphi}^2$. On the other hand fixing u with $\bar{u}$ yields

$$\varphi^2 |\nabla \bar{u}|^2,$$

which is a quadratic mass-like term weighted with a given coefficient $|\nabla \bar{u}|^2$.

Due to these complications, only few suggestions have been made to date for monolithic frameworks in quasi-static brittle phase-field fracture: quasi-monolithic formulations with a classical residual-based Newton method [221], and for fully monolithic frameworks, line-search assisted Newton methods [192], error-oriented algorithms [444], modified Newton/fixed-point methods [445], and trust-region methods [262]. A prototype computational analysis was performed in [445, Section 4] for highlighting the main challenges.

7.7.2 Quasi-monolithic schemes

We start with a terminology that is certainly neither unique nor common in the literature.

Definition 58 (Quasi-monolithic functional iteration). We call a solution scheme quasi-monolithic when the overall solution is achieved all-at-once, but certain (lower-order) terms are linearized. We lose fully implicit coupling, but we still keep the overall character of a monolithic method.

7.7.2.1 Time-lagging with previous time step solution

Ideas are based on our previous sections, namely a partial linearization such as proposed in Section 7.2, but without operator splitting. A first choice is

Proposition 56 (Time-lagging with old time step). *Linearize in the Propositions* 31, 32, *or* 33 *the first term in* $A_1(U)(\Psi)$ *by time-lagging with* $\tilde{\varphi} := \varphi^{n-1}$ *such that*

$$((1-\kappa)\tilde{\varphi}^2 + \kappa)\,\sigma(u).$$

However, this introduces a temporal discretization error of order $O(k)$ that becomes immediately visible for medium-speed growing fractures like the single edge notched shear test (Chapter 8).

7.7.2.2 Time-lagging with linear-in-time extrapolation

A better choice is a linear-in-time extrapolation (see Figure 7.2), that works for most relevant numerical examples we have considered so far surprisingly well in terms of robustness, efficiency and accuracy.

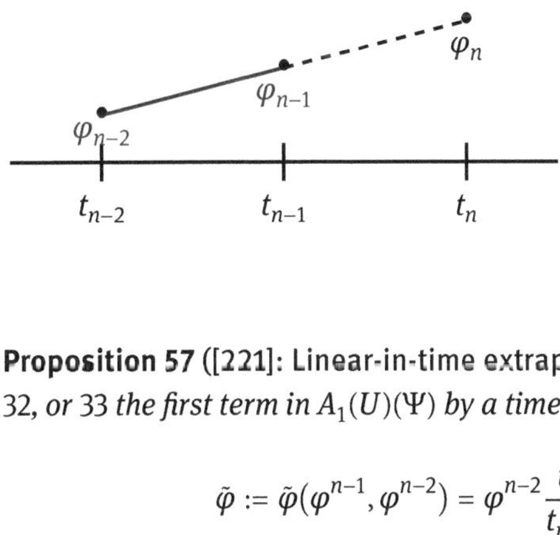

Figure 7.2: Linear-in-time extrapolation in time of $\tilde{\varphi}$.

Proposition 57 ([221]: Linear-in-time extrapolation). *Linearize in the Propositions* 31, 32, *or* 33 *the first term in* $A_1(U)(\Psi)$ *by a time-lagged extrapolation with*

$$\tilde{\varphi} := \tilde{\varphi}(\varphi^{n-1}, \varphi^{n-2}) = \varphi^{n-2}\frac{t_n - t_{n-1}}{t_{n-2} - t_{n-1}} + \varphi^{n-1}\frac{t_n - t_{n-2}}{t_{n-1} - t_{n-2}},$$

such that

$$((1-\kappa)\tilde{\varphi}^2 + \kappa)\,\sigma(u).$$

This procedure is heuristic because the proposed extrapolation requires regularity in time (see basic classes in numerical analysis on the regularity requirements for extrapolation, e. g., [381, *Section 8.4*]*), which we cannot ensure because we deal with a quasi-static problem without time derivatives. This means that the phase-field solution* φ *can have jumps in time (see, e. g.,* [323, *Section 2*] *that* $\partial_t\varphi$ *only exist as measure) and extrapolation is strictly speaking mathematically not well-defined.*

A rigorous numerical analysis of quantifying the error introduced by the extrapolation is open. Based on classical extrapolation results, we expect for the linear-in-time approximation, the order $O(k^2)$. For decreasing time step size $k \to 0$ the solutions might converge to a quasi-stationary limit. For a decoupled version under certain assumptions, this could be established in [323, Section 3.3]. For a monolithic or quasi-monolithic formulation, there are no mathematical proofs for this hypothesis. The computational verification of a time-lagging error was shown in [444, Figures 3, 11, and 12] and depends as to be expected on the crack tip velocity. Further computational results of loading step refinement are presented in Chapter 8. For steady-state fractures, the time-lagging error is negligible. For this reason, we work in the Chapters 6 and 10 simply with the previous time step solution φ^{n-1}.

7.7.2.3 Iteration-lagging/functional iteration

A third option is simply a functional iteration from Section 7.4.

Proposition 58 (Iteration-lagging). *We are given the previous iteration U^{j-1} and seek the new iterate U^j. Linearize in the Propositions 31, 32, or 33 by iteration-lagging the first term in $A_1(U)(\Psi)$ with $\tilde{\varphi} := \varphi^{j-1}$ such that*

$$((1-\kappa)\tilde{\varphi}^2 + \kappa)\,\sigma(u).$$

However, in the case of $\|\varphi^j - \varphi^{j-1}\| \to 0$ for $j \to \infty$, we approach the monolithic limit and expect therefore the same difficulties as in fully monolithic treatments.

7.7.3 An iteration on the extrapolation

We present a simple, but very effective method to enhance the approximation quality in Proposition 57. To this end, we improve the extrapolation by an additional iteration.

Algorithm 6 (Iterating on the extrapolation). We assume to be at time step t_n. Set the number of iterations $N_E \geq 1$.

1. Let φ^{n-2} and φ^{n-1} be the given two previous time step solutions.
2. Set $\varphi^{n,-2} := \varphi^{n-2}$ and $\varphi^{n,-1} := \varphi^{n-1}$.
3. Construct the linear extrapolation:

$$\tilde{\varphi}^{n,0} = \varphi^{n,-2}\frac{t_n - t_{n-1}}{t_{n-2} - t_{n-1}} + \varphi^{n,-1}\frac{t_n - t_{n-2}}{t_{n-1} - t_{n-2}}.$$

4. Set $u^{n,0} := u^{n-1}$ and $\varphi^{n,0} := \varphi^{n-1}$.
5. For $i = 1, \ldots, N_E$:
   (a) Find $(u^{n,i}, \varphi^{n,i})$ by solving the displacement phase-field system with $u^{n,i-1}$, $\varphi^{n,i-1}$, $\tilde{\varphi}^{n,i-1}$.

(b) Construct a new extrapolation:

$$\tilde{\varphi}^{n,i} = \varphi^{n,i-2}\frac{t_n - t_{n-1}}{t_{n-2} - t_{n-1}} + \varphi^{n,i-1}\frac{t_n - t_{n-2}}{t_{n-1} - t_{n-2}}.$$

(c) Increment $i \to i + 1$.

6. Set $u^n := u^{n,N_E}$ and $\varphi^n := \varphi^{n,N_E}$.

Supporting computations are presented in Chapter 8.

7.8 Newton's method: overview. Going from $\mathbb{R}$ to Banach spaces

An efficient alternative to fixed-point iterations (Section 7.3) or functional iterations (Section 7.4) is Newton's method, which is a well-known technique for solving nonlinear equations. The first classical convergence theorem for Newton's method in abstract spaces goes back to Kantorovich [255, 256] and is therefore well known as the Newton–Kantorovich theorem.

We quickly summarize in this section some well-known algorithms for which we refer to classical textbooks [131, 340, 365]. Furthermore, we refer to introduction-to-numerical-methods textbooks for similar introductions of Newton's method [381, 362]. A quick overview is given by:

– Newton–Raphson (1D), find $x \in \mathbb{R}$ via iterating $j = 0, 1, 2, \dots$ such that $x^j \to x$ via:

$$\begin{aligned} \text{Find } \delta x \in \mathbb{R}: \quad & f'(x^j)\delta x = -f(x^j), \\ \text{Update:} \quad & x^{j+1} = x^j + \delta x. \end{aligned}$$

– Newton in $\mathbb{R}^d$, find $x \in \mathbb{R}^d$ via iterating $j = 0, 1, 2, \dots$ such that $x^j \to x$ via:

$$\begin{aligned} \text{Find } \delta x \in \mathbb{R}^d: \quad & F'(x^j)\delta x = -F(x^j), \\ \text{Update:} \quad & x^{j+1} = x^j + \delta x. \end{aligned}$$

Here, we need to solve a linear equation system to compute the update $\delta x \in \mathbb{R}^d$.

– Banach spaces, find $u \in V$, with $\dim(V) = \infty$, via iterating $j = 0, 1, 2, \dots$ such that $u^j \to u$ via:

$$\begin{aligned} \text{Find } \delta u \in V: \quad & F'(u^j)\delta u = -F(u^j), \\ \text{Update:} \quad & u^{j+1} = u^j + \delta u. \end{aligned}$$

Such a problem needs to be discretized and again results in solving a linear equation system in the defect step.

– Banach spaces, applied to variational formulations, find $U \in X$, with $\dim(X) = \infty$, via iterating $j = 0, 1, 2, \ldots$ such that we find a suitable approximation $U^j \to U$ via:

$$\text{Find } \delta U \in X: \quad A'(U^j)(\delta U^j, \Psi) = -A(U^j)(\Psi), \tag{7.4}$$

$$\text{Update:} \qquad U^{j+1} = U^j + \delta U^j. \tag{7.5}$$

As before, the infinite-dimensional problem is discretized, resulting in a linear equation system in the defect step.

7.8.1 Monotonicity tests

In order to accept or reject the current update, say δU in the lastly presented scheme (7.4)–(7.5), we need to perform monotonicity tests. The classical procedure is to compare to subsequent residuals and to check whether

$$\|A(U^{j+1})(\Psi)\| < \|A(U^j)(\Psi)\|$$

holds true. If this criterion is not fulfilled, either line-search steps (damped Newton) or trust region concepts may be applied [340, 131] for expanding the convergence radius and to globalize[3] Newton's method. Following further [131, p. 144 and p. 149, Example 3.2] with illustrative computations, an alternative is to switch to an error-controlled criterion, a so-called natural monotonicity test. Rather observing the residuals, Newton updates are compared:

$$\|\delta U^{j+1}_{\text{simp}}\| < \|\delta U^j\|.$$

The simplified Newton update $\delta U^{j+1}_{\text{simp}}$ is obtained from solving

$$A'(U^j)(\delta U^{j+1}_{\text{simp}}, \Psi) = -A(U^{j+1})(\Psi),$$

with the already constructed Jacobian and a new right hand side residual using the current solution $U^{j+1} = U^j + \delta U^j$. For more information, we refer the reader to [131, Section 3.3].

7.8.2 A basic algorithm for a residual-based Newton method

In this type of methods, the main criterion is a decrease (i. e., contraction) of the residual in each step.

3 The terminology "globalization" is adopted from numerical optimization (e. g., [340]) or, in general, Newton methods (e. g., [131]) and means that the convergence radius of Newton's method is extended by, for example, line search or trust region methods.

Algorithm 7 (Residual-based Newton's method). Given some initial guess x^0. Iterate for $j = 0, 1, 2, \dots$ such that

$$\begin{aligned} \text{Find } \delta x \in \mathbb{R}^d: \quad & F'(x^j)\delta x^j = -F(x^j), \\ \text{Update:} \quad & x^{j+1} = x^j + \lambda^j \delta x^j, \end{aligned}$$

with $\lambda^j \in (0, 1]$ (see the next sections how λ^j can be determined). A full Newton step corresponds to $\lambda^j = 1$. The criterion for convergence is the contraction of the residuals measured in terms of a discrete vector norm:

$$\|F(x^{j+1})\| < \|F(x^j)\|.$$

In order to save some computational cost, close to the solution x^*, intermediate simplified Newton steps can be used. In the case of $\lambda^j = 1$, we observe

$$\theta^j = \frac{\|F(x^{j+1})\|}{\|F(x^j)\|} < 1.$$

If $\theta^j < \theta_{\max}$, e. g., $\theta_{\max} = 0.1$, then the old Jacobian $F'(x^j)$ is kept and used again in the next step $k + 1$. Otherwise, if $\theta^j > \theta_{\max}$, the Jacobian will be assembled. Finally, the stopping criterion is one of the following (a relative criterion is preferred!):

$$\begin{aligned} \|F(x^{j+1})\| &\le \mathrm{TOL}_{\mathrm{New}} \quad \text{(absolute)} \\ \|F(x^{j+1})\| &\le \mathrm{TOL}_{\mathrm{New}} \|F(x_0)\| \quad \text{(relative)} \end{aligned}$$

If fulfilled, set $x^* := x^{j+1}$ and the (approximate) root x^* of the problem $F(x) = 0$ is found.

7.8.3 Inexact Newton

For large scale nonlinear PDEs, the inner linear systems are usually not solved with a direct method (e. g., LU), but with iterative methods (CG, GMRES, multigrid). Using such iterative methods yields an inexact Newton scheme [129]. Here, we deal with two iterations:
- Outer nonlinear Newton iteration;
- Inner linear iteration.

7.8.4 Abstract schemes for monolithic formulations and their numerical solution

7.8.4.1 Variational equations

In this section, we formulate an abstract scheme for variational formulations, which can be used to discretize several, nonlinear, coupled, equations. Let us assume we are given S equations and possible inequalities are already regularized by penalization.

1. Formulate the given equations in variational forms:

$$A_1(U)(\Psi) = F_1(\Psi)$$
$$A_2(U)(\Psi) = F_2(\Psi)$$
$$\vdots$$
$$A_S(U)(\Psi) = F_S(\Psi)$$

2. Formulate compact short semilinear form: Find $U \in X$ such that

$$A(U)(\Psi) = \sum_{j=1}^{S} A_j(U)(\Psi) \quad \Psi \in X.$$

This system is time-continuous, space-continuous and nonlinear.

3. Time discretization
   (a) Classify A_j into time-dependent and stationary forms:

$$\text{Stationary terms:} \qquad A_j \Rightarrow A_S \tag{7.6}$$
$$\text{Time-dependent terms:} \qquad A_j \Rightarrow A_T, A_E, A_I \tag{7.7}$$

   where A_S are stationary terms, A_T terms with time derivatives, A_E terms that are treated explicitly in a time-stepping scheme, A_I terms that are treated implicitly in a time-stepping scheme (e. g., steady-state PDEs inside the coupled system).
   (b) Approximate the continuous time derivative in A_T with a backward difference quotient yielding $A_T(U^{n,k}) \approx A_T(U^n)$.
   (c) One-Step-θ: For $n = 1, \dots, N$:

$$\underbrace{A_T(U^{n,k})(\Psi) + \theta A_E(U^n)(\Psi) + A_I(U^n)(\Psi) + A_S(U^n)(\Psi)}_{=:A(U^n)(\Psi)}$$
$$= \underbrace{-(1-\theta)A_E(U^{n-1})(\Psi)}_{=:A(U^{n-1})(\Psi)} + \underbrace{\theta F^n(\Psi) + (1-\theta)F^{n-1}(\Psi)}_{=:F^{n,n-1}(\Psi)}$$

   Remark: Alternatively, the Fractional-Step-θ scheme (see Section 5.3) can employed.

4. Spatial discretization at t_n. For a conforming finite element method, we seek $U_h^n \in X_h \subset X$:

$$A(U_h^n)(\Psi_h) = -A(U_h^{n-1})(\Psi_h) + F_h^{n,n-1}(\Psi_h)$$

5. Nonlinear solution (Newton's method for instance). Let $M_X = \dim(X_h)$ and $\delta U_h = \sum_{i=1}^{M_X} \delta u_i \Psi_i$. Find $\delta U_h \in X_h$ for $j = 0, 1, 2, \dots$

$$\text{(i)} \quad \underbrace{A'(U_h^{n,j})(\delta U_h, \Psi_h)}_{=:AU} = \underbrace{-A(U_h^{n,j})(\Psi_h) - A(U_h^{n-1,j})(\Psi_h) + F_h^{n,n-1}(\Psi_h)}_{=:B}$$

$$\text{(ii)} \qquad U_h^{n,j+1} = U_h^{n,j} + \omega \delta U_h, \quad \omega \in [0,1].$$

6. Solve linear system:

$$AU = B,$$

for $U = (\delta u_1, \dots, \delta u_{M_X})^T \in \mathbb{R}^{M_X}$. Here, when using an iterative method, an appropriate preconditioner P^{-1} needs to be constructed such that

$$P^{-1}AU = P^{-1}B$$

where $P^{-1}A$ has a moderate condition number.

7.8.4.2 Variational inequalities: penalization

When terms are subject to inequality constraints, we present a modification of the previous algorithm using penalization.

1. Formulate the given equations in variational forms:

$$\begin{aligned}
&A_1(U)(\Psi) = F_1(\Psi)\\
&A_2(U)(\Psi) = F_2(\Psi)\\
&\vdots\\
&A_s(U)(\Psi) = F_s(\Psi)\\
&A_{s+1}(U)(\Psi - U) \geq F_{s+1}(\Psi - U)\\
&\vdots\\
&A_S(U)(\Psi - U) \geq F_S(\Psi - U)
\end{aligned}$$

2. Regularization[4] of the inequality constraints using penalty parameters $\gamma_{s+1}, \dots, \gamma_S$ yielding

$$\begin{aligned}
&A_{s+1}(U)(\Psi) - A_{\gamma_{s+1}}(U)(\Psi) = F_{s+1}(\Psi)\\
&\vdots\\
&A_S(U)(\Psi) - A_{\gamma_S}(U)(\Psi) = F_S(\Psi).
\end{aligned}$$

3. Formulate a compact short semilinear form:

$$A(U)(\Psi) = \sum_{j=1}^{s} A_j(U)(\Psi) - \sum_{j=s+1}^{S} A_{\gamma_j}(U)(\Psi).$$

This system is regularized, time-continuous, space-continuous, and nonlinear.

4. Apply the steps 3–6 from the previous section.

4 For other regularization procedures, we refer the reader to Section 5.2.

7.9 Newton-type methods applied to variational phase-field fracture

We present five Newton-type methods in the following sections:
1. a classical residual-based method (Section 7.9.1);
2. Section 7.9.2: error-oriented version with a natural monotonicity criterion;
3. Section 7.9.3: a modified Newton method in which the Jacobian is dynamically updated;
4. Section 7.9.4: error-oriented version with Jacobian modification;
5. In Section 7.9.5 a combined semismooth Newton method is constructed that combines the nonlinearity and the crack irreversibility constraint via a primal-dual active set strategy. Conceptionally, this method is similar to the first method.

We formulate most of the following algorithms in terms of phase-field fracture using $U = (u, \varphi)$ as solution variable. However, all algorithms can be applied to any other multiphysics problem. For instance, our own first Newton algorithm was originally implemented in [438] for fluid-structure interaction with $U = (v, u, p)$. The first three Newton methods were further applied in phase-field fluid-structure interaction in [441, 443] with $U = (v, u, p, \varphi)$. Implementations of basic versions can be found in
- [438]: classical residual-based Newton method.[5]
- [413, 207]: classical residual-based Newton method for forward simulations and optimization problems.[6]
- [221, 222]: combined semismooth Newton method.[7]

7.9.1 A residual-based Newton algorithm

We first present a residual-based Newton algorithm. Globalization may be achieved by a damping strategy based on a backtracking line search algorithm.

To measure the residuals and monitoring functions, we use the discrete norm $\|\cdot\| := \|\cdot\|_{l^\infty}$. At a given time instance t_n, we shall find the time step solution U^n using:

Algorithm 8 (Residual-based Newton's method). In this type of methods, the main criterion is a decrease of the residual in each step. Choose an initial Newton guess U^0, e.g., $U^0 := U^{n,0} := U^{n-1}$. For the iteration steps $j = 0, 1, 2, 3, \ldots$:

5 https://journals.ub.uni-heidelberg.de/index.php/ans/article/view/10305
6 http://www.dopelib.net
7 https://github.com/tjhei/cracks

1. Find $\delta U^j := (\delta u, \delta\varphi) \in X^D$ such that

$$A'(U^j)(\delta U^j, \Psi) = -A(U^j)(\Psi) \quad \forall \Psi \in X^0, \tag{7.8}$$

$$U^{j+1} = U^j + \lambda_j \delta U^j, \tag{7.9}$$

   for $\lambda_j = 1$.
2. The criterion for convergence is the contraction of the residuals:

$$\|A(U^{j+1})(\Psi)\| < \|A(U^j)(\Psi)\|. \tag{7.10}$$

3. If (7.10) is violated, recompute in (7.9) U^{j+1} by choosing $\lambda_j^l = 0.5$, and compute for $l = 1, \ldots, l_M$ (e. g., $l_M = 5$) a new solution

$$U^{j+1} = U^j + \lambda_j^l \delta U^j$$

   until (7.10) is fulfilled for a $l^* < l_M$ or l_M is reached. In the latter case, no convergence is obtained and the program aborts.
4. In case of $l^* < l_M$, we check next the stopping criterion:

$$\|A(U^{j+1})(\Psi)\| \le \mathrm{TOL}_{\mathrm{New}}.$$

   If this is criterion is fulfilled, set $U^n := U^{j+1}$. Else, we increment $j \to j+1$ and go to Step 1.

Remark 51 (Simplified Newton steps). Usually, when the Newton reduction rate

$$\theta_j = \frac{\|A(U^{j+1})(\Psi)\|}{\|A(U^j)(\Psi)\|},$$

was sufficiently good, e. g., $\theta_j \le \theta_{\max} < 1$ (where, e. g., $\theta_{\max} \approx 0.1$), a common strategy is to work with the "old" Jacobian matrix, but with a new right-hand side. This procedure is well established in the literature (see, e. g., [131]) and works usually very well. In phase-field fracture, we found that simplified Newton steps, do not help significantly in reducing the computational cost because more iterations are needed in comparison to building each time the Jacobian matrix.

7.9.1.1 The Jacobian matrix

To apply Newton's method for solving $A(U_h)(\Psi_h) = 0$, we need to compute the derivative of $A(U_h)(\Psi_h)$. We construct the Jacobian by evaluating the directional derivative (see also Section 2.2.4):

$$A'(U)(\delta U, \Psi) := \lim_{s \to 0} \frac{A(U + s\delta U)(\Psi) - A(U)(\Psi)}{s}$$

with $\delta U := (\delta u, \delta\varphi) \in X^0$, which represents later the Newton update. As example, let us consider Proposition 32:

$$\begin{aligned} A'(U)(\delta U, \Psi) = &\left(2\delta\varphi(1-\kappa)\varphi\sigma^+(u) + \left((1-\kappa)\varphi^2 + \kappa\right)\sigma^+(\delta u), e(w)\right) \\ &+ \left(\sigma^-(\delta u), e(w)\right) \\ &+ (1-\kappa)\left(\delta\varphi\sigma^+(u) : e(u) + 2\varphi\, \sigma^+(\delta u) : e(u), \psi\right) \\ &+ \left(\frac{1}{\varepsilon}(G_c\delta\varphi, \psi) + \varepsilon(G_c\nabla\delta\varphi, \nabla\psi)\right) \\ &+ \gamma(\delta\varphi, \psi)_{\mathcal{A}(\varphi)} \quad \forall\Psi := (w, \psi) \in X^0, \end{aligned} \tag{7.11}$$

where

$$\mathcal{A}(\varphi) = \{x \in B \mid \Xi + \gamma(\varphi(x) - \varphi(x)^{n-1}) > 0\}.$$

Remark 52. The term $\gamma(\delta\varphi, \psi)_{\mathcal{A}(\varphi)}$ is not smooth because cut-off values are used. An alternative of a smoother regularization is a fourth-order penalization as briefly discussed in Remark 29.

Proposition 59 (Directional derivative Miehe's et al. splitted stress tensor). *We are given from Section 4.5.1.1*

$$\sigma^+(u) := 2\mu e^+ + \lambda[\mathrm{tr}(e)]^+ I.$$

The directional derivative reads

$$\sigma^+(\delta u) := \sigma^+(u)'(\delta u) := 2\mu e^+(u)'(\delta u) + \lambda[\mathrm{tr}(e(u)'(\delta u))]^+ I,$$

with

$$[\mathrm{tr}(e(u)'(\delta u))]^+ = \begin{cases} \mathrm{tr}(e(\delta u)) & \textit{if } \mathrm{tr}(e) > 0 \\ 0 & \textit{if } \mathrm{tr}(e) \le 0 \end{cases}$$

and

$$e^+(\delta u) := P(\delta u)\Lambda^+ P^T + P\Lambda^+(\delta u)P^T + P\Lambda^+ P^T(\delta u).$$

The derivation of $\sigma^-(\delta u)$ is similar and omitted here. An algorithm is provided in [249].

7.9.1.2 Block structure of the Jacobian, solution vector, and right-hand side

In this section, we have a closer look into the structure of the linear system (7.8) arising at each Newton iteration. For the spatial discretization, we use the previously introduced spaces X_h^D and X_h^0, respectively, with the vector-valued basis

$$\{\psi_i \mid i = 1, \ldots, M_X\},$$

where $M_X = \dim(X_h^D)$. The basis functions are primitive (they are only nonzero in one component), so we can separate them into displacement and phase-field basis functions and sort them accordingly:

$$\psi_i = \begin{pmatrix} \chi_i^u \\ 0 \end{pmatrix}, \text{ for } i = 1, \ldots, M_{X,u},$$

$$\psi_{(M_{X,u}+i)} = \begin{pmatrix} 0 \\ \chi_i^\varphi \end{pmatrix}, \text{ for } i = 1, \ldots, M_{X,\varphi},$$

where $M_{X,u} + M_{X,\varphi} = M_X$. This is now used to transform (7.8) into a system of the form

$$M\delta U = F, \tag{7.12}$$

where M is a block matrix (the Jacobian) and F the right-hand side consisting of the residuals. The unknown solution vector is δU. The block structures are

$$M = \begin{pmatrix} M^{uu} & M^{u\varphi} \\ M^{\varphi u} & M^{\varphi\varphi} \end{pmatrix}, \qquad F = \begin{pmatrix} F^u \\ F^\varphi \end{pmatrix}, \qquad \delta U = \begin{pmatrix} \delta U^u \\ \delta U^\varphi \end{pmatrix},$$

with entries coming from (7.11):

$$\begin{aligned}
M_{i,j}^{uu} &= (((1-\kappa)\varphi^2 + \kappa)\,\sigma^+(\chi_j^u), e(\chi_i^u)) + (\sigma^-(\chi_j^u), e(\chi_i^u)),\\
M_{i,j}^{u\varphi} &= (2\chi_j^\varphi(1-\kappa)\varphi\sigma^+(u), e(\chi_i^u)),\\
M_{i,j}^{\varphi u} &= 2(1-\kappa)(\varphi\,\sigma^+(\chi_j^u) : e(u), \chi_i^\varphi),\\
M_{i,j}^{\varphi\varphi} &= (1-\kappa)(\sigma^+(u) : e(u)\chi_j^\varphi, \chi_i^\varphi)\\
&\quad + \left(\frac{1}{\varepsilon}(G_c\chi_j^\varphi, \chi_i^\varphi) + \varepsilon(G_c\nabla\chi_j^\varphi, \nabla\chi_i^\varphi)\right) + \gamma(\chi_j^\varphi, \chi_i^\varphi)_{A(\varphi)}.
\end{aligned}$$

The right-hand side consists of the corresponding residuals read

$$\begin{aligned}
F_i^u &= -A(U^j)(\chi_i^u)\\
&= (((1-\kappa)\varphi^{j2} + \kappa)\,\sigma^+(u^j), e(\chi_i^u)) + (\sigma^-(u^j), e(\chi_i^u)),\\
F_i^\varphi &= -A(U^j)(\chi_i^\varphi) = (1-\kappa)(\varphi^j\,\sigma^+(u^j) : e(u^j), \chi_i^\varphi)\\
&\quad + \left(-\frac{1}{\varepsilon}(G_c(1-\varphi^j), \chi_i^\varphi) + \varepsilon(G_c\nabla\varphi^j, \nabla\chi_i^\varphi)\right) + ([\Xi^j + \gamma(\varphi^j - \varphi^{n-1})]^+, \chi_i^\varphi).
\end{aligned}$$

7.9.1.3 Block structure for extrapolated scheme

When applying the ideas from Section 7.7.2, we obtain

$$M_{i,j}^{u\varphi} = 0$$

after differentiation with respect to φ and consequently, we deal with the triangular block structure

$$M = \begin{pmatrix} M^{uu} & 0 \\ M^{\varphi u} & M^{\varphi\varphi} \end{pmatrix}.$$

This pattern greatly facilitates the linear solution, particularly the design of preconditioners when using an iterative technique, such as GMRES, for instance. Evidence is shown in several studies for 2D and 3D problems where we could extend relatively easily the idea presented in [221] to parallel computations in 3D [285].

Since the blocks $M^{\varphi u}$ and $M^{u\varphi}$ have identical structures since the matrix M is symmetric by construction, one may try to build a symmetric approximation by

$$\begin{pmatrix} M^{uu} & 0 \\ 0 & M^{\varphi\varphi} \end{pmatrix}. \tag{7.13}$$

In some numerical tests, this option did however not improve the performance of the nonlinear solver.

7.9.2 An error-oriented Newton method

We shall address the second Newton method in this section. The idea was motivated by Deuflhard [131] based on his early work [130]. A first version for phase-field fracture was worked out in [444], but in this section extensions are suggested. The key feature of this algorithm is that it is based on a natural monotonicity test in which the ordinary Newton update is compared to a simplified Newton update (Section 7.8.1). This approach yields more flexibility in allowing for growing residuals (e. g., [131, p. 144, Figure 3.9]). In the following, we further extend the main algorithm presented in [444]:

1. As suggested by Deuflhard, we use a scaled norm employing a natural level function [131, p. 135] (see also [63]) linked to the inverse of the Jacobian matrix (in [444], we simply used the identity);
2. When the so-called regularity test fails (chosen line search parameter λ is smaller than a minimal parameter $\lambda_{\min}$), we do not abort the algorithm, but:
   (a) Try another initial guess U^0 and restart the Newton iteration;
   (b) Employ a regularization procedure similar to the Hessian modification in numerical optimization [340, Section 3.4] or also realized through a homotopy in numerical continuation [13, Chapter 1].

We briefly comment on both ideas. Usually as initial guess, we choose $U^0 := U^{n-1}$ (previous loading/time step). One example for an alternative initial guess may be a convex combination $U^0 = cU^{n-1} + (1-c)U^{n-2}$ with $c \in [0,1]$. It is obvious that this is a trial and error procedure. The second variant is more systematic and well-known in the literature as mentioned previously. Here, two further cases can be distinguished:

- Modify only the Jacobian matrix (i. e., the Hessian), but use an unperturbed right hand side in Newton's defect step. Conceptionally, this is similar to Section 7.9.3;
- Modify both: Jacobian and right hand side. Conceptionally this is similar to the partitioned stabilized scheme from Section 7.6.1.

Below, we work with the last idea and augment the given system with the stabilization terms

$$L_u(u^j - u^{j-1}, w),$$
$$L_\varphi(\varphi^j - \varphi^{j-1}, \psi),$$

where $\|u^j - u^{j-1}\| \to 0$ and $\|\varphi^j - \varphi^{j-1}\| \to 0$ for $j \to \infty$ in case of convergence. Alternatively, pure mass matrix terms can be employed:

$$L_u(u^j, w),$$
$$L_\varphi(\varphi^j, \psi),$$

where we should adapt $L_u \to 0$ and $L_\varphi \to 0$, or being sufficiently small, once convergence is detected. In contrast to the pure homotopy procedure, we propose however not to use at each time/loading step t_n the stabilization terms, but only in cases where they are needed. This is most likely the case when the crack starts growing. In most other time/loading steps, Newton's method converges as expected in a few steps (e. g., [444, 445]). To this end, the principle idea is as follows:

1. Try to solve the given physical (unperturbed) problem;
2. If Newton's method stagnates or diverges, add stabilization terms;
3. Optional: Once local and fast convergence is detected, reduce values of stabilization parameters.

The following norms are scaled measures with $P^{-1} \approx A'(\cdot)^{-1}$ (where the latter is as usually the Jacobian matrix) such that $\|P^{-1} \cdot \| := \|P^{-1} \cdot \|_{l^2}$.

The details of this algorithm are as follows:

Algorithm 9 **(Modified error-oriented Newton's method).** Set the minimal tolerance $\lambda_{\min} = 10^{-10}$ and an initial damping parameter $\lambda_0 = 1$ or $\lambda_0 \ll 1$, e. g., $\lambda_0 = 10^{-3}$. Choose an initial Newton guess $U^0 \in X^D$. For $j = 0, 1, 2, 3, \ldots$:

1. Solve for $\delta U^j := (\delta u, \delta\varphi) \in V_u^0 \times V_\varphi$:

$$A'(U^j)(\delta U^j, \Psi) = -A(U^j)(\Psi) \quad \forall \Psi \in V_u^0 \times V_\varphi. \tag{7.14}$$

2. Check if $\|\delta U^j\| \le \mathrm{TOL}_{\mathrm{New}}$. If true, the solution is found and set

$$U^* := U^j + \delta U^j.$$

Else and if $j > 0$ compute a prediction value for the damping factor:

$$\lambda_j := \min(1, \mu_j), \quad \mu_j := \frac{\|\delta U^{j-1}\| \cdot \|\delta U^j_{\text{simp}}\|}{\|\delta U^j_{\text{simp}} - \delta U^j\| \cdot \|\delta U^j\|},$$

where δU^j_{simp} is available from the previous iteration $j-1$ by solving the simplified system (7.17).

3. If

$$\lambda_j < \lambda_{\min} \tag{7.15}$$

then
 - Choose another initial guess $U^0 \in X^D$
 - Or choose stabilization parameters $L_j > 0$ in order to increase the definiteness properties of the given system. Here, $L_j(U^j - U^{j-1}, \Psi)$. These L_j can be chosen as constants or adapted during the Newton iteration j. For instance, we employ increasing L_j at the beginning and decreasing L_j later once fast convergence, e. g., $\theta_{j-1} < 0.01$, is detected.

 Then, go to Step 1.
4. If $\lambda_j > \lambda_{\min}$ continue and compute the trial iterate

$$U^{j+1} = U^j + \lambda_j \delta U^j \tag{7.16}$$

and evaluate the new residual $A(U^{j+1})(\Psi)$. Solve the simplified linear system using the old Jacobian: Find $\delta U^{j+1}_{\text{simp}} := (\delta u_{\text{simp}}, \delta\varphi_{\text{simp}}) \in V^0_u \times V_\varphi$:

$$A'(U^j)(\delta U^{j+1}_{\text{simp}}, \Psi) = -A(U^{j+1})(\Psi) \quad \forall \Psi \in V^0_u \times V_\varphi. \tag{7.17}$$

5. Compute the monitoring functions:

$$\theta_j := \frac{\|\delta U^{j+1}_{\text{simp}}\|}{\|\delta U^j\|}, \quad \mu'_j := \frac{0.5\|\delta U^j\| \cdot \lambda_j^2}{\|\delta U^{j+1}_{\text{simp}} - (1-\lambda_j)\delta U^j\|}, \quad \lambda'_j := \min\left(\mu'_j, \frac{1}{2}\lambda_j\right) \tag{7.18}$$

6. If $\theta_j \geq 1$ (no convergence of the updates) set

$$\lambda_j := \lambda'_j$$

and go to Step 3 and continue from there.
Else: set $\lambda'_j := \min(1, \mu'_j)$.
7. If $\theta_j < 1$, we have convergence and continue:
 (a) If $\lambda'_j = \lambda_j = 1$, check if

$$\|\delta U^{j+1}_{\text{simp}}\| \leq \text{TOL}_{\text{New}}$$

 then stop and the solution is found:

$$U^* := U^{j+1} + \delta U^{j+1}_{\text{simp}}.$$

(b) Else if $\lambda_j' \geq c\lambda_j$, e. g., $c = 4$: replace λ_j by λ_j' and go to Step 4
(c) Else accept U^{j+1} (computed in (7.16)) as new iterate. Then increment $j \to j+1$ and go to the beginning to Step 1.

7.9.3 A modified Newton method with Jacobian modification

In this section, we present our third Newton-type method. The key idea is a dynamic update inside the Jacobian matrix that automatically switches between full Newton steps and fixed-point iterations. Our original work was published in [445] and is inspired by [238, 296]. We also mention the paper [306] that was published at the same time as our method.

A hint for a dynamic Jacobian modification can be found in Section 7.9.1.3 in which one block in the Jacobian is zero due to an extrapolation in the phase-field variable, yielding a robust and efficient method.

Rather than employing line search in Step 3 in Algorithm 8, we introduce a control parameter $\omega \in [0,1]$ inside the Jacobian, which decides whether a full Newton system ($\omega = 1$), a fixed-point system (see Section 7.3) with $0 < \omega < 1$ or even $\omega = 0$ is solved. The choice of this parameter is heuristic, but the key idea is simple.

As shown in Section 7.9.1.2, formally, the Jacobian reads at each Newton step j:

$$M = \begin{pmatrix} M^{uu} & M^{u\varphi} \\ M^{\varphi u} & M^{\varphi\varphi} \end{pmatrix}.$$

The critical block is $M^{u\varphi}$, particularly

$$(2\chi_j^{\varphi}(1-\kappa)\varphi\sigma^{+}(u), e(\chi_i^{u})), \tag{7.19}$$

as it was already identified in [221] and Section 7.7.2. Thus the goal is to design a procedure in which this block is dynamically activated or disabled during a Newton iteration. Incorporating ω brings us to

$$M = \begin{pmatrix} M^{uu} & \omega M^{u\varphi} \\ M^{\varphi u} & M^{\varphi\varphi} \end{pmatrix} = \begin{pmatrix} M^{uu} & 0 \\ M^{\varphi u} & M^{\varphi\varphi} \end{pmatrix} + \omega \begin{pmatrix} 0 & M^{u\varphi} \\ 0 & 0 \end{pmatrix}. \tag{7.20}$$

7.9.3.1 Computing the control parameter ω

The choice of ω is done in a dynamic way dependent on the previous two Newton residuals. Thus, at each Newton step j, the parameter $\omega := \omega_j$ is updated if applicable. The residual and reciprocal residual reductions are defined by, respectively,

$$Q_{j+1} = \frac{\|A(U^{j+1})(\Psi)\|}{\|A(U^{j})(\Psi)\|}, \quad Q_{j+1}^{\text{rec}} = \frac{\|A(U^{j})(\Psi)\|}{\|A(U^{j+1})(\Psi)\|}. \tag{7.21}$$

If $Q_{j+1} < 1$, the new residual is smaller and we classify this step as a "good" step as it fulfills the residual-based monotonicity criterion (Section 7.8.1). Moreover, if $Q_{j+1} \to 0$, the better the current step. On the other hand, if $Q_{j+1} \geq 1$, the new residual is larger than the old one and we have the situation in which a residual-based Newton method would fail (assuming globalization techniques such as line search do not do their job) and, for example, an error-oriented version as we presented in Section 7.9.2 may perform better.

We summarize the key ideas and construction of the control parameter ω in the following.

Definition 59 (Computing ω_{j+1}). At the Newton step j, let $0 \leq \omega_j \leq 1$ be given and let $S \in \mathbb{R}_+ \cup \{0\}$. We define

$$\omega := \omega_{j+1} = S\omega_j. \tag{7.22}$$

Proposition 60 (Motivation of S). *The scaling parameter S is motivated as follows:*
1. *S must yield $\omega_{j+1} \in [0, 1]$.*
2. *$S \gg 1$ should yield $\omega_{j+1} \to 1$ (full Newton).*
3. *$S \to 0$ should yield $\omega_{j+1} \to 0$ (fixed-point scheme).*

Proposition 61 (A specific realization of S). *Given Proposition 60, the scaling parameter S can be realized, using the residuals Q_{j+1} and Q_{j+1}^{rec} defined in (7.21), as follows:*
1. *For $Q_{j+1} \to 0$ (and $Q_{j+1}^{\text{rec}} \to \infty$), the matrix M defined in (7.20) has good properties and we can work with a full Newton step, i. e., $S \gg 1$ yielding $\omega_{j+1} = 1$.*
2. *On the other hand, for $Q_{j+1}^{\text{rec}} \to 0$ (and $Q_{j+1} \to \infty$), the matrix M becomes ill-conditioned, and $S \to 0$ yielding $\omega_{j+1} \ll 1$ should be employed.*
3. *These observations yield the following possible realization of S:*

$$S := \left(\frac{a}{\exp\left(Q_{j+1}^{\text{rec}}\right)} + \frac{b}{\exp\left(Q_{j+1}\right)} \right). \tag{7.23}$$

4. *The control parameter a is related to a fixed-point step with small ω_{j+1}, and thus $a < 1$ should be chosen.*
5. *The control parameter b is related to a full Newton step with $\omega_{j+1} = 1$, and thus $b \geq 1$ should be chosen.*
6. *Both control parameters will be explained and specified further in Section 7.9.3.3.*

Corollary 5 (Further properties of S). *The scaling parameter S has the following properties:*
1. *S is bounded from below by zero: since $a, b, Q_{j+1}, Q_{j+1}^{\text{rec}} \geq 0$, we have $S \geq 0$ yielding $\omega_{j+1} \geq 0$.*
2. *S is not bounded from above. Consequently, it may easily happen that $\omega_{j+1} > 1$ in (7.22). Therefore, we use a simple projection:*

$$\text{Set} \quad \omega_{j+1} := 1 \quad \text{if} \quad \omega_{j+1} > 1.$$

7.9.3.2 The modified Newton algorithm

Let us consider as before Proposition 32 (the other two follow trivially). Weighting the first block with ω yields the Jacobian:

$$\begin{aligned}A'_\omega(U)(\delta U,\Psi) = {} & (\omega 2\delta\varphi(1-\kappa)\varphi\sigma^+(u) + ((1-\kappa)\varphi^2+\kappa)\,\sigma^+(\delta u), e(w)) \\ & + (\sigma^-(\delta u), e(w)) \\ & + (1-\kappa)(\delta\varphi\sigma^+(u) : e(u) + 2\varphi\,\sigma^+(\delta u) : e(u), \psi) \\ & + \left(\frac{1}{\varepsilon}(G_c\delta\varphi,\psi) + \varepsilon(G_c\nabla\delta\varphi,\nabla\psi)\right) \\ & + \gamma(\delta\varphi,\psi)_{A(\varphi)}.\end{aligned} \tag{7.24}$$

Algorithm 10 **(Modified Newton's method with Jacobian modification).** Choose an initial Newton guess U^0 and an initial guess for the control parameter, i. e., $\omega_0 = 1$. For the iteration steps $j = 0, 1, 2, 3, \ldots$:

1. Find $\delta U^j := (\delta u, \delta\varphi) \in X^0$ such that

$$A'_\omega(U^j)(\delta U^j,\Psi) = -A(U^j)(\Psi) \quad \forall \Psi \in X^0, \tag{7.25}$$

$$U^{j+1} = U^j + \delta U^j. \tag{7.26}$$

2. Compute:

$$\omega := \omega_{j+1} = S\omega_j, \tag{7.27}$$

   with S determined by (7.23).

3. Check the stopping criterion

$$\|A(U^{j+1})(\Psi)\| \le \mathrm{TOL}_{\mathrm{New}}.$$

   If this criterion is fulfilled, set $U^n := U^{j+1}$. Else, we increment $j \to j+1$ and go to Step 1.

Remark 53. In this algorithm, we do not have any convergence monitor and it can happen that Newton's method diverges. Thus, we also check in Step 3 whether

$$\|A(U^{j+1})(\Psi)\| < \mathrm{TOL}^{\mathrm{up}}_{\mathrm{New}}, \quad \mathrm{TOL}^{\mathrm{up}}_{\mathrm{New}} = 10^{12},$$

otherwise we stop the algorithm because of divergence. In [445], we see that for backtracking line-search such a behavior is indeed detected, but the modified Newton method (without convergence monitor) yields excellent performance. We notice that $\mathrm{TOL}^{\mathrm{up}}_{\mathrm{New}}$ seems high, but there we found numerical examples [445, Figures 26 and 29], where the residual goes up to 10^7 but where Newton's method will still converge nonetheless.

7.9.3.3 Choosing the parameters *a* and *b*

The choices of a and b are heuristic. As outlined in Proposition 61, the parameter a controls the influence of block $M^{u\varphi}$. The parameter b controls the rate for returning to full Newton steps in case sufficient performance of the solver is detected. Therefore, we propose the following bounds:

$$0 \le a < 1 \quad \text{and} \quad 1 \le b < \infty.$$

Let us discuss the idea in more detail. If $Q_{j+1} \ll 1$, we had a good reduction and we can use a higher ω_{j+1} in the next step. Formula (7.23) yields

$$\lim_{Q_{j+1}^{\text{rec}} \to \infty} \lim_{Q_{j+1} \to 0} S \to b \quad \Rightarrow \quad \omega_{j+1} = b\omega_j \quad \Rightarrow \quad \omega_{j+1} \ge \omega_j.$$

On the other hand, if $Q_{j+1} > 1$ or even $Q_{j+1} \gg 1$ (thus $Q_{j+1}^{\text{rec}} \to 0$), we want to eliminate the irregular terms in the Jacobian matrix and rather work with a Newton-like method in which the Jacobian is approximated by minimizing the influence of the term (7.19). Here,

$$\lim_{Q_{j+1}^{\text{rec}} \to 0} \lim_{Q_{j+1} \to \infty} S \to a \quad \Rightarrow \quad \omega_{j+1} = a\omega_j \quad \Rightarrow \quad \omega_{j+1} < \omega_j.$$

Since we cannot ensure a priori that S is bounded from above due to the construction, the requirement $0 \le \omega_{j+1} \le 1$ may be violated. In this case, a projection is used (see Corollary 5).

Possible choices of a and b:
- Choice 1: $a = 0.001$ and $b = 10$ fastly tries to remove the entire block $M^{u\varphi}$ and moderately goes back to full Newton;
- Choice 2: $a = 0.1$ and $b = 2$ tries moderately to remove the influence of block $M^{u\varphi}$ and moderately goes back to full Newton;
- Choice 3: $a = 0$ and $b = 0$ resulting in $S = 0$ from which we obtain $\omega_j = 0$ for all j and thus never work with the block $M^{u\varphi}$.

Obviously, the smaller $a \ll 1$ is chosen, the faster we obtain a Newton-like (fixed-point) scheme. Second, the larger we choose $b \ge 1$, the faster we go back to a full Newton scheme.

Remark 54 (Numerical results). For more choices of a and b and their consequences in numerical examples, we refer to [445].

7.9.4 An error-oriented Newton method with Jacobian modification

In order to enhance the convergence tests of the modified Newton scheme, we propose a fourth variant in which the Jacobian is updated in the error-oriented version if the convergence monitor in step 5 is not sufficiently good.

We propose the following combination:

1. Replace Q_{j+1} in (7.21) by θ_j defined in (7.18), respectively replace Q_{j+1}^{rec} by $1/\theta_j$;
2. Evaluate (7.23) to compute ω with (7.22);
3. Use the modified Jacobian (7.24) in the error-oriented algorithm in (7.14) and (7.17).

Thus, Step 6 of Algorithm 9 is modified to:

6. If $\theta_j \geq 1$ (no convergence of the updates), set

$$\lambda_j := \lambda_j'$$

and compute ω with (7.22). Then go to Step 3 and continue from there.

The idea is two-fold. First, we improve robustness when re-evaluating a rejected step. Second, we then keep the new ω for the next full cycle to improve upfront the following steps. As further modification, also in Step 7 of Algorithm 9, the parameter ω is each time updated, depending on the value of θ_j.

7.9.5 A combined Newton algorithm and using a primal-dual active set strategy

In this section, we present our fifth and last Newton-type method. Here, we do not use penalization (simple or augmented Lagrangian) for treating the crack irreversibility constraint, but a primal-dual active set (PDAS) method. The latter can be related to a semismooth Newton method (see our discussions in Section 5.2.3). The idea is to combine this approach with the Newton procedure necessary for the nonlinearities inside the PDEs. Conceptionally, the resulting algorithm is close to Section 7.9.1, namely a residual-based monotonicity criterion with line search steps is adopted.

Since we deal with a nonlinear problem, we formulate the concepts from Section 5.2.3 to the Newton system. As before: given an initial guess U^0, find δU for $j = 0, 1, 2, 3, \ldots$:

$$\begin{aligned} A'(U^j)(\delta U, \Psi) &= -A(U^j)(\Psi) \quad \forall \Psi \in X^0, \\ U^{j+1} &= U^j + \omega \delta U. \end{aligned}$$

In order to show the conceptional equivalence to Section 5.2.3, let us define

$$\begin{aligned} G &:= A'(U^j)(\cdot, \Psi), \\ F &:= -A(U^j)(\Psi). \end{aligned}$$

Before we can state the Lagrange multiplier system, we need to have a closer look on the update δU. Linearizing the irreversibility constraint $\varphi \leq \varphi^{n-1}$ at the point φ, we obtain $\delta\varphi \leq 0$. Consequently, we have $\delta U \leq 0$ on the space $0 \times V_\varphi$. First, this allows us to write a minimization problem for the phase-field part:

$$\min_{\delta U \in \tilde{V}_\varphi} \frac{1}{2}(\delta U, G\delta U) - (F, \delta U),$$

where $\tilde{V}_\varphi := \{\delta U \in 0 \times V_\varphi \mid \delta U \leq 0\}$. Similar to Proposition 30, we obtain a Lagrange multiplier system, but now on the Newton level.

Proposition 62 (Lagrange multiplier system on Newton level). *Let $\delta U = (\delta u, \delta\varphi) \in X^0 = V_u^0 \times V_\varphi$. Find $(\delta U, \Xi) \in X^0 \times V_\Xi$ such that*

$$A'(U)(\delta U, \Psi) + \langle \Xi, \psi \rangle = -A(U)(\Psi) \quad \forall \Psi = (w, \psi) \in X^0,$$
$$\langle \Xi - \chi, \delta\varphi \rangle \geq 0 \quad \forall \chi \in V_\Xi.$$

The complementarity system to the last inequality reads

$$\delta\varphi \leq 0 \quad \text{in } B \times I,$$
$$\Xi \geq 0 \quad \text{in } B \times I,$$
$$\langle \Xi, \delta\varphi \rangle = 0 \quad \text{in } B \times I.$$

We obtain as in Section 5.2.3

$$C(\delta\varphi, \Xi) = 0,$$

where

$$C(\delta\varphi, \Xi) = \Xi - [\Xi + c\delta\varphi]^+. \tag{7.28}$$

We can now formulate a primal-dual active set algorithm according to Section 5.2.3 (based on the work [231, Section 2]). For phase-field fracture, this was worked out in [221, Section 3.3]. The arising linear equation systems in Proposition 62 have the following block structure:

$$\begin{pmatrix} G & B \\ B^T & 0 \end{pmatrix} \begin{pmatrix} \delta U \\ \Xi \end{pmatrix} = \begin{pmatrix} F \\ 0 \end{pmatrix}$$

Here, the matrix B represents the duality pairing $\langle \Xi, \psi \rangle$. On the spatially discretized level for $\langle \Xi_h, \psi_h \rangle$, the trick is to employ a quadrature rule for Ξ_h in such a way, that B becomes diagonal and Ξ_h can be explicitly eliminated from the block system. An example for such a numerical quadrature is Gauss–Lobatto. This yields the active set $\mathcal{A}^j$ at the Newton iteration j for index i (e. g., DoF):

$$(B^{-1})_{ii}(F - G\delta U^j)_i + c(\delta U^j)_i > 0. \tag{7.29}$$

If

$$(B^{-1})_{ii}(F - G\delta U^j)_i + c(\delta U^j)_i \leq 0, \tag{7.30}$$

then the index i is in the inactive set $\mathcal{N}^j$. These preparations will be merged with the nonlinear Newton solver yielding the following algorithm [221]:

Algorithm 11. Given the initial guess U^0. For $j = 0, 1, 2, 3, \dots$:
1. Assemble residual $R(U_h^j) := -A(U_h^j)(\Psi)$
2. Compute active set $\mathcal{A}_j = \{i \mid (B^{-1})_{ii}(R_j)_i + c(\delta U_h^j)_i > 0\}$
3. Assemble matrix $G = A'(U_h^j)(\delta U_h^j, \Psi_h)$ and right-hand side $F = -A(U_h^j)(\Psi_h)$
4. Eliminate rows and columns in $\mathcal{A}_j$ from G and F to obtain $\widetilde{G}$ and $\widetilde{F}$
5. Solve linear system $\widetilde{G}\delta U_j = \widetilde{F}$, i. e., find $\delta U_h^j \in X_h^0$ with

$$A'(U_h^j)(\delta U_h^j, \Psi_h) = -A(U_h^j)(\Psi_h) \quad \forall \Psi_h \in X_h^0 \tag{7.31}$$

6. Find a step size $0 < \lambda \leq 1$ using line search to get

$$U_h^{j+1} = U_h^j + \lambda \delta U_h^j,$$

   with $\widetilde{R}(U_h^{j+1}) < \widetilde{R}(U_h^j)$.
7. Check both stopping criteria simultaneously:

$$\mathcal{A}_{j+1} = \mathcal{A}_j \qquad \text{and} \qquad \widetilde{R}(U_h^j) < \mathrm{TOL}_{\mathrm{New}}.$$

Remark 55. It is important to distinguish between the full residual $R(U_h^j)$ and the reduced residual $\widetilde{R}(U_h^j)$. The latter is the residual on the inactive set, which can be computed by eliminating the active set constraints from the former.

Remark 56. In optimization it is well known that active set and primal-dual active set methods may suffer from cycling and stalling of the algorithm, e. g., [293, 214]. Solutions to cope with these issues have been proposed therein. Our implementation in [222] to avoid cycling is based on [120].

Remark 57 (Implementation). For a C++ implementation of this algorithm, we refer the reader to Chapter 13, Software 1 [222].

7.10 Linear solution inside Newton's method of monolithic phase-field fracture systems

We explain in this section the structure of the linear equation system when both solution variables u and φ are treated simultaneously, i. e., in a monolithic fashion. We recall:

$$A'(U_h^{n,j})(\delta U_h^n, \Psi) = -A(U_h^{n,j})(\Psi) + F(\Psi),$$
$$U_h^{n,j+1} = U_h^{n,j} + \lambda \delta U_h^n.$$

The linear equation system in matrix form reads:

$$M\delta U = B \tag{7.32}$$

where M has been defined before, B is the discretization of the residual:

$$B \sim -A(U_h^{n,j})(\Psi) + F(\Psi)$$

and the solution vector δU is given by

$$\delta U = (\delta u_1, \ldots, \delta u_{M_u}, \delta\varphi_{M_u+1}, \ldots, \delta\varphi_{M_\varphi})^T.$$

7.10.1 Fully monolithic approximation

Let us come back to the block system from Section 7.9.1.2 and briefly develop the principle idea of constructing a preconditioner. We are given

$$\begin{pmatrix} M_{uu} & M_{u\varphi} \\ M_{\varphi u} & M_{\varphi\varphi} \end{pmatrix} \begin{pmatrix} \delta u \\ \delta\varphi \end{pmatrix} = \begin{pmatrix} B_u \\ B_\varphi \end{pmatrix},$$

where B_u and B_φ are the residual parts corresponding to the first and second problem, respectively. Since in general such matrix systems are solved with iterative solvers, the block form allows for a better view on the structure and construction of preconditioners. For instance, starting again from (7.32), a preconditioner is a matrix P^{-1} such that

$$P^{-1} M \delta U = P^{-1} B,$$

such that the condition number of $P^{-1}M$ is moderate. Obviously, the ideal preconditioner would be the inverse of A, e. g., $P^{-1} := A^{-1}$. In practice one tries to build P^{-1} in such a way that

$$P^{-1} M = \begin{pmatrix} I & * \\ 0 & I \end{pmatrix}$$

and where P^{-1} is a lower triangular block matrix:

$$P^{-1} = \begin{pmatrix} P_1^{-1} & 0 \\ P_3^{-1} & P_4^{-1} \end{pmatrix}.$$

The procedure is as follows (see lectures for linear algebra in which the inverse is explicitly constructed):

$$\begin{aligned} M &= \begin{pmatrix} M_{uu} & M_{u\varphi} \\ M_{\varphi u} & M_{\varphi\varphi} \end{pmatrix} \begin{pmatrix} I & 0 \\ 0 & I \end{pmatrix} \\ &= \begin{pmatrix} I & M_{uu}^{-1} M_{u\varphi} \\ M_{\varphi u} & M_{\varphi\varphi} \end{pmatrix} \begin{pmatrix} M_{uu}^{-1} & 0 \\ 0 & I \end{pmatrix} \end{aligned}$$

$$= \begin{pmatrix} I & M_{uu}^{-1}M_{u\varphi} \\ 0 & \underbrace{M_{\varphi\varphi} - M_{\varphi u}M_{uu}^{-1}M_{u\varphi}}_{=S} \end{pmatrix} \begin{pmatrix} M_{uu}^{-1} & 0 \\ -M_{\varphi u}M_{uu}^{-1} & I \end{pmatrix}$$

$$= \begin{pmatrix} I & M_{uu}^{-1}M_{u\varphi} \\ 0 & I \end{pmatrix} \underbrace{\begin{pmatrix} M_{uu}^{-1} & 0 \\ -S^{-1}M_{\varphi u}M_{uu}^{-1} & S^{-1} \end{pmatrix}}_{=P^{-1}}$$

where $S = M_{\varphi\varphi} - M_{\varphi u}M_{uu}^{-1}M_{u\varphi}$ is the so-called Schur complement. The matrix P^{-1} is used as an (exact) preconditioner for M. Indeed we double-check:

$$\begin{pmatrix} M_{uu}^{-1} & 0 \\ -S^{-1}M_{\varphi u}M_{uu}^{-1} & S^{-1} \end{pmatrix} \begin{pmatrix} M_{uu} & M_{u\varphi} \\ M_{\varphi u} & M_{\varphi\varphi} \end{pmatrix} = \begin{pmatrix} I & M_{uu}^{-1}M_{u\varphi} \\ 0 & I \end{pmatrix}.$$

Tacitly, we assumed in the entire procedure that S and M_{uu} are invertible.

Using P^{-1} in a Krylov subspace method, we only have to perform matrix-vector multiplications such as

$$\begin{pmatrix} X_{\text{new}} \\ Y_{\text{new}} \end{pmatrix} = \begin{pmatrix} P_1^{-1} & 0 \\ P_3^{-1} & P_4^{-1} \end{pmatrix} \begin{pmatrix} X \\ Y \end{pmatrix}.$$

Now we obtain

$$X_{\text{new}} = P_1^{-1}X, \tag{7.33}$$

$$Y_{\text{new}} = P_3^{-1}X + P_4^{-1}Y. \tag{7.34}$$

7.10.2 A block-diagonal preconditioner for the extrapolated scheme

For the quasi-monolithic scheme (Section 7.7.2), we obtained in Section 7.9.1.3 the block-triangular matrix

$$\begin{pmatrix} M^{uu} & 0 \\ M^{\varphi u} & M^{\varphi\varphi} \end{pmatrix}.$$

Here, a simple preconditioner is the block-diagonal version

$$P^{-1} = \begin{pmatrix} (P^{uu})^{-1} & 0 \\ 0 & (P^{\varphi\varphi})^{-1} \end{pmatrix}.$$

This variant was used in [221], but not further exploited in detail therein. In [222], performance tests are provided, yielding excellence results. Let us have a closer look into the single entries. Here, the prototype problems introduced in Section 2.3 provide useful insight. The main entries, namely the diagonal terms are elliptic. For such structures, it is simple to approximate the inverse matrices. One possibility is an algebraic multigrid (AMG) solver from Trilinos [224]. Approximatively, we have

$$P_{uu} \approx -\nabla \cdot (\varphi^2 \nabla u),$$

$$P_{\varphi\varphi} \approx -\varepsilon\Delta\varphi - \frac{1}{\varepsilon}(1-\varphi) + |\nabla u|^2\varphi,$$

where both are of elliptic type and are therefore "nice" terms to deal with. For the more advanced models based on Proposition 20, we observe the following linear solver behaviors:

1. For Sneddon's configuration [396] in which a steady-state example is solved with a given, fixed pressure, we observed between 17–52 GMRES[8] iterations per Newton step using the above mentioned block-diagonal preconditioner [222].
2. Similar iteration numbers were observed for the single edged notched shear test in the code from [221]; but these numbers were not explicitly mentioned therein.
3. For phase-field fracture in nonlinear porous media in which also a parabolic pressure equation is solved, we observed in average 15–100 GMRES iterations [425]. Therein, the above mentioned block-diagonal preconditioner was simply extended by an additional diagonal block for the pressure equation part.
4. Similar GMRES iteration numbers of 15–100 iterations for fluid-filled fractures were observed in [287, 322].

7.10.3 A matrix-free geometric multigrid solver

Recently, we have developed a matrix-free geometric multigrid (GMG) method in [250]. Technical details are omitted here and can be found in [250] and [249]. This solver shows a good performance with 5–50 GMRES iterations in average per Newton step for the *2D* L-shaped panel test (original configuration proposed in [454] and extended to a cyclic version in [17]). For a 3D version, we observe 5–10 GMRES iterations per Newton step. In [249], a comparison between the AMG implementation of [221, 222] and the matrix-free GMG solver from [250] is made. For the single edge notched shear test, 5–20 iterations in average are necessary using the AMG implementation,[9] while the GMG solver performs slightly better with 5–6 iterations.

7.10.4 Parallel solution on uniform and adaptively refined meshes

The principle framework for parallel solutions of the previously explained solvers uses deal.II [26], MPI,[10] Trilinos [224], and p4est [97]. Detailed explanations of illustrative implementations are provided in step-40, 41, 42[11] of deal.II [26]; see also the main corresponding publication [185]. Basic versions of our programming code [222] are as

8 GMRES = generalized minimal residual method.

9 Thanks to both Katrin Mang and Daniel Jodlbauer for running these tests again for me in the end of April 2020.

10 MPI (Message Passing Interface): for instance, https://computing.llnl.gov/tutorials/mpi/

11 https://www.dealii.org/

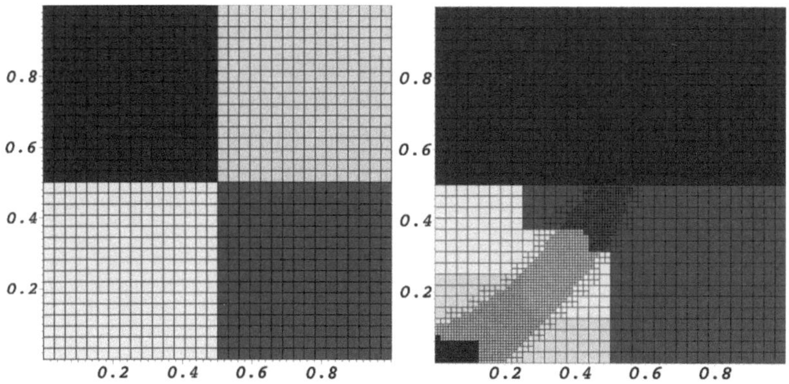

Figure 7.3: Single edge notched shear test (see also Chapter 8): Parallel computing on 4 processors for the uncracked configuration and the final state. The different subdomains are associated with different processors. Depending on mesh refinement, the workload for each processor is adjusted dynamically at each time step. The original code was developed in [221] and recently extended in [222].

well open-source (see Chapter 13). For the single edge notched shear test (Chapter 8), a sketch of processor distribution for locally refined meshes is provided in Figure 7.3.

Based on the formulations derived in [221], we have studied two linear parallel solutions in more detail:

1. GMRES with parallel AMG [222] (code open source; see Chapter 13);
2. GMRES with a matrix-free GMG method [250, 249].

The former was further developed in [285, 322, 425]. The parallel performance was studied in detail in [222], where we observed a good performance in terms of parallel scalability – including adaptively refined meshes. Our matrix-free implementation was studied exclusively on uniformly refined meshes so far, but showed a slightly better performance than the AMG method [249]. Therein, weak and strong scalability tests are presented. Matrix-free solutions are an active research field and the closest work in comparison to [249] is the very recently published study [126] for finite-strain hyperelasticity. We also mention another direction, namely [468], in which a GPU (graphical processing unit) based implementation was achieved. To the best of our knowledge, our own studies [222, 249] are the only ones investigating in detail the parallel performance of (the inner) linear solvers for phase-field fracture problems.

7.11 The augmented Lagrangian loop and an inexact version

In order to enforce the penalty parameter γ, when working with Propositions 19 or 20 (and their corresponding discrete versions), simple penalization leads to ill-conditioned systems as described earlier. One possibility is to start with a moderate or small γ, while increasing during a simulation run. As we already encountered, a better choice is to work with the Lagrange multiplier function Ξ. Rather than updating directly γ, the function Ξ is updated, e. g., [244]. At each time point t_n, the augmented Lagrangian loop constitutes the outer loop wherein at each step the Newton solver is adopted. Moreover, we propose an (heuristic) adaptive criterion in order to reduce the computational cost. This leads to an inexact augmented Lagrangian/Newton loop.

7.11.1 Augmented Lagrangian penalization

The iteration for updating Ξ reads as follows:

Algorithm 12 (Inexact augmented Lagrangian loop with inner Newton solver [444]). At each t_n for $n = 1, 2, 3, \ldots, N$ let $\Xi_{h,0}$ be given, e. g., $\Xi_{h,0} = 0$. Moreover, let $\gamma > 0$ be fixed and given for all t_n. At each time t_n iterate for $j = 0, 1, 2, \ldots$

1. Given Ξ_h^j, we seek $U_h^{n,j+1} = (u_h^{n,j+1}, \varphi_h^{n,j+1})$ by solving Proposition 31 or Proposition 32 in the augmented Lagrangian fashion with the help of a nonlinear iterative method (choose one of this chapter).
2. Update
$$\Xi_h^{n,j+1} = [\Xi_h^{n,j} + \gamma(\varphi_h^{n,j} - \varphi_h^{n-1})]^+.$$
3. Check the stopping criterion
$$\{\|u_h^{n,j+1} - u_h^{n,j}\|_{L^2}, \|\Xi_h^{n,j+1} - \Xi_h^{n,j}\|_{L^2}\} \le \mathrm{TOL}_{\mathrm{AL}}, \quad \mathrm{TOL}_{\mathrm{AL}} > 0. \tag{7.35}$$
4. (a) If the stopping criterion is satisfied, set $U_h^n := U_h^{n,j^*}$ where j^* is the final j that satisfies (7.35).
   (b) Else increment $j \to j + 1$ and go to Step 1.

Remark 58. Compare this algorithm to Section 7.6.1 in which Ξ is updated along with the L stabilization parameters inside iterative coupling.

7.11.2 An adaptive Newton stopping criterion for an inexact augmented Lagrangian method

In order to decrease the computational cost, we adaptively determine the stopping tolerance of Newton's method depending on the augmented Lagrangian norm. Such strategies are in particular well known for (adaptive) inexact Newton methods, e. g., [129, 99, 131], in which the linear equations are only solved approximately at each stage. Related techniques using adaptive (or inexact) augmented Lagrangian realizations are discussed, for instance, in [353]. In such adaptive inexact methods, the accuracy of the inner method should be chosen such that the convergence pattern of the outer loop remains unperturbed, i. e., $TOL_{New} < TOL_{AL}$. Therefore, $\alpha < 1$ is a necessary choice. From our practical observations, the largest α should be chosen as 10^{-3} in these types of problems. For computational results, we refer to [444, Section 5].

Proposition 63 ([444]). *For $j = 0$ set Ξ_h^j and, e. g., $TOL_{New} = 10^{-8}$. For $j = 0, 1, 2, \ldots$ use Algorithm 12 and compute in Step 4b*

$$\Delta := \|\Xi_h^{j+1} - \Xi_h^j\|_{L^2}.$$

For each further augmented Lagrangian iteration step j, we use in the inner Newton loop (part of Step 1) as stopping criterion

$$TOL_{New} := \alpha\Delta,$$

where $\alpha = 10^{-3} < 1$.

7.12 Comparing primal-dual active set and augmented Lagrangian in Newton-type algorithms

In the previous sections of this chapter, we designed Newton methods for basically two major algorithms to treat the crack irreversibility constraint:
- Augmented Lagrangian (AL) iterations;
- Primal-dual active set (PDAS) method (semismooth Newton).

As previously noticed in Section 5.2.3 (see in the literature for instance [244]), both algorithms share important similarities.

Indeed both algorithms work sufficiently well for phase-field fracture and yield simulation results that others have obtained in the literature as well. A rigorous comparison of both approaches for phase-field fracture in terms of numerical analysis and computational analysis remains open to date. In theory, PDAS should beat AL because it can be interpreted as a semismooth Newton method [231] with better convergence under certain assumptions. In our work so far, we made the following observations:
- Pro PDAS/con AL: The primal-dual active set method has no penalization parameter γ, but still a parameter c, which however is less influential in computations.
- Con PDAS: might run into cycling; a simple, nonrobust solution to cope with this issue was suggested in [222] based on [120].
- Con PDAS/con AL: both methods need iterations to converge. While the Newton method for the nonlinear PDE converges in a fast way using the extrapolation $\tilde{\varphi}$ (see Section 7.7.2), the PDAS or AL algorithms often need several iterations more. Here, a clear picture could not be made, which of both algorithms beats the other, and further research is necessary.
- Pro AL/con PDAS: the augmented Lagrangian loop is simpler to implement.

Summarizing these observations, a clear picture of which algorithm should be preferred for phase-field fracture computations cannot be made today. What is true, however, is the fact, that the PDAS algorithm is published as open source (see Chapter 13 [Software No. 1]) and jointly developed and tested with other colleagues (Timo Heister, Sanghyun Lee, Daniel Jodlbauer, Katrin Mang). Therefore, the programming code is much more sophisticated and debugged than the in-house AL implementations [430, 444, 445, 88].

7.13 Updating the strain history field

We finally come back to the alternative Proposition 33, which directly results into two equations without inequality constraints. Efficient procedures were explained by the original authors in [313]. Based on the methods considered so far in this chapter, we propose two models:

- Monolithic treatment of φ in the u-equation;
- Using the φ extrapolation in the u-equation.

Either option is then solved with one of the first four Newton methods from Section 7.9. In particular, when the second model is employed, we may use the suggested extrapolation (see Section 7.7) or the iteration on the extrapolation presented in Section 7.7.3. For the strain history, we apply the same schemes: we can update either at each time step (yielding a relatively large approximation error in time) or we can iterate per time step a few times, which reduces the temporal error. The procedure to update the strain history field is as follows.

Algorithm 13. Initialize $\mathcal{H}(x, t_0) = \mathcal{H}^0$. At time point t_n:

1. Given $\mathcal{H}^{n-1}$ solve Proposition 33;
2. Evaluate

$$\mathcal{H} = \mathcal{H}^n = \begin{cases} \psi_E^+(e), & \psi_E^+(e(u(x_q, t_n))) > \mathcal{H}^{n-1} \\ \mathcal{H}^{n-1} & \text{otherwise} \end{cases}$$

   for each quadrature point $x_q \in B$. We recall the positive strain energy $\psi_E^+ := \psi_E^+(e(u))$ defined in Section 4.5.3.
3. Increment time $n \mapsto n + 1$.

Remark 59. The strain history function $\mathcal{H}$ needs to be evaluated at each quadrature point in the inner FEM loop and their history needs to be stored. This procedure is not simple and we adopted implementational ideas from step 18 in deal.II [26].

Remark 60. If the iteration on the extrapolation (Section 7.7.3) is used, then $\mathcal{H}^n$ can be updated as well at each extrapolation iteration step.

8 Simulations II: single edge notched shear test

We consider in this chapter the single edge notched shear test, which was mentioned before and which has often been studied in the literature. Phase-field modeling references of this test case include [317, 313, 67, 17, 221, 444, 445, 88, 301, 154, 249].

8.1 Governing frameworks

The governing CVISs/PDEs are given in Proposition 20 and 25 (SH),[1] respectively. In the former, AL penalization (Section 5.2.2.1), or the PDAS method (Sections 5.2.3 and 7.9.5) are applied.

8.2 Goals and test cases

We describe the computational setting and present several computational results from which most are not published in our own journal papers and are therefore novel findings. A list of the methods we shall compare is provided in Table 8.1. The focus is then on:

- imposing the crack irreversibility condition with AL (Proposition 32), SH (Proposition 33), and the PDAS strategy (Section 7.9.5)
- relaxing the nonlinearity in the first term of the u-equation by using the extrapolation from Section 7.7.2
- iterating on the extrapolation (Section 7.7.3)
- iterative coupling, L-scheme and dynamic L-scheme (Section 7.6.1)
- refining the loading-step size (related to temporal convergence)

The simulations are analyzed in terms of

- the time step at which the crack reaches the lower boundary;
- the load-displacement curves (i. e., evolution of traction forces versus loading increments) on the top boundary in shear direction;
- Nonlinear iteration numbers.

8.3 Results elsewhere in this monograph

Furthermore, findings of this numerical test can be found in:

- Figure 1.6: Section 1.3.3;
- Figure 7.3: Partitioning of the computational domain using parallel computing [221, 222];

1 SH = strain history, AL = augmented Lagrangian, PDAS = primal-dual active set.

https://doi.org/10.1515/9783110497397-008

Table 8.1: Summary of all test cases. SH(j) = strain history with j iterations (together with extrapolation iterations), AL = augmented Lagrangian, PDAS = primal-dual active set, k time step size (loading step size). In case 17, different updating parameters $a = 5, 10, 20$ are compared.

Case	PFF var.	Iter. on extra.	Crack irr.	k
1	φ	–	SH(0)	$1e{-}4$
2	φ	–	AL	$1e{-}4$
3	$\tilde{\varphi}$	0	SH(0)	$1e{-}4$
4	$\tilde{\varphi}$	0	AL	$1e{-}4$
5	$\tilde{\varphi}$	3	SH(0)	$1e{-}4$
6	$\tilde{\varphi}$	3	AL	$1e{-}4$
7	$\tilde{\varphi}$	3	SH(3)	$1e{-}4$
8	$\tilde{\varphi}$	5	SH(5)	$1e{-}4$
9	$\tilde{\varphi}$	0	SH(0)	$5e{-}5$
10	$\tilde{\varphi}$	0	SH(0)	$2.5e{-}5$
11	$\tilde{\varphi}$	0	SH(0)	$1.25e{-}5$
12	$\tilde{\varphi}$	0	SH(0)	$1e{-}5$
13a	$\tilde{\varphi}$	0	PDAS	$1e{-}4$
13b	$\tilde{\varphi}$	0	PDAS	$1e{-}4, \ldots, 1e{-}5$
14	$\tilde{\varphi}$	3	PDAS	$1e{-}4$
15	φ^{j-1}	–	Partitioned scheme	$1e{-}4$
16	φ^{j-1}	–	L-scheme, $L = 1e{-}2$	$1e{-}4$
17	φ^{j-1}	–	L-scheme, L dyn., Sec. 7.6.1	$1e{-}4$

- Figure 7.3 and Figure 9.5: Predictor-corrector mesh adaptivity (the mesh grows with the crack path) [221];
- Open-source codes (Chapter 13) on https://github.com/tjhei/cracks [222] and [207] on www.dopelib.net in `dopelib/Examples/PDE/InstatPDE/Example8`.

8.4 Configuration

The geometric and material properties are the same as suggested in [313, Section 5.1/5.2, p. 2773]. The domain is given by $B = (10\,\text{mm}, 10\,\text{mm})^2$. The geometry and boundary conditions are displayed in Figure 8.1. In particular the initial domain already has a slit (fracture). The initial mesh is 4 times uniformly refined, leading to 1024 mesh cells, yielding 2210 DoFs for the solid, 1105 DoFs for the phase-field variable, and 3315 DoFs in total. Here, $h = 0.044\,\text{mm}$. For mesh refinement studies, we refer to the published work, e. g., [221].

The characteristic feature of this test is that an initial crack is prescribed in the geometry rather than with phase-field and that the crack will slowly develop, followed by faster growth. It is important to consider the correct boundary conditions and splitting of the stress tensor (Section 4.5). As further discussed in [17], however, Miehe's et al.

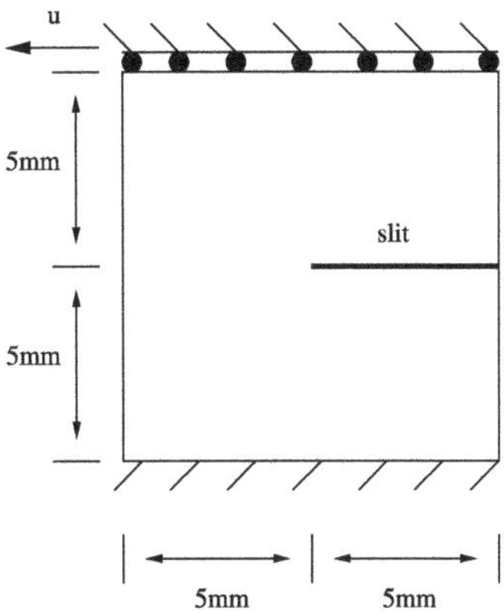

Figure 8.1: Single edge notched shear test (SENS). We prescribe the following conditions: On the left and right boundaries $\Gamma_{\text{left}} \cup \Gamma_{\text{right}}$, we have $u_y = 0$ mm and traction-free conditions in x-direction. On the bottom part Γ_{bottom}, we use $u_x = u_y = 0$ mm and on Γ_{top}, we prescribe $u_y = 0$ mm and u_x as stated in (8.1). Finally, the lower part $\Gamma_{\text{slit,bottom}}$ of the slit is fixed in y-direction, i. e., $u_y = 0$ mm. We notice that the initial crack is described in the geometry by doubling the degrees of freedom on the respective faces. Consequently, the initial phase-field is $\varphi^0 = 0$ in the entire domain.

splitting [313] does not release all stresses once the fracture reaches the bottom part of the specimen.

8.5 Boundary conditions

Let us decompose the boundary into

$$\partial B = \Gamma_{\text{left}} \cup \Gamma_{\text{right}} \cup \Gamma_{\text{bottom}} \cup \Gamma_{\text{top}} \cup \Gamma_{\text{slit,bottom}} \cup \Gamma_{\text{slit,top}}.$$

We increase the displacement on Γ_{top} over time, namely we apply a time-dependent non-homogeneous Dirichlet condition:

$$u_x = t\bar{u}, \quad \bar{u} = 1\,\text{mm/s}, \tag{8.1}$$

where t denotes the total time. For phase-field, we prescribe homogeneous Neumann conditions (traction-free) on the entire boundary ∂B. All boundary conditions read in detail:

$$\begin{aligned}
\sigma(u)ne_x &= 0 \quad \text{on } \Gamma_{\text{left}} \times I\\
u_y &= 0 \quad \text{on } \Gamma_{\text{left}} \times I\\
\sigma(u)ne_x &= 0 \quad \text{on } \Gamma_{\text{right}} \times I\\
u_y &= 0 \quad \text{on } \Gamma_{\text{right}} \times I\\
u &= 0 \quad \text{on } \Gamma_{\text{bottom}} \times I\\
u_x &= t\bar{u} \quad \text{on } \Gamma_{\text{top}} \times I\\
u_y &= 0 \quad \text{on } \Gamma_{\text{top}} \times I\\
\sigma(u)ne_x &= 0 \quad \text{on } \Gamma_{\text{slit,bottom}} \times I\\
u_y &= 0 \quad \text{on } \Gamma_{\text{slit,bottom}} \times I\\
\sigma(u)n &= 0 \quad \text{on } \Gamma_{\text{slit,top}} \times I\\
\varepsilon\partial_n\varphi &= 0 \quad \text{on } \partial B \times I,
\end{aligned}$$

where $e_x = (1,0)^T$ denotes the unit vector in y direction in the Cartesian coordinate system.

8.6 Initial conditions

The initial phase-field is prescribed as

$$\varphi(0) = 1 \quad \text{in } B \times \{0\},$$

because the initial crack is directly represented in this example inside the geometry. This is simply done by doubling the number of degrees of freedom in the initial coarse mesh and furthermore yields the “internal” boundaries $\Gamma_{\text{slit,bottom}}$ and $\Gamma_{\text{slit,top}}$. For the displacements, no initial conditions need to be prescribed.

8.7 Parameters

Specifically, we use $\mu = 80.77\,\text{kN/mm}^2$, $\lambda = 121.15\,\text{kN/mm}^2$, and $G_c = 2.7\,\text{N/mm}$. Furthermore, we set $\kappa = 10^{-12}$ [mm] and $\varepsilon = 2h$.

The time/loading step size is chosen as $k = 10^{-4}\,\text{s}$. The time interval is $I = (0, 0.014\,\text{s})$. Since we deal with an incremental problem without time-derivatives in the main equations, we have some liberty in choosing the relationship of loading step size k and the increments $\bar{u}$ on Γ_{top}:

1. $u_x = t\bar{u}$ with increments of 10^{-4} [mm]: time t is advanced in increments of $k = 10^{-4}\,\text{s}$, while setting $\bar{u} = 1\,\text{mm/s}$;
2. $u_x = t\bar{u}$ with increments of 10^{-4} [mm]: time t is advanced in increments of $k = 1\,\text{s}$, while setting $\bar{u} = 10^{-4}\,\text{mm/s}$.

It is obvious that in time-dependent problems with time derivatives in the main equations, these two choices yield different results.

8.8 Quantities of interest

To verify the solution, we observe the crack path and in particular the time instants when the crack starts growing and when it reaches the lower boundary. Second, we evaluate the surface load vector on $\Gamma_{\text{top}} := \{(x,y) \in B \mid 0\text{mm} \le x \le 10\text{mm},\ y = 10\text{mm}\}$ as

$$\tau = (F_x, F_y) := \int_{\Gamma_{\text{top}}} \sigma(u) n \, ds,$$

with the normal vector n, and we are particularly interested in F_x.

8.9 Discussion of our findings

8.9.1 General comparisons

The crack pattern in terms of φ at various times are shown in Figure 1.6 (Section 1.3.3) and is in good agreement to other results reported in the literature, e. g., [313, 67, 221, 444, 445, 17]. The first important observation can be found in Figure 8.2, which shows in the left and right subfigures similar load-displacement curves, independently of the nonlinear solver; here, two different Newton methods. Further load-displacement curves are shown in the Figures 8.3, and 8.4. An analysis is provided in the figure captions. Newton iterations numbers are shown and discussed in the Figures 8.5, 8.6, and 8.7. In the figures, the "time" on the x-axis corresponds to a loading increment u_x of $k\bar{u} = 10^{-4}$ mm on Γ_{top}.

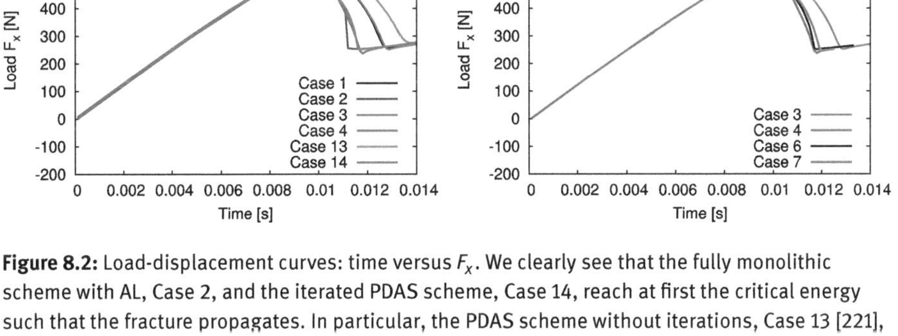

Figure 8.2: Load-displacement curves: time versus F_x. We clearly see that the fully monolithic scheme with AL, Case 2, and the iterated PDAS scheme, Case 14, reach at first the critical energy such that the fracture propagates. In particular, the PDAS scheme without iterations, Case 13 [221], has a significant time-lagging error. For the latter, see also Section 8.9.3 below.

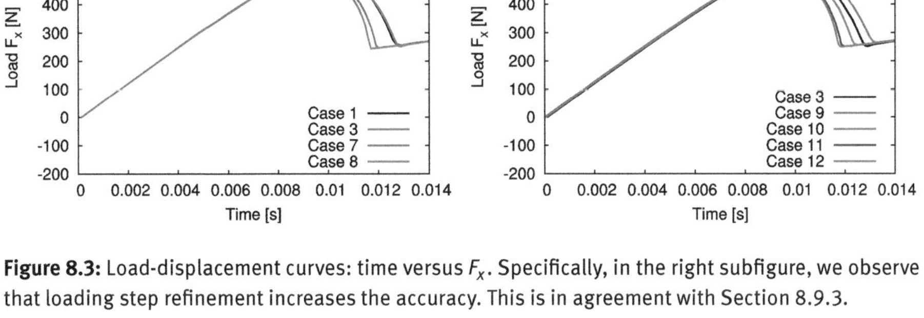

Figure 8.3: Load-displacement curves: time versus F_x. Specifically, in the right subfigure, we observe that loading step refinement increases the accuracy. This is in agreement with Section 8.9.3.

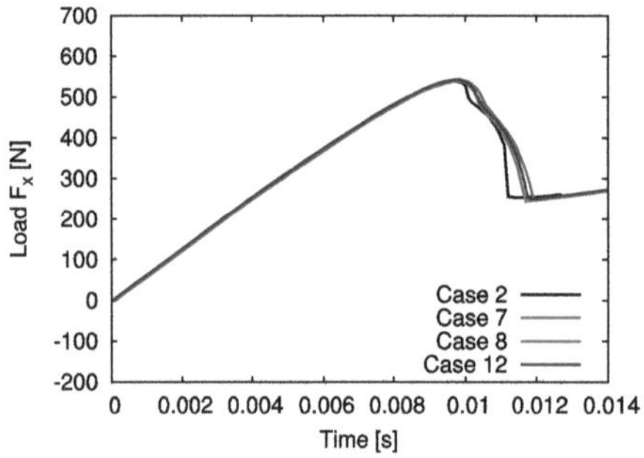

Figure 8.4: Load-displacement curves: time versus F_x. Here, we finally compare some strain history function results to the fully monolithic scheme.

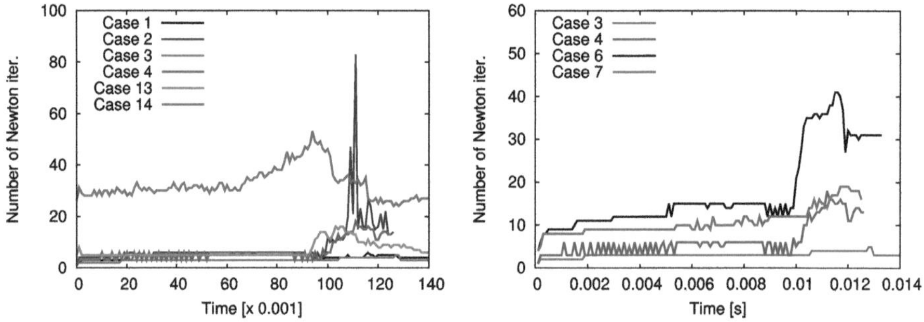

Figure 8.5: Newton iterations per time step. When the fracture starts growing the number of Newton steps increases. This is true for all schemes. The monolithic AL scheme, Case 2, has a high maximum number, which is a drawback of this scheme. Surprisingly, the monolithic strain history scheme has a very low amount of iterations. In average all schemes have less than 10 Newton iterations.

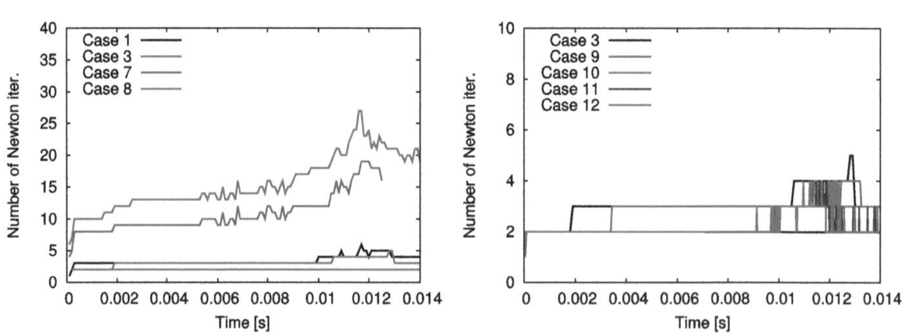

Figure 8.6: Newton iterations per time step. Here, we observe in some more detail and see that the strain history formulations have a very low number of iterations.

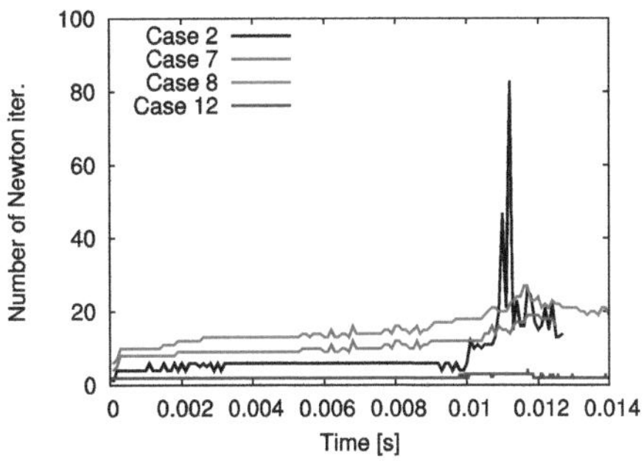

Figure 8.7: Newton iterations per time step. Comparison of strain history formulations with the fully monolithic AL scheme. As a conclusion, we confirm [313] that the strain history formulation is a very competitive scheme in terms of accuracy and computational cost.

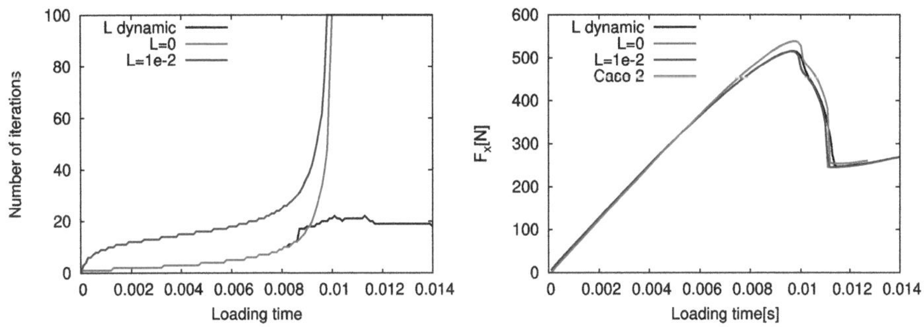

Figure 8.8: Comparison of dynamic *L* updates, the weighted version, constant *L* and the fully monolithic version, Case 2. Left: number of iterations. Right: load-displacement curves.

8.9.2 Stabilized *L*-scheme with updating parameters

In the following, we show findings for the L-scheme (Section 7.6.1), which are summarized in Figure 8.8. The numerical solutions for different strategies for choosing L are practically identical, only the number of iterations being different. Here, $L = 0$ and $L = 10^{-2}$ denote tests in which $L = L_u = L_\varphi$ are taken constant throughout the entire computation. The dynamic update recently proposed in [154] is denoted by `L dynamic`.

We observe a significant reduction in the computational cost when using the dynamic L-schemes. The maximum number of iterations is 21. This number is reduced to 12 iterations using $a = 20$ while the accuracy only slightly changes (see Figure 8.9).

8.9.3 Loading step refinement

We finally compare in Table 8.2 Case 2 (fully monolithic AL scheme) with Case 13 (quasi-monolithic PDAS). In the latter, loading step refinements[2] are done in order to

2 These tests were run by Katrin Mang. Thanks!

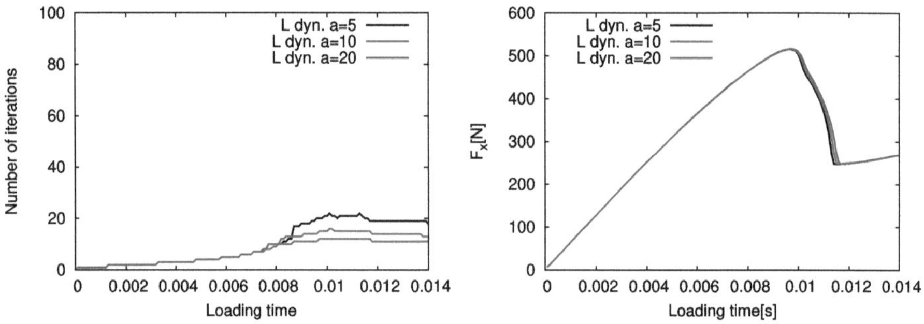

Figure 8.9: Comparison of different a for the dynamic L-scheme.

Table 8.2: Loading step refinement. In the last row, the fully monolithic reference solution from Case 2 is shown.

N	Time, crack starts growing	Max. loading $\max_{t\in[0,T]} F_x$
140	0.0101	551.784
160	0.009975	550.277
180	0.00995556	549.2
200	0.00994	548.138
280	0.0098	546.365
700	0.00974	544.732
1400	0.00968	553.503
140	0.0096	538.596

study whether the quasi-monolithic scheme yields closer solutions to the fully monolithic scheme. In general, convergence is to be expected, however for our schemes, we have no rigorous convergence proofs so far.

The computations are done on a uniform refined mesh with 3315 DoFs (Q_1^c elements for u and φ) per time step and end time for all tests is $T = 0.014$ s. For instance, $N = 140$ corresponds to a loading step size of $k = 10^{-4}$ s, while $\bar{u} = 1$.

The crack with $N = 1400$ incremental steps, corresponding to an incremental size of $k = 10^{-5}$ s, starts propagating at a very similar time point as the monolithic approach; see also Figure 8.2.

9 Numerical modeling part III: adaptive concepts

In this chapter, we concentrate on adaptive numerical modeling and address various technologies.

9.1 General motivation of adaptivity

The goal of numerical simulations is to obtain a space/time discrete solution U_{hk} with sufficient accuracy at minimal computational cost. However, often, not the entire numerical solution U_{hk} is of interest, but only a part. Such parts can be evaluations on subdomains, boundaries, interfaces, or point values of the entire solution, only specific components, primary function values or derivative information. These can be expressed through a so-called goal functional $J : X \to \mathbb{R}$ with $U_{hk} \mapsto J(U_{hk})$. Often these are motivated by requests from engineering, physics, or other applications, and represent a (physical) quantity of interest (QoI). The aim is then to estimate the error:

$$J(U) - J(U_{hk}).$$

More precisely, for $h \to 0$ and $k \to 0$, we aim for

$$|J(U) - J(U_{hk})| \to 0.$$

In practice, we use

$$|J(U) - J(U_{hk})| < \mathrm{TOL}.$$

This objective can be achieved by the following possibilities:
- Brute-force computing on uniform meshes using black-box discretizations and black-box solvers: often hopeless in coupled problems;[1]
- Parallel computing;
- Adaptive discretization;
- Design of adaptive stopping criteria for linear and nonlinear solvers;
- Model adaptivity;
- Model order reduction and reduced basis methods;
- Adaptive multiscale modeling.

In the remainder of this chapter, we concentrate on adaptive methods with an emphasis on spatial mesh adaptivity. We shall also briefly discuss an adaptive global-local approach and a heuristic method for time step error control.

1 Brute-force with a direct solver was used to compute reference solutions in Section 9.3.1 and Section 10.1. In both tests, it is clear, that we did not get far by using "standard" methods. Scientifically, this is unsatisfying, but serves the purpose for some prototype computations.

https://doi.org/10.1515/9783110497397-009

9.1.1 Specific comments to spatial and temporal adaptivity for phase-field fracture

There are several studies that have considered spatial mesh refinement. These include anisotropies introduced by the mesh [333], residual-based adaptive finite elements [94, 95], anisotropic adaptive mesh refinement [30], prerefined meshes when the crack path is known a priori [67], our predictor-correction approach [221, 285], and also other's work [35], goal-oriented techniques [442], and mesh refinement in multiscale phase-field methods [350]. We are aware of attempts and studies that investigate the dependence of the crack path versus mesh refinement. In our initial studies on mesh refinement, e. g., [221, 285] we often compared solutions obtained on uniformly refined meshes and locally refined meshes. Therein, mesh influences played a minor role. Moreover, perturbed meshes [430, Figures 12 and 13] were used and also rotated fractures in [280, Section 7.1]. For rigorous studies on temporal error control, less work exists and we refer to some studies referenced in Section 1.5[point 12].

9.1.1.1 Relation to obstacle problems

Phase-field fracture shares similarities with the obstacle problem as discussed in Section 2.3 and Section 4.4.3.4, including references to classical books. In error estimation, we can learn principle techniques from these settings. First, for a priori finite element error estimates for the obstacle problem, we refer to [84, 85]. A recent overview article of various general aspects is [456]. For first a posteriori error estimates employing duality arguments, we refer to [254]. Goal-oriented a posteriori error estimation is discussed in detail in [402, 367, 60, 61, 404, 403, 368, 389, 230]. The key ingredient in these papers is the construction of the adjoint problem for which for instance [60] employed a generalization, proposed by Natterer [332], of the well-known Aubin–Nitsche trick.

9.1.2 Further comments and recent developments

In the Sections 5.5.3.5 and 5.5.3.6, we provided extensive discussions about the phase-field model and discretization error estimates. In this context, we refer again to error control and computational investigations in related fields such as Allen–Cahn phase-field models, Ginzburg–Landau-type equations, and functions of bounded variations [41, 39, 40, 463, 163, 164] as well as multiscale problems in which small scales interact with the spatial discretization [1].

In the following, we concentrate on the discretization error only. Adaptive stopping criteria for numerical solvers will only be slightly touched. For such techniques, specifically using goal-oriented techniques, we refer for instance to [77, 344] (model and discretization errors), [370, 150] (discretization and nonlinear iteration error),

[50, 307] (discretization and linear error). We also want to mention the work on adaptive inexact Newton solvers with equilibrated flux reconstructions [156] and the extension to goal-oriented methods [295]. The recent Acta Numerica paper from J. T. Oden [345] is recommended; see Section 9.2 in [345].[2] Most recent developments in goal-oriented methods are in time step control [309, 159], multiscale problems [294], stochastic estimations [305], machine-learning enhanced solutions [83], and theoretical work towards optimal convergence rates [162, 49]. Monographs summarizing important techniques for a posteriori error control for residual-based, functional type, and goal-oriented methods are [4, 37, 335, 371, 213, 428, 403].

9.1.3 Efficiency, reliability, basic adaptive algorithm

Following [52, 51], [37], [363], we now provide details for a posteriori error estimates not only for norms, but for a general differentiable functional of interest $J : X \to \mathbb{R}$. Norm-based error estimation, $\|U - U_{hk}\|$ can be expressed in terms of $J(\cdot)$. We work with U as solution variable and X or V as function spaces and understand that $U := u$ or $U := \varphi$ or $U := (u, \varphi)$ to name a few realizations. Moreover, we consider now both spatial and temporal discretization in U_{hk}, but later mainly concentrate on spatial discrete solutions U_h or time-discrete solutions U_h^n.

When designing a posteriori error estimates, we are in principle interested in the following relationship:

$$C_1\eta \le |J(U) - J(U_{hk})| \le C_2\eta \tag{9.1}$$

where $\eta := \eta(U_{hk})$ is the error estimator and C_1, C_2 are positive constants. Moreover, $J(U) - J(U_{hk})$ is the true error.

Definition 60 (Efficiency and reliability). A good estimator $\eta := \eta(U_{hk})$ should satisfy two bounds:

1. An error estimator η of the form (9.1) is called efficient when

   $$C_1\eta \le |J(U) - J(U_{hk})|,$$

   which means that the error estimator is bounded by the error itself.
2. An error estimator η of the form (9.1) is called reliable when

   $$|J(u) - J(U_{hk})| \le C_2\eta.$$

   Here, the true error is bounded by the estimator.

2 Just as remark to [345] for a general education point-of-view: it is worth it to read Sections 1 and 2 in [345].

Remark 62. For general goal functionals $J(\cdot)$, it is much simpler to derive reliable estimators, e. g., [52, 37], rather than proving their efficiency. For general goal functionals, it can happen that $J(U) - J(U_{hk}) = 0$. For this reason, showing efficiency remains a questionable task. Efficiency estimates, using a mesh-dependent norm, with common worst-case bounds of the true error and the estimator were shown in [380]. For stronger results on efficiency, assuming a saturation assumption, we refer the reader to very recent work [152, 151].

Remark 63 (AFEM). Finite element frameworks working with a posteriori error estimators applied to local mesh adaptivity, are called adaptive FEM.

Definition 61 (Basic algorithm of AFEM). The basic algorithm for AFEM is:
1. Solve the PDE on the current space-time mesh $\mathcal{T}_h$;
2. Estimate the error via a posteriori error estimation to obtain η;
3. Mark the elements by localizing the error estimator;
4. Refine/coarsen the elements with the highest/lowest error contributions using a certain refinement strategy.

9.1.4 Measuring the quality of the error estimator η

As a quality measure of how well the estimator approximates the true error, we use the (reciprocal) effectivity index I_{eff}.

Definition 62 ([34]: Effectivity index). The effectivity index is defined as

$$I_{\text{eff}} := \left| \frac{\eta}{J(U) - J(U_{hk})} \right|. \tag{9.2}$$

Problems with good I_{eff} satisfy asymptotically $I_{\text{eff}} \to 1$ for $h \to 0$ and $k \to 0$. We say that
- for $I_{\text{eff}} > 1$, we have an overestimation of the error,
- for $I_{\text{eff}} < 1$, we have an underestimation of the error.

In the effectivity index, localized error indicators η_i might cancel out because the absolute sign is taken outside of the sum $\sum_i \eta_i =: \eta$ (the index i will become clear below).

In order to measure the quality of the localization process, we further define the indicator index in which the absolute sign is applied to each single summand.

Definition 63 ([380]: Indicator index). The indicator index is defined as

$$I_{\text{ind}} := \frac{\sum_i |\eta_i|}{|J(U) - J(U_{hk})|}. \tag{9.3}$$

The index i is the localization index of each mesh element or degree of freedom associated with the localization procedure.

9.2 Goal-oriented a posteriori error estimation for steady-state phase-field fracture

In this section, we briefly recall the dual-weighted residual (DWR) method for estimating the spatial discretization error in terms of the goal functional $J(\cdot)$, in combination with adaptive stopping criteria for the nonlinear solver. This general theorem was first proposed by Rannacher/Vihharev [370] and is based on earlier work by Becker/Rannacher [52] (see also [37]). In [150], this methodology was extended to multigoal-oriented error estimation. From now on, we work with U_h and disregard temporal errors, but we will revisit those in the last section of this chapter. To the best of our knowledge, a full space-time adaptive method for phase-field fracture has not yet been developed to date.

9.2.1 Problem statements and setups

Importantly, we have regularized formulations in mind such as Propositions 27, 28, or 25 and their corresponding finite element discretizations. Designing the DWR method directly for variational inequalities using Propositions 19 or 20 would require techniques as for instance explained by Suttmeier [403] (see also Section 9.1.1.1 for more references). To this end, we propose to work with

Formulation 20. For the loading steps $n = 1, 2, 3, \dots, N$, given φ^{n-1}, find $U := U^n = (u^n, \varphi^n) \in X$ with $\varphi(0) = \varphi_0$, such that

$$A(U)(\Psi) := \bar{A}(U)(\Psi) + (\gamma[\varphi - \varphi^{n-1}]_+, \psi) = 0 \quad \forall \Psi = (w, \psi) \in X,$$

where $\bar{A}(U)(\Psi)$ denotes the main parts of Propositions 19 or 27 yet without penalization.

For adjoint-based error estimation using the DWR method, we need a second problem, the so-called adjoint. For this reason, we denote the previous semilinear form as primal problem. In the following, we concentrate on a single time point t_n. This is important since we avoid computing a backward-in-time running adjoint problem by evaluating the adjoint at each $t_n, n = 1, \dots, N$. We have no theory that this procedure is justified for phase-field fracture. We only have hints such as [368, 369] in which for quasi-stationary problems, rather than computing the dual problem backward in time, the time-evolving problems are split into a sequence of stationary problems (see again [367]). This means that time-dependent problems are discretized in time with a backward Euler scheme. In [368], Lemma 6.2, and Theorem 6.1, it has been shown that in the case of the quasi-stationary Prandtl–Reuss model in solid mechanics, the error in each time/loading step can accumulate at most linearly and does not become dominant, which is the argument to neglect a full backward-time dual problem. Specifically, meshes are kept fixed during a Newton iteration and are only refined during time steps. Therefore, the global spatial error is approximated through a sequence of stationary problems.

Definition 64 (Primal problem). The primal problem reads as follows: Find $U \in X$ such that

$$A(U)(\Psi) = 0 \quad \forall \Psi \in X. \tag{9.4}$$

If there is a right-hand side $F(\Psi)$, for convenience, we simply assume it to be included in $A(U)(\Psi)$.

Proposition 64 (Adjoint problem). *Given the goal functional $J(U)$, the adjoint problem reads: Find $Z \in X$ such that*

$$A'(U)(\Phi, Z) = J'(U)(\Phi) \quad \forall \Phi \in X. \tag{9.5}$$

Proof. The adjoint problem arises naturally in Proposition 65 (below) as part of the Euler–Lagrange system for the goal functional minimization problem. □

Example 12. In the following, we demonstrate the calculation of the adjoint for the penalized obstacle problem (Section 2.4.1) since this is again very close to our problem statements:

$$-\Delta u - \gamma[g - u]^+ = f.$$

The adjoint is derived as follows:
1. $A(u)(\phi) = (\nabla u, \nabla \phi) - \gamma([g - u]^+, \phi) - (f, \phi)$ (Primal problem, weak form)
2. $A'(u)(\delta u, \phi) = (\nabla \delta u, \nabla \phi) + \gamma(\delta u, \phi)_{B(u)}$ (Tangent problem)
3. $A'(u)(\phi, \delta u) = (\nabla \phi, \nabla \delta u) + \gamma(\phi, \delta u)_{B(u)}$ (Switch test and trial function)
4. $A'(u)(\phi, z) = (\nabla \phi, \nabla z) + \gamma(\phi, z)_{B(u)}$ (Adjoint, weak form)

with $B(u) = \{x \in B \mid g(x) > u(x)\}$. We notice that the tangent problem is nothing else than the Jacobian matrix required for Newton's method; see Section 7.9.

9.2.2 Brief recapitulation of the DWR method for nonlinear problems

Based on the last statements, we recapitulate adjoint-based error estimation for nonlinear PDEs. Therein, the variational inequality constraint is already regularized as in Formulation 20 such that we deal with PDE systems rather than CVISs. We recall the main theorem from [370], which was later applied to multiple goal functionals in [150] (see also Section 9.4). Motivated by the goal outlined in Section 9.1 and using the primal and adjoint problems, we design an optimization problem.

Formulation 21. Given $U_h \in X_h$. Let $J(U)$ and $A(U)(\Psi)$ be sufficiently differentiable. Solve

$$\min_{U \in X}(J(U) - J(U_h)) \quad \text{s. t. } A(U)(\Psi) = 0.$$

Proposition 65. *Let $U \in X$ solve (9.4) and let $Z \in X$ solve (9.5) and let $\tilde{U}$ and $\tilde{Z}$ be approximations in X. Then*

$$J(U) - J(\tilde{U}) = \frac{1}{2}\rho(\tilde{U})(Z - \tilde{Z}) + \frac{1}{2}\rho^*(\tilde{U},\tilde{Z})(U - \tilde{U}) + \rho(\tilde{U})(\tilde{Z}) + \mathcal{R}^{(3)}$$

with

$$\begin{aligned}\rho(\tilde{U})(\cdot) &:= -A(\tilde{U})(\cdot)\\ \rho^*(\tilde{U},\tilde{Z})(\cdot) &:= J'(\tilde{U})(\cdot) - A'(\tilde{U})(\cdot,\tilde{Z})\end{aligned}$$

and some remainder term $\mathcal{R}^{(3)}$ of third order. Furthermore, we have the primal and adjoint errors:

$$e = U - \tilde{U}, \quad e^* = Z - \tilde{Z}.$$

Proof. Let us formulate the Lagrangian of the optimization problem in Formulation 21 first:

$$\mathcal{L}(U,Z) = J(U) - A(U)(Z).$$

We are then interested in the minimization of $\mathcal{L}(U,Z)$. The corresponding Euler–Lagrange system[3] reads:

$$\begin{aligned}\mathcal{L}'_U(U,Z)(\delta U) &= J'(U)(\delta U) - A'(U)(\delta U,Z) = 0,\\ \mathcal{L}'_Z(U,Z)(\delta Z) &= -A(U)(\delta Z) = 0.\end{aligned} \tag{9.6}$$

This shows that the primal problem $\mathcal{L}'_Z(U,Z)(\cdot)$ is still nonlinear, while the adjoint problem $\mathcal{L}'_U(U,Z)(\cdot)$ is linearized. Now, we define

$$x = (U,Z) \in X \times X, \quad \tilde{x} = (\tilde{U},\tilde{Z}) \in X \times X$$

and observe the Lagrangian

$$\mathcal{L}(x) - \mathcal{L}(\tilde{x}) = \int_0^1 \mathcal{L}'(\tilde{x} + se)e\,ds, \quad e = x - \tilde{x}. \tag{9.7}$$

3 As in the previous chapters, the Euler–Lagrange system are the first variations.

The key trick goes back to basics in numerical analysis. We solve (9.7) with the trapezoidal rule

$$\int_0^1 f(s)\mathrm{ds} = \frac{1}{2}[f(0) + f(1)] + \underbrace{\frac{1}{2}\int_0^1 f''(s)\, s(s-1)\mathrm{ds}}_{=\mathcal{R}^{(2)}} \tag{9.8}$$

and after applying (9.8) to (9.7) we observe with $f(s) := \mathcal{L}'(\tilde{x} + se)e$

$$\mathcal{L}(x) - \mathcal{L}(\tilde{x}) = \frac{1}{2}\left[\mathcal{L}'(x)e + \mathcal{L}'(\tilde{x})e\right] + \mathcal{R}^{(3)}. \tag{9.9}$$

By using the Lagrangians,

$$\begin{aligned} J(U) &= \mathcal{L}(U,Z) + A(U)(Z) \\ J(\tilde{U}) &= \mathcal{L}(\tilde{U},\tilde{Z}) + A(\tilde{U})(\tilde{Z}) \end{aligned}$$

we finally obtain

$$\begin{aligned} J(U) - J(\tilde{U}) &= \mathcal{L}(U,Z) + A(U)(Z) - \mathcal{L}(\tilde{U},\tilde{Z}) - A(\tilde{U})(\tilde{Z}) \\ &= \mathcal{L}(U,Z) - \mathcal{L}(\tilde{U},\tilde{Z}) + \underbrace{A(U)(Z)}_{=0} - A(\tilde{U})(\tilde{Z}) \\ &= \mathcal{L}(U,Z) - \mathcal{L}(\tilde{U},\tilde{Z}) - A(\tilde{U})(\tilde{Z}) \\ &\overset{(9.9)}{=} \frac{1}{2}\left[\mathcal{L}'(x)e + \mathcal{L}'(\tilde{x})e\right] - A(\tilde{U})(\tilde{Z}) + \mathcal{R}^{(3)} \\ &= \frac{1}{2}(-A(\tilde{U})(e^*) + J'(\tilde{U})(e) - A'(\tilde{U})(e,\tilde{Z})) - A(\tilde{U})(\tilde{Z}) + \mathcal{R}^{(3)} \\ &=: \frac{1}{2}\rho + \frac{1}{2}\rho^* + \rho(\tilde{U})(\tilde{Z}) + \mathcal{R}^{(3)}. \end{aligned}$$

In the next to last row, we have used

$$\begin{aligned} \mathcal{L}'(x)e + \mathcal{L}'(\tilde{x})e = J'(U)(e) &- A'(U)(e,Z) - A(U)(e^*) \\ &+ J'(\tilde{U})(e) - A'(\tilde{U})(e,\tilde{Z}) - A(\tilde{U})(e^*), \end{aligned}$$

which follows trivially from the chain rule (Section 2.2.5) because

$$\mathcal{L}'(x)(\cdot) = \mathcal{L}'_U(U,Z)(\cdot) + \mathcal{L}'_Z(U,Z)(\cdot)$$

where we now employ (9.6). □

Corollary 6 (Full error estimator/identity). *The full error estimator from Proposition 65 is given by*

$$\eta = J(U) - J(\tilde{U}) = \frac{1}{2}\rho(\tilde{U})((Z - \tilde{Z})) + \frac{1}{2}\rho^*(\tilde{U},\tilde{Z})((U - \tilde{U})) + \rho(\tilde{U})(\tilde{Z}) + \mathcal{R}^{(3)}.$$

This estimator is impractical because $U, \tilde{U}, Z, \tilde{Z}, \mathcal{R}^{(3)}$ must be known. Furthermore, it is still in global form and cannot be employed to extract localized information.

For our later results, we introduce more notation.

Definition 65. The full error estimator is split into the error parts

$$\eta = \eta_p + \eta_a^* + \eta_m + \mathcal{R}^{(3)}$$

with

$$\begin{aligned}
\eta_p &:= \frac{1}{2}\rho(\tilde{U})((Z - \tilde{Z})), && \text{(Primal estimator)},\\
\eta_a^* &:= \frac{1}{2}\rho^*(\tilde{U},\tilde{Z})((U - \tilde{U})), && \text{(Adjoint estimator)},\\
\eta_m &:= \rho(\tilde{U})(\tilde{Z}), && \text{(Iteration error)}.
\end{aligned}$$

9.2.3 Localization of the error estimator

The error estimator η in Corollary 6 is yet given in a global form with general approximations $\tilde{U}$ and $\tilde{Z}$. For local mesh adaptivity, we need to localize η to single elements. Traditionally, an element-based version using partial integration [52] or a patchwise weak formulation [77] were employed. In [380], a DoF-based estimator was constructed that also works with the weak form. Specifically, the latter feature is attractive for nonlinear, coupled problems. The reasons are two-fold:

- Deriving the strong form is error-prone; specifically because of integration by parts, which introduces face terms;
- Implementing the strong form is again error-prone, needs additional debugging and is computationally more costly since second-order operators must be evaluated.

To this end, we introduce a partition-of-unity (PU) [310, 33], which can be realized in terms of another finite element function. The procedure is therefore easy to realize in existing codes.

Definition 66 (PU – partition-of-unity). The PU is given by

$$V_{\mathrm{PU}} := \{\psi_1, \ldots, \psi_M\}$$

with $\dim(V_{\mathrm{PU}}) = M$. The PU has the property

$$\sum_{i=1}^{M} \psi_i \equiv 1. \tag{9.10}$$

Remark 64. The PU can be simply chosen as a low order finite element space with linear or bilinear elements, i. e.,

$$V_{\mathrm{PU}} = V_h^{(1)}.$$

To understand the idea, we recall that in the classical error estimator the face terms are essential since they gather information from neighboring cells measuring the discrete smoothness in edge terms. Indeed these terms are important as shown in [104]. When we work with the variational form, no integration by parts is necessary. Therefore, the information of the neighboring cells is missing. Using the PU, we touch different cells per PU-node, and consequently we gather known information from neighboring cells.

Proposition 66 (Localized error estimator/identity). *Let the previous PU be given. The localized form of the error estimator η in Corollary 6 is*

$$J(U) - J(\tilde{U}) = \eta = \sum_{i=1}^{M} \frac{1}{2}\rho(\tilde{U})((Z - \tilde{Z})\psi_i) + \frac{1}{2}\rho^*(\tilde{U}, \tilde{Z})((U - \tilde{U})\psi_i) + \rho(\tilde{U})(\tilde{Z}) + \mathcal{R}^{(3)}.$$

As before, the last part determines the deviation of the approximate solution $\tilde{U}$ in comparison to the "exact" discrete solution U_h, which will be used in practice. These can be iteration errors due iterative linear or nonlinear solution procedures. Since they act globally on the entire solution, no localization with the PU function is required here.

Proof. Inserting ψ_i into Corollary 6 and using property (9.10), we immediately establish the result. □

Neglecting the remainder term yields a computable form.

Corollary 7 (Localized error estimator). *It holds*

$$J(U) - J(\tilde{U}) \approx \eta = \sum_{i=1}^{M} \frac{1}{2}\rho(\tilde{U})((Z - \tilde{Z})\psi_i) + \frac{1}{2}\rho^*(\tilde{U}, \tilde{Z})((U - \tilde{U})\psi_i) + \rho(\tilde{U})(\tilde{Z}).$$

Definition 67. The error estimator in Corollary 7 is composed by the following parts:

$$\eta = \eta_{\text{dis}} + \eta_m := \sum_{i=1}^{M} (\eta_i + \eta_i^*) + \eta_m,$$

where η_{dis} denotes the discretization error and η_m the nonlinear iteration error. Specifically, we have

$$\begin{aligned} \eta_i &:= \frac{1}{2}\rho(\tilde{U})((Z - \tilde{Z})\psi_i), \\ \eta_i^* &:= \frac{1}{2}\rho^*(\tilde{U}, \tilde{Z})((U - \tilde{U})\psi_i), \\ \eta_m &:= \rho(\tilde{U})(\tilde{Z}). \end{aligned}$$

9.2.4 Finite element approximations and comments on adjoint-based error estimation

In the computational realization, it is immediately clear and well known that Z and $\tilde{Z}$ as well as U and $\tilde{U}$ are approximated through discrete unknowns from spaces such as X_h. Here, it is important that $Z_h \approx Z$ and $\tilde{Z}_h \approx \tilde{Z}$ come from different discrete spaces since otherwise $Z_h - \tilde{Z}_h \equiv 0$ and $U_h - \tilde{U}_h \equiv 0$; see [52, 37]. A discrete version of the previous localized form is

Proposition 67 (Practical error estimator). *Let $U_h := U_h^{(1)} \in X_h^{(1)}$ be a low-order solution, $U_h^{(2)} \in X_h^{(2)}$ a higher-order solution, and $Z_h := Z_h^{(1)} \in X_h^{(1)}$, $Z_h^{(2)} \in Z_h^{(2)}$ the low- and higher-order adjoint solutions, respectively. The practical localized PU error estimator reads*

$$J(U) - J(U_h) \approx \eta = \sum_{i=1}^{M} \frac{1}{2}\rho(U_h)((Z_h^{(2)} - Z_h)\psi_i) + \frac{1}{2}\rho^*(U_h, Z_h)((U_h^{(2)} - U_h)\psi_i) + \rho(U_h)(Z_h),$$

where we now redefine the previous notation and obtain as error parts

$$\eta = \eta_h + \eta_m := \sum_{i=1}^{M}(\eta_i + \eta_i^*) + \eta_m$$

with

$$\begin{aligned}
\eta_i &:= \frac{1}{2}\rho(U_h)((Z_h^{(2)} - Z_h)\psi_i),\\
\eta_i^* &:= \frac{1}{2}\rho^*(U_h, Z_h)((U_h^{(2)} - U_h)\psi_i),\\
\eta_m &:= \rho(U_h)(Z_h).
\end{aligned}$$

Despite that we explicitly used the orders $r = 1$ and $r = 2$ for the primal and adjoint problems, the estimator also applies for higher orders $r > 2$.

9.2.4.1 Comments on adjoint-based error estimation

Let us briefly comment on adjoint-based error estimation. The appealing features are that

1. a specific functional of interest $J(U)$ or even multiple goal functionals (Section 9.4) can be quantitatively measured;
2. the error estimator can be used for local mesh adaptivity and adaptive solver control.[4]

4 Further applications as model error control are listed in Section 9.1.2.

The obvious disadvantage is that a second solution $Z_h^{(2)} \in X_h^{(2)}$ must be computed. However, as it becomes clear in the proof of Proposition 65 and the construction, the adjoint problem is always linear—even for nonlinear primal problems and nonlinear goal functionals. Here, the computational cost may become less significant for computing the additional adjoint problem. For time-dependent problems, the adjoint is however running backwards in time and often has jump terms due to discontinuous Galerkin temporal discretizations. For these problems, the adjoint must be carefully designed and implemented [388, 55, 56, 269].

We also notice that from the theoretical point of view, convergence of the adaptive scheme for arbitrary goal functionals could not be shown yet. Recent advances toward convergence properties for general goal-oriented error estimation, not necessarily with the help of the DWR method, include for instance [162, 49].

9.2.5 A PU-DWR a posteriori error estimator for quasi-static phase-field fracture

In the following, we present a primal error estimator ρ (thus neglecting ρ^*) for a steady-state phase-field fracture. Therein, we assume that we only compute one single time step from t_0 to t_1. Such configurations comprise Sneddon's pressurized fracture (Chapter 12) and a phase-field slit (Chapter 10) as we shall study later in detail. The extension to a quasi-static procedure by solving the adjoint problem at each time point t_n is not difficult and a proof of concept is provided in [442, Section 6.3]. The further extension to a space-time setting in which the adjoint equation is running backwards in time has not been addressed for phase-field fracture problems as mentioned before.

Proposition 68 (Primal PU-DWR error estimator for phase-field fracture). *Let $U = (u, \varphi) \in X^D$ and $U_h = (u_h, \varphi_h) \in X_h^D$ be the continuous and discrete primal solutions, respectively. Let $Z = (z_u, z_\varphi) \in X^0$ and $Z_h \in X_h^0$ be the continuous and discrete adjoint solutions. With the superscript (2), we indicate higher-order solutions. For the finite element approximation of the penalized phase-field problem in Proposition 27, we have for the primal error part $\rho(U_h)(\cdot)$ the a posteriori error estimate at the loading point t_1:*

$$|J(U) - J(U_h)| \le \eta := \left| \sum_{i=1}^{M} \eta_i \right| \le \sum_{i=1}^{M} |\eta_i|$$

with

$$\begin{aligned}
\eta_i &= -A(U_h)((Z_h^{(2)} - i_h Z_h^{(2)})\psi_i) \\
&= -(g(\varphi_h)\sigma(u_h), \nabla((z_{u,h}^{(2)} - i_h z_{u,h}^{(2)})\psi_i)) \\
&\quad - ((1-\kappa)\varphi_h \sigma(u_h) : e(u_h), ((z_{\varphi,h}^{(2)} - i_h z_{\varphi,h}^{(2)})\psi_i)) \\
&\quad + \left(\frac{G_c}{\varepsilon}(1-\varphi_h), ((z_{\varphi,h}^{(2)} - i_h z_{\varphi,h}^{(2)})\psi_i) \right) - (G_c \varepsilon \nabla \varphi_h, \nabla((z_{\varphi,h}^{(2)} - i_h z_{\varphi,h}^{(2)})\psi_i)) \\
&\quad - \gamma([\varphi_h - \varphi_h^0]^+, ((z_{\varphi,h}^{(2)} - i_h z_{\varphi,h}^{(2)})\psi_i)).
\end{aligned}$$

Proof. Again, we apply Proposition 65 (by only considering the primal residual), thus no factor 1/2, and use

$$\rho(U_h) := -A(U_h)(Z - \Phi_h),$$

where $\Phi_h = (\phi_h^u, \phi_h^\varphi)$. We then introduce the interpolations $i_h^u : V \to V_h$ and $i_h^\varphi : W \to W_h$ and set $\Phi_h := i_h Z$ with $i_h := (i_h^u, i_h^\varphi)$ and obtain

$$\rho(U_h) = -A(U_h)(Z - i_h Z).$$

We now replace Z by a higher-order finite element solution (or higher-order interpolation), i. e., $Z \approx Z_h^{(2)}$. Then

$$\rho(U_h) = -A(U_h)(Z_h^{(2)} - i_h Z_h^{(2)}).$$

This part is now fully computable. Next, we insert the PU for restricting the error $\rho(U_h)$ to single degrees of freedom. Here, we notice that the function ψ_i is scalar-valued as it only marks the DoFs at which the error estimator is evaluated. Then

$$\rho(U_h) = \sum_i \underbrace{(-1)A(U_h)((Z_h^{(2)} - i_h Z_h^{(2)})\psi_i)}_{=:\eta_i}.$$

Finally, we take absolute values and the triangle inequality and obtain the assertion. □

Remark 65. Due to the crack irreversibility in the phase-field variational inequality, the problem is not purely stationary as it seems because φ and φ^{n-1} enter. For steady-state phase-field fracture as considered in Chapter 10, we impose the initial condition φ^0 and compute one single loading step, i. e., $N = 1$, only.

Formally, we immediately obtain the full version by considering the adjoint residual $\rho^*(U_h, Z_h)(\cdot)$ and the iteration error as well:

Proposition 69 (PU-DWR error estimator for phase-field fracture). *Let $U = (u, \varphi) \in X^D$ and $U_h = (u_h, \varphi_h) \in X_h^D$ be the continuous and discrete primal solutions, respectively. Let $Z = (z_u, z_\varphi) \in X^0$ and Z_h be the continuous and discrete adjoint solutions, respectively. With the superscript (2) we indicate higher-order solutions. For the finite element approximation of the penalized phase-field problem in Proposition 27, we have the a posteriori error estimate*

$$|J(U) - J(U_h)| \le \eta := |\eta_h| + |\eta_m| \le \sum_{i=1}^M |\eta_i + \eta_i^*| + |\eta_m|,$$

where η_h is defined in Proposition 67. The indicators η_i are composed by the terms as in Proposition 68, but now with the factor 1/2 (Proposition 65) such that

$$\eta_i = -\frac{1}{2}A(U_h)((Z_h^{(2)} - i_h Z_h^{(2)})\psi_i),$$

and the iteration error

$$\eta_m = -A(U_h)(Z_h),$$

and the adjoint error part:

$$\eta_i^* = \frac{1}{2}(J'(U_h)((U_h^{(2)} - i_h U_h^{(2)})\psi_i) - A'(U_h)((U_h^{(2)} - i_h U_h^{(2)})\psi_i, Z_h)).$$

Remark 66 (Possible future extensions). Based on the theoretical and algorithmic work for quasi-linear problems [150, 152], the previous result may be extended in four directions:

- Efficiency and reliability results for the previous problem based on [152];
- Extension to space-time adaptivity in which the inequality constraint must be considered as well. Ideas might be taken from [357, 389, 404, 403]. Moreover, the thesis [55] is useful in which a space-time DWR for the Navier–Stokes equations has been carefully worked out. An open source framework for linear space-time problems was published in [269]. Finally, recent general advances in space-time methods are compiled in [273];
- Multiple goal functional evaluations based on [150];
- Practical framework for large-scale problems of phase-field fracture combining the previous suggestions and including balancing discretization, nonlinear iteration, and linear iteration errors.

9.2.6 Mesh refinement and marking strategies

9.2.6.1 Basic algorithm

We now have an error value on each element $K_m \in \mathcal{T}_h$ or each PU-DoF i. It remains to set-up a strategy that tells us which elements shall be refined to enhance the accuracy in terms of the goal functional $J(\cdot)$.

Let an error tolerance TOL be given. Mesh adaptation is realized using extracted local error indicators from the a posteriori error estimate on the mesh $\mathcal{T}_h$. A cellwise assembly reads:

$$|J(U) - J(U_h)| \le \eta := \sum_{K \in \mathcal{T}_h} \eta_K.$$

To construct η_K, the DoF-wise indicators η_i can easily be summed up on each element. Alternatively, the PU allows for a direct DoF-wise assembly:

$$|J(U) - J(U_h)| \le \eta := \sum_{i=1}^{M} \eta_i.$$

This information is used to adapt the mesh using the following strategy.

Algorithm 14.

1. Compute the primal solution U_h and the (higher-order) adjoint solution Z_h on the present mesh $\mathcal{T}_h$.
2. Determine the cell indicator η_K at each cell K.
   Alternatively, determine the DoF-indicator η_i at each PU-DoF i.
3. Compute the sum of all indicators $\eta := \sum_{K \in \mathcal{T}_h} \eta_K$.
   Alternatively, $\eta := \sum_i \eta_i$.
4. Check, if the stopping criterion is satisfied: $|J(U) - J(U_h)| \leq \eta \leq \text{TOL}$, then accept U_h within the tolerance TOL. Otherwise, proceed to the following step.
5. Mark all cells K_i that have values η_{K_i} above the average $\frac{\alpha\eta}{M_{\text{el}}}$ (where M_{el} denotes the total number of cells of the mesh $\mathcal{T}_h$ and $\alpha \approx 1$).
   Alternatively, all PU-DoFs are marked which are above, say, the average $\frac{\alpha\eta}{M_{\text{el}}}$. Other marking strategies are listed in the next subsection.

9.2.6.2 Marking strategies

In order to refine the mesh we need to flag the elements to be refined. In the following, three methods will be introduced.

9.2.6.2.1 Averaging strategy

Select indicators according to

$$\eta_i > \frac{\alpha\eta}{M_{\text{el}}},$$

where $\alpha \approx 1$ is often a reasonable value.

9.2.6.2.2 Fixed error reduction (fixed fraction)

For fractions X, Y with $1 \quad X > Y$, determine indices $M_*, M^* \subset \{1, \ldots, M_{\text{el}}\}$ such that

$$\sum_{i=1}^{M^*} \eta_i \approx X\eta, \quad \sum_{i=M_*}^{M_{\text{el}}} \eta_i \approx Y\eta.$$

Then the elements $K_1, \ldots, K_{M^*}$ are refined and the elements $K_{M^*}, \ldots, K_{M_{\text{el}}}$ are coarsened. For example, $X = 0.3$, $Y = 0.01$. Refining/coarsening according to a reduction of the error estimate is also known as bulk criterion or Dörfler marking [142].

9.2.6.2.3 Fixed-rate (fixed number)

Here, we refine the $X \cdot M_{\text{el}}$ elements and coarsen $Y \cdot M_{\text{el}}$ elements with the highest and lowest error indicators, respectively.

Remark 67. We emphasize that the tolerance TOL should be well above the tolerances of the numerical solvers. Just recall that $\text{TOL} = 0.01$ would mean that the goal functional is measured up to a tolerance of 1 %.

When the DoF-based estimator is adopted, the error indicators η_i are nodewise contributions of the error. Mesh adaptivity can be carried out in two ways:

- in a nodewise fashion: if a node i is picked for refinement, all elements touching this node will be refined;
- alternatively, one could also first assemble for each $K \in \mathcal{T}_h$ elementwise indicators by summing up all indicators belonging to nodes of this element and then carry out adaptivity in the usual elementwise way.

On adaptive meshes with hanging nodes, the evaluation of the PU indicator is straightforward: First, the PU is assembled by employing the basis functions $\psi_i \in V_{PU}$ for $i = 1, \dots, M$. In a second step, the contributions belonging to hanging nodes are condensed in the usual way by distribution to the neighboring indicators.

9.2.6.3 Mesh refinement strategies for quasi-stationary (incremental) problems

If the previous error estimators shall be evaluated for a sequence of time/loading steps $n = 1, 2, 3, \dots, N$ for $N > 1$, then the following incremental strategy can be employed:

Algorithm 15. For the loading steps $n = 1, 2, \dots, N$, we perform at each t_n:

1 Compute the primal solution U_h on the current mesh $\mathcal{T}_h$;
2 Compute the higher-order adjoint solution Z_h on the current mesh $\mathcal{T}_h$;
3 Compute error estimator (Proposition 68 or 69) and determine the local indicators η_i at each PU-nodal point.
4 Compute the sum of all indicators $\eta := \sum_i \eta_i$, respectively, $\eta := \sum_i \eta_i + \eta_i^*$.
5 Check if the stopping criterion is satisfied: $|J(U) - J(U_h)| \leq |\eta| \leq \mathrm{TOL}$, then accept U_h within the tolerance TOL. Otherwise, proceed to the following step.
6 Mark all nodes i that belong to PU-nodal indicators η_i above the average $\frac{\alpha\eta}{M_{\mathrm{el}}}$ (where M_{el} denotes the total number of degrees of freedom of the PU and $\alpha \approx 1$). All cells that touch this node will then be refined.
7 Adapt the mesh.
8 Increment $n \mapsto n + 1$ and go to Step 1.

The marking strategy in Step 6 can be augmented with phase-field based refinement to have both accurate crack paths and goal functionals.

Rather than fixing the finest mesh level in order to keep the computational cost reasonable, another strategy for keeping the number of cells constant is based on the following strategy:

1. Order all nodes according to their size η_i.
2. A fixed portion of nodes (i. e., their touching cells) with the smallest contributions to the total indicator η is marked to be coarsened.
3. In the second step, the nodes (i. e., cells) with the largest η_i are refined such that the desired number of cells $M_{\max}$ is approximately achieved.

This strategy has been successfully applied for Prandtl–Reuss models in perfect plasticity [367, 369].

9.3 Comments on a posteriori error estimation in numerical simulations

A posteriori error estimation, local mesh adaptivity, verification of the developed methods, and code validation are achieved in three steps (with increasing level of difficulty). These steps hold true for classical norm-based error-estimation $\|U - U_h\|$ as well as goal-oriented error estimation $J(U) - J(U_h)$.

1. Does the solution make sense?
   - If possible test your code with an acknowledged benchmark configuration and verify whether $J(U_h)$ matches the benchmark values in a given range and on a sequence of at least three meshes. This first step can be performed with uniform and adaptive mesh refinement. In time-dependent problems, please compute at least with three different time step sizes.
   - If no benchmark configuration is available, study a simple, prototype configuration and observe whether the solution makes sense.
2. Do the true error $J(U) - J(U_h)$ and the error estimator η decrease under mesh refinement?
   - Compute $J(U)$ either on a uniformly-refined superfine mesh or even analytically (i. e., a manufactured solution). Compute the error $J(U) - J(U_h)$ and observe whether the error is decreasing.
   - If a priori estimates are available, see if the orders of convergence are as expected. But be careful, often goal functionals are nonlinear, for which rigorous a priori error estimates might not be available.
3. Compare η and $J(U) - J(U_h)$ in terms of the effectivity index I_{eff}. Do we asymptotically obtain values around $I_{\text{eff}} \approx 1$? For nonlinear problems, one easily observes $0.1 \le I_{\text{eff}} \le 10$, which might be still okay depending on the problem.

In particular the last step No. 3 is often difficult when not all theoretical requirements (smoothness of data and boundary conditions, regularity of the goal functional, smoothness of domains and boundaries) are fulfilled. Often, the adjoint error part ρ^* and the remainder term $\mathcal{R}^3$ are neglected, which result in a loss of accuracy of the error estimator for nonlinear problems. For computational illustrations, we refer the reader to the next section and further to Chapter 10.

9.3.1 Example: Stationary Navier–Stokes 2D-1 benchmark

We demonstrate the previous developments with a well-known benchmark in fluid mechanics: the 2D-1 benchmark [391]. Despite the fact that fluid flow has not been in the core of this book so far, there are two good reasons:
- There exist widely-accepted reference values;
- Fluid flow is a key part in Chapter 11.

The underlying equations are the stationary incompressible Navier–Stokes equations [410, 361, 201], which we briefly touched in Section 3.3 as an internally-coupled problem. The goals are the accurate measurements of drag, lift and a pressure difference around an obstacle. We notice upfront that we only develop the primal form of the error estimator ρ and neglect the adjoint part ρ^*.

9.3.1.1 Equations

Formulation 22 (Stationary incompressible Navier–Stokes). Let $\Omega \subset \mathbb{R}^2$ with boundary $\partial\Omega = \Gamma_{\text{wall}} \cup \Gamma_{\text{in}} \cup \Gamma_{\text{out}}$. Find vector-valued velocities $v : \Omega \to \mathbb{R}^2$ and a scalar-valued pressure $p : \Omega \to \mathbb{R}$ such that

$$\begin{aligned} \rho(v \cdot \nabla)v - \operatorname{div}(\sigma) &= \rho f \quad \text{in } \Omega, \\ \operatorname{div}(v) &= 0 \quad \text{in } \Omega, \\ v &= 0 \quad \text{on } \Gamma_{\text{wall}}, \\ v &= g \quad \text{on } \Gamma_{\text{in}}, \\ \nu\partial_n v - pn &= 0 \quad \text{on } \Gamma_{\text{out}}, \end{aligned}$$

where $f : \Omega \to \mathbb{R}^2$ is a force, $g : \Gamma_{\text{in}} \to \mathbb{R}^1$ a Dirichlet inflow profile, n the normal vector, ρ the density and

$$\sigma = \sigma(v,p) = -pI + \rho\nu(\nabla v + \nabla v^T), \quad [\sigma] = \text{Pa} = \text{N/m}^2$$

is the Cauchy stress tensor for a Newtonian fluid. Moreover, I is the identity matrix and ν the kinematic viscosity.

The Reynolds number Re is given by the dimensionless expression

$$\text{Re} = \frac{LU}{\nu}$$

where L is a characteristic length and U a characteristic velocity.

Remark 68 (Stokes). Neglecting the convection term $(v\cdot\nabla)v$ for viscous flow, we obtain the linear Stokes equations.

For a variational formulation, we define the spaces (for details, see [410, 361, 201]):

$$V_v := \{v \in H^1(\Omega)^2 \mid v = 0 \text{ on } \Gamma_{\text{in}} \cup \Gamma_{\text{wall}} \cup S\},$$
$$V_p = L^2(\Omega)/\mathbb{R}.$$

Here, V_v is a vector-valued space. The pressure is only defined up to a constant. For existence of the pressure, the inf-sup condition is adopted [410, 361, 201].

On the outflow boundary Γ_{out}, we apply a natural boundary condition:

$$\rho v \partial_n v - pn = 0.$$

This type of condition implicitly normalizes the pressure [229, 360] to a unique solution. Thus, the pressure space can be written as

$$V_p = L^2(\Omega).$$

Remark 69. Applying $\rho v \partial_n v - pn = 0$ on Γ_{out} is not consistent with the total stress $\sigma \cdot n = -pI + \rho v(\nabla v + \nabla v^T) \cdot n$. The symmetric part $\rho v \nabla v^T \cdot n$ needs to be subtracted on Γ_{out}. For more details see [229, 440]. For a better concentration on the main parts in this section, we neglect this term in the following.

As variational formulation, we obtain the following.

Proposition 70. *Find $v \in \{g + V_v\}$ and $p \in V_p$ such that*

$$(\rho(v \cdot \nabla)v, \phi) + (\sigma, \nabla\phi) = (\rho f, \phi) \quad \forall \phi \in V_v,$$
$$(\operatorname{div}(v), \psi) = 0 \quad \forall \psi \in V_p.$$

The finite element spaces must satisfy a discrete inf-sup condition [201]. A stable FE pair is the so-called Taylor–Hood element $Q_2 \times Q_1$; here defined on quadrilaterals. The velocities are biquadratic and the pressure bilinear. Hence we seek $(v_h, p_h) \in V_h^2 \times V_h^1$ with

$$V_h^s := \{u_h \in C(\bar{\Omega}) | \, u_h|_K \in Q_s(K), \forall K \in \mathcal{T}_h\}.$$

This is a conforming finite element space, i. e., $V_h^s \subset H^1(\Omega)$.

Proposition 71. *Find $v_h \in \{g + V_h^2\}$ and $p_h \in V_h^1$ such that*

$$(\rho(v_h \cdot \nabla)v_h, \phi_h) + (\sigma, \nabla\phi_h) = (\rho f, \phi_h) \quad \forall \phi_h \in V_h^2,$$
$$(\operatorname{div}(v_h), \psi_h) - 0 \quad \forall \psi_h \in V_h^1.$$

A semi-linear form is formulated as follows.

Proposition 72. *Find $U_h := (v_h, p_h) \in \{g + V_h^2\} \times V_h^1$ such that*

$$a(U_h)(\Psi_h) = (\rho(v_h \cdot \nabla)v_h, \phi_h) + (\sigma, \nabla\phi_h) - (\rho f, \phi_h)$$
$$+ (\operatorname{div}(v_h), \psi_h) = 0 \quad \forall \Psi_h = (\phi_h, \psi_h) \in V_h^2 \times V_h^1.$$

9.3.1.2 Functionals of interest

The goal functionals are drag, lift, and pressure difference, respectively,

$$J_1(U) = \int_S \sigma \cdot n e_1 \, ds, \quad J_2(U) = \int_S \sigma \cdot n e_2 \, ds, \quad J_3(U) = p(x_0, y_0) - p(x_1, y_1)$$

with $x_0, x_1, y_0, y_1 \in \Omega$ and where e_1 and e_2 are the unit vectors in x and y direction. We are interested in the drag and lift coefficients:

$$c_D = \frac{2J_1}{\rho V^2 D}, \quad c_L = \frac{2J_2}{\rho V^2 D}$$

with $D = 0.1\,\mathrm{m}$ (the cylinder diameter) and the mean velocity

$$V = V(t) = \frac{2}{3} v_x(0, H/2)$$

where v_x is defined below and with the height of the channel $H = 0.41\,\mathrm{m}$.

9.3.1.3 A duality-based a posteriori error estimator (primal part)

We develop an error estimator based only on the primal part ρ. Since the problem is nonlinear, the adjoint part ρ^* (Proposition 65) should be evaluated as well, but is not considered in the following. The adjoint bilinear form is given by the following.

Proposition 73. *The adjoint residual is computed as*

$$\begin{aligned} a'_U(U_h)(\Psi_h, Z_h) &= (\rho(\phi_h \cdot \nabla) v_h + \rho(v_h \cdot \nabla)\phi_h, z_h^v) \\ &\quad + (-\psi_h I + \nu\rho(\nabla\phi_h + \nabla\phi_h^T), \nabla z_h^v) + (\mathrm{div}(\phi_h), z_h^p). \end{aligned}$$

Proposition 74 (Adjoint problem). *The adjoint problem depends on the specific goal functional $J_i(U)$, $i = 1, 2, 3$. For example, for the drag, it holds: Find $Z_h = (z_h^v, z_h^p) \in V_h^4 \times V_h^2$ such that*

$$a'_U(U_h)(\Psi_h, Z) = J_1(\Psi_h) \quad \forall \Psi_h = (\phi_h, \psi_h) \in V_h^4 \times V_h^2$$

with $a'_U(U_h)(\Psi_h, Z)$ defined in Proposition 73 and

$$J_1(\Psi_h) = \int_S (-\psi_h I + \rho\nu(\nabla\phi_h + \nabla\phi_h^T)) \cdot n e_1 \, ds, \quad \Psi_h = (\phi_h, \psi_h).$$

Recall that the adjoint solution must be approximated with a higher-order method. Therefore, we choose the space $V^4 \times V^2$.

The primal PU-DWR error estimator reads:

Proposition 75. *It holds*

$$|J(U) - J(U_h)| \leq \eta := \left|\sum_i \eta_i\right| \leq \sum_i |\eta_i|,$$

with

$$\eta_i = -a(U_h)((Z_h^{(4/2)} - i_h Z_h^{(4/2)})\psi_i), \quad Z_h^{(4/2)} = (z_h^v, z_h^p), \quad \psi_i \in V_{PU}$$

where $Z_h^{(4/2)}$ is a higher-order discrete solution with polynomials degree $r = 4$ for the velocity part and $r = 2$ for the adjoint pressure. Furthermore, the PU is realized as $V_{PU} := V_h^1$ and where $a(U_h)(\cdot)$ has been defined in Proposition 72.

Remark 70. In the error estimator the nonhomogeneous Dirichlet data g on Γ_{in} are neglected and assumed to be small. Also, the outflow correction condition (see Remark 69) is neglected.

9.3.1.4 2D-1 configuration

The configuration is displayed in [391, Figure 1] and Figure 9.1.

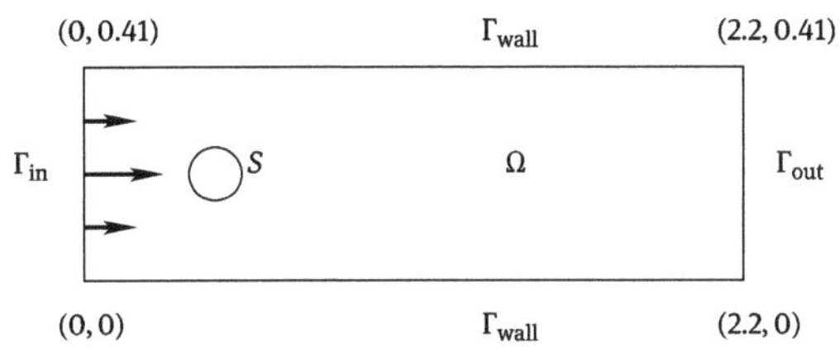

Figure 9.1: 2D-1 benchmark configuration. The cylinder with boundary S has the origin in $(0.2, 0.2)$.

9.3.1.5 Boundary conditions

We previously specified all boundaries except the inflow:

$$g = (g_1, g_2)^T = (v_x(0, y), v_y)^T = \left(4 v_m y \frac{H - y}{H^2}, 0\right)^T$$

with $v_m = 0.3\,\mathrm{m/s}$. The resulting Reynolds number is $\mathrm{Re} = 20$ (use D and V as defined above).

9.3.1.6 Parameters and right-hand side data

We use $\rho = 1\,\mathrm{kg/m^3}$, $\nu = 10^{-3}\,\mathrm{m^2/s}$ and $f = 0$.

9.3.1.7 Step 1: Verification of benchmark values

In Table 3 in [391], the following bounds are given:
- Drag, c_D: 5.57–5.59.
- Lift, c_L: 0.0104–0.0110.
- Pressure difference, Δp: 0.1172–0.1176.

We obtain on level 3 with 1 828 elements and with 15868 velocity DoFs and 2 044 pressure DoFs the results shown in Table 9.1.

Table 9.1: Our numerically computed reference results.

Functional	Value
Drag	5.5754236905876873e+00
Lift	1.0970590684824442e−02
Press. diff	1.1745258793930979e−01

9.3.1.8 Step 2 and Step 3: Observing $J(u) - J(u_h)$ and I_{eff}

The reference value $J_1(U)$ for the drag is

$$J_1(U) := 5.5787294556197073, \tag{9.11}$$

and was obtained on a four times (Level 4) uniformly refined mesh. We obtain the results shown in Table 9.2.

Table 9.2: 2D-1 benchmark with PU-DWR: errors and effectivity indices for the drag evaluation.

Level	$J_1(U) - J_1(U_h)$	η	I_{eff}
0	3.51e−01	1.08e−01	3.07e−01
1	9.25e−02	2.08e−02	2.25e−01
2	1.94e−02	6.07e−03	3.12e−01
3	3.31e−03	2.45e−03	7.40e−01

We observe that the true error and the estimated error η both decrease. The effectivity indices are less than 1 and indicate an underestimation of the true error (clearly seen in columns no. 2 and 3 as well). Because of our several assumptions and relatively coarse meshes, the results are more or less satisfying w. r. t. Step 3. Regarding Steps 1 and 2, our findings are excellent.

In[5] Figure 9.2, the true error for local DWR-based refinement and uniform refinement are compared. Adaptive mesh refinement clearly yields a better constant of more

5 These findings were provided by Jan Philipp Thiele. Thanks!

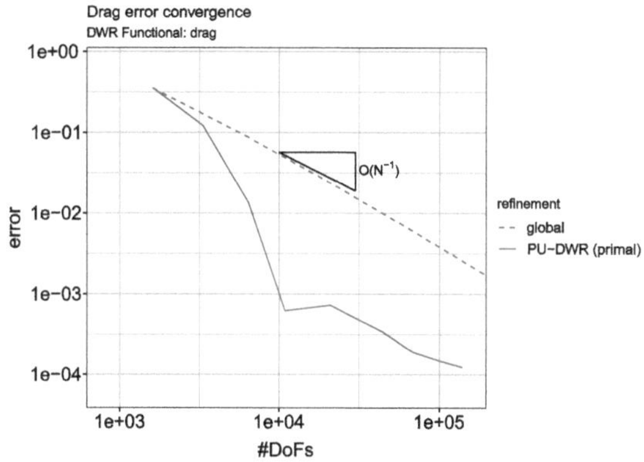

Figure 9.2: True error $J_1(U) - J_1(U_h)$ for local and global refinement.

than one order. The convergence order however seems the same as uniform mesh refinement. At this place, we also refer to the literature for other results considering stationary flow or the 2D-1 benchmark. Nabh computed in 1998 further reference values [331]. Then, in [47], the DWR method was first applied to the 2D-1 benchmark. In [52, Section 8] details of the DWR method for stationary flow problems are provided. In [78], the 3D flow benchmark versions from [391] were considered. Results using multigoal-oriented error estimation are provided in [152, Figure 12, Table 2] and [148, Figure 5]. Finally, in [374], the FSI-1 benchmark from [240] was considered in which the 2D-1 benchmark composes the flow part. Some of these studies provided error curves as Figure 9.2. On coarse meshes, the convergence orders seem comparable. Only on very fine meshes, local mesh adaptivity seems to outperform uniform mesh refinement with respect to the convergence order.

9.3.1.9 Graphical solution plots

The resulting mesh and solution plots of all three variables (v_x, v_y) and p are shown in Figure 9.3. The corresponding adjoint solutions are shown in Figure 9.4.

9.4 Multigoal-oriented a posteriori error estimation

We now extend single goal-oriented error estimation to multiple goal functionals. This is particularly interesting for multiphysics problems in which quantities of interest in each physical subproblem might be of interest.

9.4.1 Examples of multiple goal functionals

As an example, in phase-field fracture, we might simultaneously control the crack energy, the total crack volume, and point values, respectively:

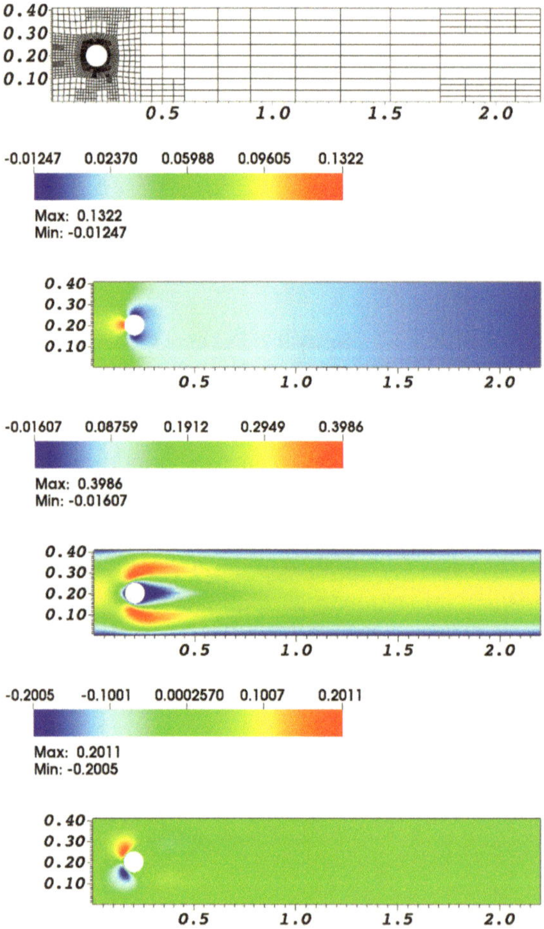

Figure 9.3: 2D-1 fluid benchmark: adaptively refined mesh with respect to $J_1(U)$, the pressure solution, x-velocity, y-velocity.

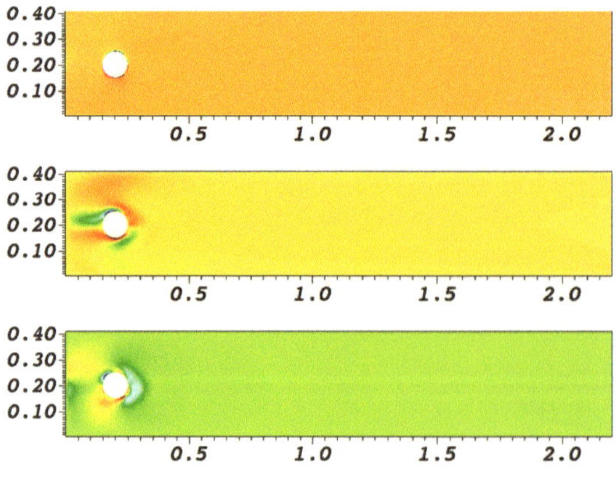

Figure 9.4: 2D-1 fluid benchmark: adjoint solutions of the pressure solution, x-velocity, y-velocity.

$$J_1(\varphi) = E_S(\varphi) = \frac{1}{2}\int_B G_c\Big(\frac{1}{\varepsilon}(1-\varphi)^2 + \varepsilon|\nabla\varphi|^2\Big)\,dx,$$

$$J_2(u,\varphi) = \int_B u\cdot\nabla\varphi\,dx,$$

$$J_3(u) = u(x), \quad x\in B.$$

The bulk energy, global L^2-norms, or stress evaluations are further reasonable goal functionals. In multiphysics problems involving fluids: drag/lift forces, pressure values, or the vorticity are further quantities of interest.

A naive approach would be to apply Proposition 65 by using these $S \in \mathbb{N}$ goal functionals with S adjoint problems such that we control

$$|J_i(U) - J_i(U_h)| \le \mathrm{TOL}_i, \quad \text{for } i = 1,\dots,S.$$

For $S \gg 1$, it is clear that this approach becomes very expensive. Hartmann and Houston [216] proposed to consider the sum of the relative errors

$$\sum_{i=1}^{S} \frac{|J_i(U) - J_i(U_h)|}{|J_i(U_h)|}, \tag{9.12}$$

while assuming $J_i(U_h) \neq 0$. This eventually leads to a combined error functional.

9.4.2 Constructing a combined goal functional

Based on [216, 215], we ourselves have been putting tremendous efforts in the last four years to further develop multigoal-oriented error estimation [153, 150, 152, 147, 149, 148]. Recent developments from others include [348, 15, 257, 423]. As model problem, we often used the p-Laplace equation (e. g., [203, 115, 3, 291, 132, 415]), which shares similarities with phase-field fracture in the sense that both are quasi-linear. At the same time, our implementation works for stationary coupled problems [152, 148] such as the stationary Navier–Stokes equations (Section 9.3.1). All results will be summarized in an upcoming PhD thesis by Bernhard Endtmayer [146].

In the following, we are given S goal functionals $J_i(\cdot)$, $i = 1,\dots,S$. Motivated by (9.12), these single functionals are combined to one goal functional $J_c(\cdot)$ creating a linear combination resulting in

$$J_c(U) := \sum_{i=1}^{S} w_i J_i(U), \tag{9.13}$$

with some weights $w_i \in \mathbb{R}$. Generalizing for a function $\Phi \in X$, it holds

$$J_c(\Phi) := \sum_{i=1}^{S} w_i J_i(\Phi). \tag{9.14}$$

One crucial aspect is the sign of w_i. Assuming $w_i = 1$ for $i = 1,\dots,S$, and $J_i = -1$ for all odd indices and $J_i = 1$ for even indices, yields $J_c \equiv 0$. Thus, we must include a sign computation and use as weights

$$w_i := \frac{\operatorname{sign}(J_i(U) - J_i(U_h))\omega_i}{|J_i(U_h)|}, \tag{9.15}$$

where $\omega_i > 0$. Then

$$J_c(\Phi) = \sum_{i=1}^{S} \omega_i \frac{\operatorname{sign}(J_i(U) - J_i(U_h))}{|J_i(U_h)|} J_i(\Phi). \tag{9.16}$$

If the weights ω_i are similar, the relative errors of all single functionals are of the same order. Setting $\Phi = U - U_h$, we obtain immediately in the linear case the error representation

$$J_c(U - U_h) = \sum_{i=1}^{S} \omega_i \frac{\operatorname{sign}(J_i(U) - J_i(U_h))}{|J_i(U_h)|} J_i(U - U_h). \tag{9.17}$$

A first sign computation as used in (9.15) was proposed in [216] by solving an adjoint problem as in Section 9.2 and an additional discrete-error problem (adjoint-to-adjoint). As a result, at each step of an adaptive algorithm, three problems must be solved:
1. Primal problem (possibly nonlinear): U_h;
2. Adjoint problem (linear): Z_h;
3. Discrete error problem: $e_h := U - U_h$.

By solving these systems, we obtain for $\Phi := U$ in (9.16):

$$J_c(U) = \sum_{i=1}^{S} \omega_i \frac{\operatorname{sign}(J_i(e_h))}{|J_i(U_h)|} J_i(U). \tag{9.18}$$

This combined goal functional is fully computable and can be used as right-hand side in the adjoint problem. For general nonlinear problems, we recall however that we need the Fréchet derivative according to Proposition 65. Then, it holds as in the following.

Proposition 76 (Adjoint problem with combined goal functional). *Let (9.18) be given and $J_c'(U)(\Phi)$ its Fréchet derivative at $U \in X$ into direction $\Phi \in X$. Then we seek $Z \in X$ such that*

$$A'(U)(\Phi, Z) = J_c'(U)(\Phi) \quad \forall \Phi \in X. \tag{9.19}$$

Details on the practical evaluation of $J_c'(U)(\Phi)$ can be found in [150, *Section 4*].

Proof. The adjoint problem follows from Proposition 65, while using (9.18) as right-hand side. □

In [153], we improved the efficiency of the above algorithm. Rather than solving the primal problem and the discrete error problem, we solve the primal problem twice: with low-order and higher-order elements such that we obtain U_h and $U_h^{(2)}$. On the first view, there is no benefit because we still need to solve three problems:

1. Primal problem (possibly nonlinear): U_h;
2. Adjoint problem (linear): Z_h;
3. Higher-order primal problem (possibly nonlinear): $U_h^{(2)}$.

However, for Proposition 65, we need the higher-order primal solution for the adjoint error part ρ^* nonetheless, which is now serving for two purposes. Second, from an algorithmic point of view, the two primal problems can be computed independently, while the discrete error problem depends on U_h and can only be solved once this solution is known. Third, in hierarchical spaces, it is sufficient, to compute the primal problem only once with higher-order elements and restrict the solution to the lower-order space like a geometric multigrid method. To this end, we save one problem solve.

These ideas yield an error representation for the sign computation. We recall our result for linear problems [153].

Proposition 77 (Higher-order solutions versus discrete error problem). *For linear problems, the solutions of the following three problems can be linked. Find $U_h \in X_h$, $U_h^{(2)} \in X_h^{(2)}$ and $e_h^{(2)} \in X_h^{(2)}$ such that*

$$
\begin{aligned}
A(U_h)(\Psi_h) &= F(\Psi_h) && \text{(Low-order primal problem)},\\
A(U_h^{(2)})(\Psi_h^{(2)}) &= F(\Psi_h^{(2)}) && \text{(High-order primal problem)},\\
A(e_h^{(2)})(\Psi_h^{(2)}) &= F(\Psi_h^{(2)}) - A(U_h)(\Psi_h^{(2)}) && \text{(Discrete error problem)}.
\end{aligned}
$$

Then, for $X_h \subset X_h^{(2)}$, the solution $e_h^{(2)}$ obtained by solving the discrete error problem can be equivalently obtained by simply constructing

$$e_h^{(2)} = U_h^{(2)} - U_h.$$

Proof. For the proof, we refer the reader to [153]. □

With Proposition 77, it holds

$$J_c(U) = \sum_{i=1}^{S} \omega_i \frac{\operatorname{sign}(J_i(U_h^{(2)} - U_h))}{|J_i(U_h)|} J_i(U). \tag{9.20}$$

In [150], we applied this procedure to nonlinear problems. We have recently proven in [151] that the opposite works as well under certain conditions, namely computing the low-order finite element primal solution and prolongating this solution into the higher-order space. The idea itself is not new and was proposed in early works [52, 37] with some proofs of effectivity in [380]. The novelty of [151] are rigorous two-sided estimates and a novel class of resulting adaptive algorithms.

Remark 71 (Justification of the multigoal technology vs. uniform mesh refinement). We emphasize that one may argue instead of constructing a joint goal functional and the overhead computing an additional problem for the sign computation, simple uniform mesh refinement would be more efficient. In [153, 150, 149], we have shown that this is often not the case and the proposed multigoal-oriented method pays off in terms of computational cost and memory requirements. Moreover, we also have an a posteriori estimator for error quantification, which simple (uniform) mesh refinement does not provide.

9.4.3 A multigoal-oriented a posteriori error estimate for steady-state phase-field fracture

We assume as in Section 9.2 a quasi-static problem for the loading steps $n = 1, 2, 3, \ldots, N$. The main result for one step $N = 1$ is the following:

Proposition 78. *Let $J_i(U)$, $i = 1, \ldots, S$ be given and let $J_c(U)$ be the combined goal functional defined in (9.20). Let $U = (u, \varphi) \in X^D$ and $U_h = (u_h, \varphi_h) \in X_h^D$ be the continuous and discrete primal solutions, respectively. Let $U_h^{(2)} = (u_h^{(2)}, \varphi_h^{(2)}) \in V_{u,h}^{(2)} \times V_{\varphi,h}^{(2)}$ be the higher-order primal solution, respectively. Let $Z = (z_u, z_\varphi) \in X^0$ be the continuous adjoint solution with respect to the combined goal functional. Let Z_h and $Z_h^{(2)}$ be the corresponding discrete adjoint solutions. For the finite element approximation of the penalized phase-field problem in Proposition 27, we have the a posteriori error estimate*

$$|J_c(U) - J_c(U_h)| \leq \eta := \left|\sum_{i=1}^{M} \eta_i\right| \leq \sum_{i=1}^{M} |\eta_i|$$

with

$$\begin{aligned}
\eta_i &= -A(U_h)((Z_h^{(2)} - i_h Z_h^{(2)})\psi_i) \\
&= -(g(\varphi_h)\sigma(u_h), \nabla((z_u^{(2)} - i_h z_u^{(2)})\psi_i)) \\
&\quad - ((1-\kappa)\varphi_h \sigma(u_h) : e(u_h), ((z_\varphi^{(2)} - i_h z_\varphi^{(2)})\psi_i)) \\
&\quad + \left(\frac{G_c}{\varepsilon}(1 - \varphi_h), ((z_\varphi^{(2)} - i_h z_\varphi^{(2)})\psi_i)\right) - (G_c \varepsilon \nabla\varphi_h, \nabla((z_\varphi^{(2)} - i_h z_\varphi^{(2)})\psi_i)) \\
&\quad - \gamma([\varphi_h - \varphi_h^0]^+, ((z_\varphi^{(2)} - i_h z_\varphi^{(2)})\psi_i)).
\end{aligned}$$

Proof. The proof is similar to Proposition 68, except that the adjoint problem is given by (9.19) using the discrete combined functional $J_c'(U_h)(\Phi)$. Algorithmically, we notice that therein the sign computations enter for which we adopt Proposition 77 and use $e_h = U_h^{(2)} - U_h$ in (9.20). Based on Proposition 68, we then obtain the error indicators η_i. □

Furthermore, we obtain based on Proposition 69:

Corollary 8. *Let the assumptions from Proposition 78 be given. Then the full error estimator reads*

$$|J_c(U) - J_c(U_h)| \leq \eta := |\eta_h| + |\eta_m| \leq \sum_{i=1}^{M} |\eta_i + \eta_i^*| + |\eta_m|,$$

with $\eta_h = \sum_{i=1}^{M} (\eta_i + \eta_i^*)$.

In both statements, we mainly adapted the findings presented in [150] to a steady-state phase-field fracture problem. Numerical studies for testing the performance in practice are left for future research. Furthermore, the extension to quasi-static or time-dependent versions including consistent adjoints for the VI are also of future interest.

9.4.4 Adaptive algorithm for multigoal-oriented discretization error estimation

A first algorithm for the discretization error in Proposition 78 is as follows:

Algorithm 16.

1. Solve two primal problems: Compute the low-order primal solution U_h and higher-order solution $U_h^{(2)}$, respectively.
2. Construct combined functional J_c as in (9.20).
3. Solve the adjoint problem: Compute the adjoint solution $Z_h^{(2)}$ by solving the adjoint problem $A'(U_h)(\Phi_h, Z_h^{(2)}) = J_c'(U_h)(\Phi_h)$ on a larger finite element space than for U_h.
4. Estimate:
   - Determine the indicator η_i at each node i by using Proposition 78.
   - Compute the sum of all indicators $\eta := \sum_i \eta_i$.
   - Check, if the stopping criterion is satisfied:

     $$\eta < \mathrm{TOL}_c,$$

     where $\mathrm{TOL}_c := \inf_{1\leq i\leq S}\{\frac{\omega_i \mathrm{TOL}_i}{|J_i(U_h)|}\}$. If this criterion is satisfied, stop the computation since $J_c(U_h)$ has been computed with a desired accuracy. Otherwise, proceed to the following step.
5. Mark all cells K_i that touch nodes that have values η_i above the average $\frac{\alpha\eta}{M_{\mathrm{el}}}$ (where M_{el} denotes the total number of cells of the mesh $\mathcal{T}_h$ and $\alpha \approx 1$).
6. Refine the mesh.
7. Go back to 1.

Remark 72. The reason for the special choice of TOL_c is to ensure that for all functionals J_i holds that $|J_i(U)-J_i(U_h)| < \mathrm{TOL}_i$ in which we tacitly assume that the exact adjoint solution would be known.

Remark 73. Of course, the above construction carries immediately over to higher order problems with $U_h^{(r)}$, $U_h^{(r+1)}$, $Z_h^{(r)}$, $Z_h^{(r+1)}$ for the finite element order r. In [153], we have computed the solutions for some linear problems up to $r = 8$.

9.4.5 Adaptive algorithm for balancing discretization and nonlinear iteration errors

We now extend the previous algorithm to include the nonlinear iteration error η_m as well, such that discretization and nonlinear iteration errors are balanced with respect to the goal functional $J_c(\cdot)$. The basic structure of the following algorithms is similar to those presented in [370, 156, 150]. Based on [150, Section 5], we design the following algorithms. Let l be the refinement level and j as usual the Newton iteration index. The error estimators η_h and η_m (computed in Corollary 8) on mesh level l are denoted by η_h^l and η_m^l, respectively.

Algorithm 17 (Adaptive Newton algorithm for multiple-goal functionals on level l).

1: Start with some initial guess $U_h^{l,0} \in X_h^l$ and $j = 0$.

2: Find $Z_h^{l,0}$ such that

$$A'(U_h^{l,0})(\Phi_h, Z_h^{l,0}) = (J_c^{(0)})'(U_h^{l,0})(\Psi_h) \quad \forall \Phi_h \in X_h^l,$$

with $(J_c^{(0)})'$ defined in (9.20), constructed with the higher-order solution $U_h^{l,(2)}$ and the low-order solution $U_h^{l,0}$.

3: **while** $|A(U_h^{l,j})(Z_h^{l,j})| =: \eta_m^l > 10^{-2}\eta_h^{l-1}$ **do**

4: For $\delta U_h^{l,j}$, solve

$$A'(U_h^{l,j})(\delta U_h^{l,j}, \Psi_h) = -A(U_h^{l,j})(\Psi_h) \quad \forall \Psi_h \in X_h^l.$$

5: Update: $U_h^{l,j+1} = U_h^{l,j} + \lambda \delta U_h^{l,j}$ with $\lambda \in (0,1]$.

6: $j = j + 1$.

7: For $Z_h^{l,j}$, solve

$$A'(U_h^{l,j})(\Phi_h, Z_h^{l,j}) = (J_c^{(j)})'(U_h^{l,j})(\Phi_h) \quad \forall \Phi_h \in X_h^l,$$

with $(J_c^{(j)})'$ constructed with $U_h^{l,(2)}$ and $U_h^{l,j}$ as in (9.20)

8: **end while**

Remark 74. In [370], the stopping criterion $10^{-1}\eta_h^l$ (current mesh level l) was proposed. To avoid the computation of η_h^l in every iteration, we use $10^{-2}\eta_h^{l-1}$ as noticed in [150, Section 5.4].

9.4.6 Algorithm for multigoal-oriented error estimation

We obtain the following algorithm for multigoal-oriented error estimation. Therein, at each refinement level l, four problems are solved: two solutions in Step 3, one higher-order primal solution in Step 2, and one (higher-order, linear) adjoint solution in Step 5.

Algorithm 18.

1: Start with some initial guesses $U_h^{0,(2)}$ and U_h^0, set $l = 1$ and set $\mathrm{TOL}_{\mathrm{dis}} > 0$.
2: Solve the (nonlinear) higher order primal problem for $U_h^{l,(2)} \in X_h^{l,(2)}$ with the initial guess $U_h^{l-1,(2)}$ and some Newton method from Chapter 7.
3: Solve the (nonlinear) low-order primal problem and (linear) adjoint problem using Algorithm 17 with the initial guess U_h^{l-1} on the discrete X_h^l.
4: Construct the combined functional J_c as in (9.20).
5: Solve the adjoint problem for J_c on $X_h^{l,(2)}$.
6: Construct the error estimator η in Corollary 8.
7: Check $|\eta| < \mathrm{TOL}_{\mathrm{dis}}$, e. g., $\mathrm{TOL}_{\mathrm{dis}} = \frac{1}{M_{\mathrm{el}}} \sum_i \mathrm{TOL}_i$ (average). If yes:, solution $U_h^l \in X_h^l$ with sufficient accuracy in $J_c(U_h)$ found $\to$ stop.
8: Mark all elements K_i with, e. g., $\eta_i > \frac{\alpha\eta}{M_{\mathrm{el}}}$, $\alpha \approx 1$.
9: Refine marked elements: $\mathcal{T}_h^l \mapsto \mathcal{T}_h^{l+1}$ and $l = l + 1$.
10: Go to 2.

9.5 Predictor-corrector adaptivity

In this section, we exploit predictor-corrector adaptivity for two technologies:
1. Adaptive mesh refinement around the crack path;
2. Adaptive choice of fine scale problems within a non-intrusive global/local approach.

Predictor-corrector adaptivity is motivated by the fact that the unknown crack path is predicted in a first step, and the new (accurate) solution on the current crack path is obtained in a second step.

9.5.1 Local mesh refinement around growing fractures

This section is concerned with local mesh adaptivity with a focus on the unknown crack path. As motivated in Section 5.5 (see specifically Remark 42) phase-field approaches require fine meshes around the fracture in order to capture the gradient behavior with sufficient accuracy.

Therefore, the goal is as follows: we want to exclusively refine the fracture zone in order to work with a small regularization parameter ε such that $h \lesssim \varepsilon$ holds true. As briefly mentioned in Section 5.5, ε should not be changed locally during a simulation. If we link for instance $\varepsilon = 2h$ and perform local mesh refinement, we have different ε in the domain B. Neither is this a desirable robust procedure, nor do we want to change ε locally since this basically means that we change the model.

One challenge is the a priori unknown crack path (see Section 1.3). For this reason, we first predict a new path and then recompute the accurate solution on the desired fine mesh. This was first proposed in [221] and extended to three dimensions in [448, 285]. Here, we choose $\varepsilon > h$ at the beginning of the computation. Our aim is to work with reasonable coarse meshes outside the fracture region. In those parts of the domain, $\varepsilon > h$ may be violated. Again we refer to Remark 42 for a justification.

To this end, we arrive at a four-step procedure (see also Figure 9.5):

Algorithm 19 (Predictor-corrector mesh adaptivity). At time t_n:
1. Given the current solution (u^n, φ^n) at t_n;
2. Predictor step: Solve for $(\tilde{u}, \tilde{\varphi})$ and predict the fracture path at t_{n+1};
3. Refine the mesh in regard to $(\tilde{u}, \tilde{\varphi})$ and go back to the solution (u^n, φ^n) afterward;
4. Corrector step: Solve on the new mesh for (u^{n+1}, φ^{n+1}) at t_{n+1}.

Increment $n \to n + 1$ and go to Step 1.

It has been shown in [221] (two-dimensional) and [448, 285] (three-dimensional) that this procedure is efficient and robust. As a refinement criterion we simply take a threshold value of the phase-field variable, i. e., once $\varphi < \eta_C$ with $0 < \eta_C < 1$, e. g., $\eta_C \sim 0.6$, we flag a mesh element for refinement. Specifically, on each element $K \in \mathcal{T}_h$, we loop over all quadrature points $x_q, q = 1, \ldots, N_q$. Once $\varphi(x_q) < \eta_C$ for at least one x_q, we flag the element K for refinement.

In the following, we present in Figure 9.6 some findings, that were obtained with the code published in [221]. The graphical outcome is shown in Figure 1.6 at the beginning of this monograph. For the single edge notched shear test its configuration, we refer the reader to Chapter 8. For the following findings, we used an Intel(R) Core(TM) i5-3320M CPU @ 2.60 GHz machine with two processors.

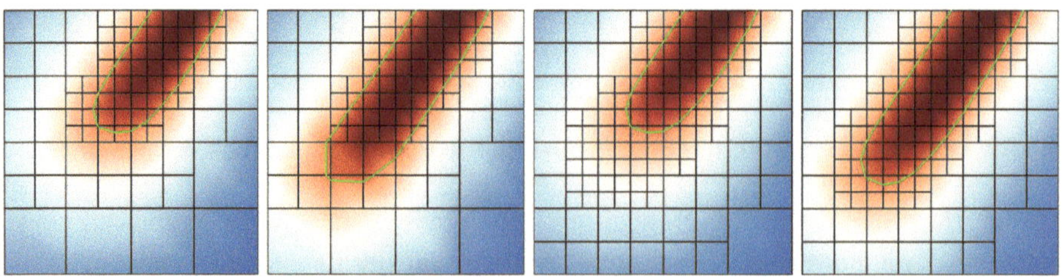

Figure 9.5: Mode of operation of predictor-corrector mesh adaptivity: (a) old state at t_n. (b) advance in time, crack leaves fine mesh. (c) refine and go back in time (interpolate old solution). (d) advance in time t_{n+1} on new mesh. Repeat until mesh does not change anymore. As a refinement indicator, we choose $\eta := \varphi < C = 0.2$. Reprinted from [221] with permission from Elsevier.

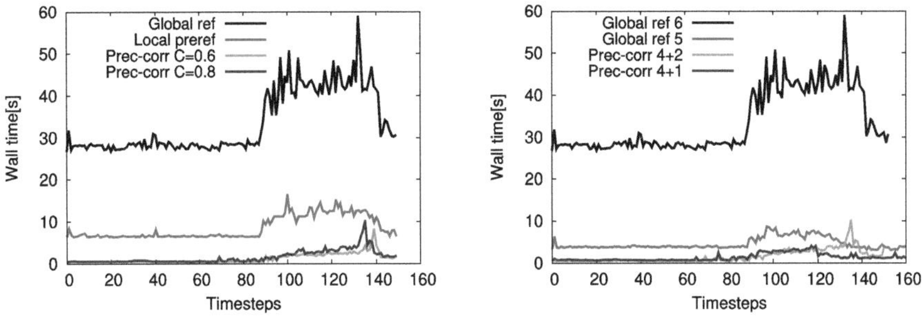

Figure 9.6: Wall time comparisons for predictor-corrector mesh refinement for the single edge notched shear test. Various refinement types and levels are compared. Here, C denotes the threshold phase-field value for mesh refinement. These graphics are produced from the raw data obtained in [221]. Additional comparisons can be found in [221].

9.5.2 An adaptive nonintrusive global-local procedure

In this section, we present an adaptive predictor-corrector scheme for a nonintrusive global-local approach. Therein, local fine-scale problems are chosen dynamically depending on the fracture path. Nonintrusive techniques for solid problems were proposed in [191] and applied to phase-field fracture in [194]. Therein, a global macroscale problem is solved with a simple approximate model, say linearized elasticity.

Formulation 23 (Macroscale/global problem). Find $u \in V_u^D$ such that

$$A_M(u)(w) = l(w) \quad \forall w \in V_u^0,$$

with $A_M(u)(w) := (\sigma(u), \nabla w)$ and $l(w) := (f, w)$ and for instance the linearized stress tensor $\sigma(u) = 2\mu e(u) + \lambda\, \mathrm{tr}(e)I$. We notice that no phase-field solution φ is sought.

On the microscale, the full phase-field model with nonlinear constitutive models is considered, i. e., Proposition 20. Here, we have the CVIS.

Formulation 24 (Microscale/local problem). Find $U = (u, \varphi) \in X^D$ such that

$$A_m(U)(\Psi - \Phi) \geq 0 \quad \Phi \in X^0.$$

This configuration has some similarities with the coupling of elasticity and poroelasticity discussed in Section 3.4. Therein as well, an enhanced model (local problem) is coupled with a simplified, linearized, PDE (global problem) in the surrounding area.

As we shall see in the following, there is however an important difference. Recapitulating Section 1.3, we identified Section 3.4 as a type 1 situation where the interface is fixed in space at all times. In contrast, the current situation is of type 3 with a propagating fracture with an unknown fracture path, which will require that the interface between the fine-scale problem and the macroscale problem must be adjusted during a computation. Consequently, the local and global subdomains are time-dependent and change during a simulation. This is challenging in terms of an algorithmic realization and implementation because the interface conditions between the local and global subdomains must be updated during time. A flexible choice is provided by using Lagrange multipliers. For details, we refer the reader to [194].

9.5.2.1 Predictor-corrector adaptive global-local scheme

The question remains how to determine the local and global subdomains for growing fractures. Here, we base ourselves on the previous predictor-corrector Algorithm 19. The mode of operation for the global-local approach is illustrated in Figure 9.7. The basic adaptive global-local algorithm is provided in the following and worked out in [341]. The numerical analysis is open and not yet addressed.

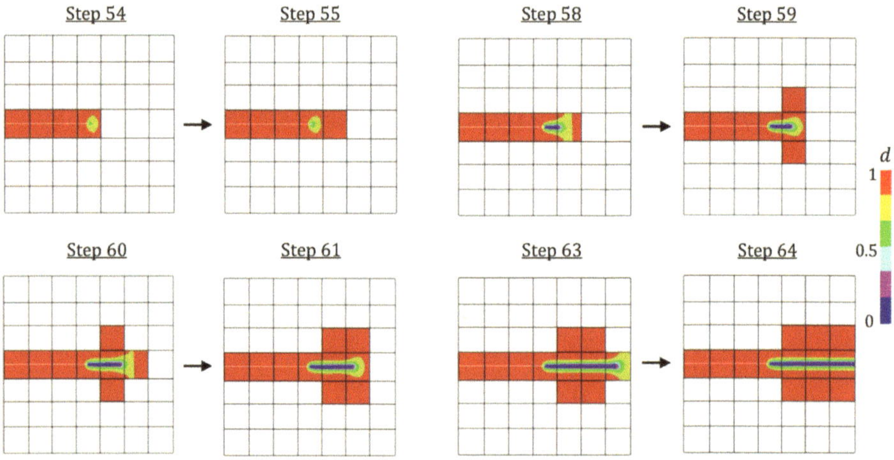

Figure 9.7: Illustration of the predictor-corrector adaptive global-local approach for eight different time points. Reprinted from [341] with permission from Elsevier.

From [341, Section 4], we extract the key algorithm and rewrite the main steps in the notation used in this monograph.

Algorithm 20 (Adaptive global-local scheme). The adaptive global-local scheme consists of the following steps:

1. Solve the coupled global-local problem at time t_n:

$$\begin{aligned} \text{Find } \tilde{u} \in V_u : \quad & A_M(\tilde{u})(w) = 0 \quad w \in V_u \\ \text{Find } \tilde{U} \in V_u \times V_\varphi : \quad & A_m(\tilde{U})(\Psi) = 0 \quad \Psi \in V_u \times V_\varphi. \end{aligned}$$

2. Predictor step:

- If $\tilde{\varphi} < \eta_C$ at an internal edge ∂K of an global element K_M and a local element K_m of the current decomposition $\mathcal{T}_n$, then $K_M \to K_m$ to obtain a new global-local decomposition $\mathcal{T}_{n+1}$
- Else increment time $n \to n+1$ and go to Step 1.

3. Interpolate global solution u onto the newly assigned local mesh element K_m. We notice that u is now known on the new local element K_m, but φ is not.
4. Corrector step: Solve on the new global-local decomposition $\mathcal{T}_{n+1}$:

$$\begin{aligned} \text{Find } u_{n+1} \in V_u : &\quad A_M(u_{n+1})(w) = 0 \quad w \in V_u, \\ \text{Find } U_{n+1} \in V_u \times V_\varphi : &\quad A_m(U_{n+1})(\Psi) = 0 \quad \Psi \in V_u \times V_\varphi. \end{aligned}$$

Increment time $n \to n+1$ and go to Step 1.

9.6 Adaptive time step control – a heuristic estimator

In this section, we present an algorithm for adaptive time step control. Since the presented phase-field model is only quasi-static it is questionable if the algorithm works in all cases because of the missing regularity in time. However, it makes sense to discuss this topic since in many fracture propagation problems we basically deal with three temporal regimes (see Chapter 4):

- $G < G_c$ and the fracture does not yet propagate (larger time steps/loading steps possible)
- $G = G_c$ fracture propagation (possibly smaller time steps necessary);
- $G < G_c$ or a completely damaged material (again larger time steps or even finishing the simulation due to total damage/fracture).

Our aim is to control again the accuracy of a goal functional $J(U)$. The principle idea, e. g., [362], is based on ODE theory and utilizes the local truncation error. Thus the only assumptions we have to make are enough regularity of the truncation error and discrete stability of the underlying difference scheme used for discretizing the continuous time derivatives.

The following description is adapted from a heuristic estimator developed in [159] in which time step control has been developed for the Fractional-Step-θ scheme for FSI multiphysics computations.

The entire approach is a combination of known results from the literature, which we briefly recapitulate to acknowledge the original authors. We follow an idea proposed by [420] but extend his criterion with additional terms, which are inspired by [210, 252]. The result is a simple method but the price to pay are additional solves of the problem per time step. Specifically, per time step, we compute the problem once with $2k$, compute the goal functional $J(U)$, and then compute two consecutive steps with time step size k, after we compare $J(U_k)$ and $J(U_{2k})$ by evaluating the absolute error $\text{abs}_{\text{err}} := |J(U_{2k}) - J(U_k)|$.

Proposition 79. *Let the error tolerance* TOL_{TS} *be given. We evaluate*

$$\text{abs}_{\text{err}} := |J(U_{2k}) - J(U_k)| \tag{9.21}$$

and then

$$\theta = \gamma * \left(\frac{\text{TOL}_{\text{TS}}}{\text{abs}_{\text{err}}}\right)^K, \tag{9.22}$$

where $\gamma \approx 1$ *is a safety factor (in* [210], $\gamma = 0.9$*),* $K = \frac{1}{15}$ *as in* [210]. *We then compute the new time step size as*

$$k_{\text{new}} = \begin{cases} k_{\text{old}} & 1 \leq \theta \leq 1.2, \\ k_{\text{old}} * \theta & \textit{otherwise.} \end{cases} \tag{9.23}$$

Moreover, we check whether $k_{\min} \leq k_{\text{new}} \leq k_{\max}$. *If* $k_{\text{new}} < 0.5k_{\text{old}}$, *i. e., the new time step would be much smaller than the old one, we redo the current time step* [420].

Remark 75. The check (9.23) prevents high time step oscillations and just keeps the old time step if the difference is not too large. Several illustrating computations have been presented in [210].

Remark 76. The last check $k_{\text{new}} < 0.5k_{\text{old}}$ is expensive since the entire calculation of the solution of this time step has to be repeated. However, in practice this happens rarely, but is necessary since a dramatic decrease of the time step size indicates that the current solution is by far not accurate enough.

Algorithm 21. The adaptive time step control for One-Step-θ schemes reads at t_n:
1. Set $k_{\text{old}} := k_{n-1}$
2. Perform
   - one computation with $2k_{\text{old}}$, evaluate $J(U_{2k})$ at $t_n + 2k_{\text{old}}$;
   - compute two steps with k_{old}, evaluate at the end of the 2nd step $J(U_k)$ at $t_n + k_{\text{old}} + k_{\text{old}}$;
3. Evaluate (9.21) and compute (9.22);
4. Compute the new k_{new} with (9.23);
5. If $k_{\text{new}} < 0.5k_{\text{old}}$, set $k_{\text{old}} := k_{\text{new}}$ and go to Step (ii);
6. Otherwise accept k_{new} and set $k_n = k_{\text{new}}$;
7. Increment $n \to n + 1$ and go to Step (i).

This heuristic time control estimator was used by us in two multiphysics publications:

- Time step control for nonlinear, nonstationary fluid-structure interaction [159];
- Time step control for fluid-structure interaction coupled to phase-field fracture [443].

The application of this time step control procedure to pure phase-field fracture is largely open because we mainly focused on quasi-static processes so far in which time step control remains questionable because of the missing time regularity. Some first results are presented in this work in the next chapter. For dynamic fracture, time step control does make more sense, but a rigorous computational analysis has not been performed yet.

10 Simulations III: goal-oriented error control for a phase-field slit and time control for a screw

10.1 A PU-DWR method for a slit modeled by phase-field

In this section, we continue to work with the slit configuration proposed in Chapter 6. The configuration and problem data are described therein. In extension, we now study some results obtained from a goal-oriented error estimator and local mesh adaptivity. Related studies with different goal functionals were published in [442].

10.1.1 Configuration, boundary conditions, initial condition, and parameters

All general information is provided in Chapter 6.

10.1.2 Goal functionals and reference values

As goal functionals, we choose the displacement value in the upper half of the domain:

$$J^D(u) = \int_{\{x\geq 0\}} u\,dx \quad \text{(Example 1)},$$

and the point value

$$J^P(u) = u(0.75, 0.75) \quad \text{(Example 2)}.$$

The reference values are computed for fixed ε on a ten times refined grid:

$$J^D(u_{\text{ref}}) := 0.34587686191839784,$$

and

$$J^P(u_{\text{ref}}) := 0.40136760427090823.$$

We employ two different marking strategies (Section 9.2.6.2): First, the fixed-rate strategy with $X = 0.1$ and $Y = 0.0$ (no coarsening) and second, the averaging strategy with $\alpha = 10$. While experimenting with this test case, we found that the refinement strategies play an important role in getting reasonable effectivity indices. In total, we therefore conduct four studies.

https://doi.org/10.1515/9783110497397-010

10.1.3 Goal-oriented a posteriori error estimator

We use the error estimator stated in Proposition 68 and notice that U and Z are single-valued for the displacements, i. e., we work with a classical Poisson-type problem. In total, we have $U = (u, \varphi)$ with $u : B \to \mathbb{R}$ and $\varphi : B \to \mathbb{R}$, and $Z = (z_u, z_\varphi)$ with $z_u : B \to \mathbb{R}$ and $z_\varphi : B \to \mathbb{R}$.

10.1.4 Example 1a/b: upper half domain goal functional

We obtain the results shown in Table 10.1 (Example 1a) and 10.2 (Example 1b). The primal and adjoint solutions are displayed in the Figures 10.1 and 10.2.

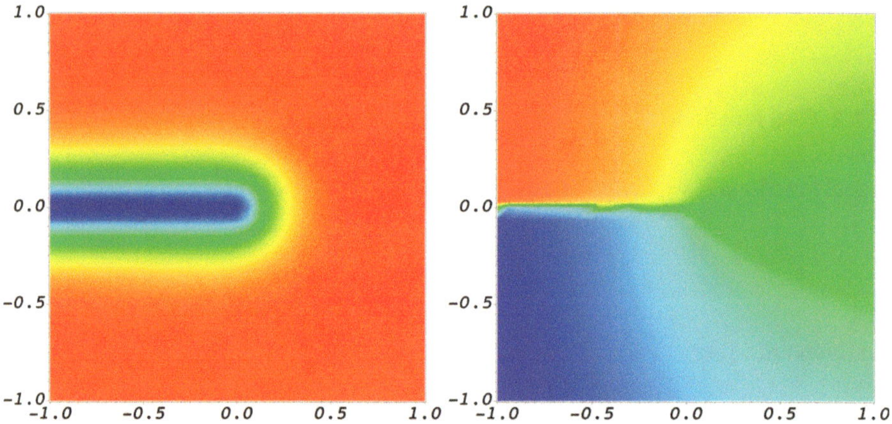

Figure 10.1: Examples 1/2: Plots of the (primal) phase-field solution and the primal displacements.

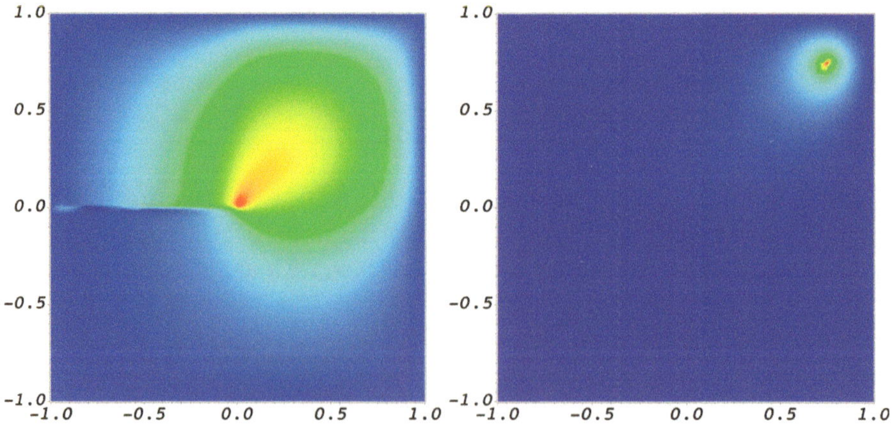

Figure 10.2: Plots of the adjoint displacements: Example 1 (left) and Example 2 (right).

10.1.4.1 Example 1a

The total error is decreased by about one and a half orders of magnitude. The error indicator η does not show a good performance; however, by its weak behavior, the effectivity indices I_{eff} become excellent. For the definition of the indicator index (last column) and its estimator (fourth column), we refer to [380] and Section 9.1.4. Satisfying values for I_{ind} are observed. Despite the good I_{eff}, the results should be carefully questioned because the adaptively refined meshes seem somewhat arbitrary as displayed in Figure 10.3. Some future improvements for further verification are: quality of the reference value, influence of the penalization parameter γ to approximate the variational inequality constraint, the nonlinearities in the coupled system, which would require a nonlinear DWR version including the adjoint error part ρ^* (Proposition 65).

Table 10.1: Example 1a: PU-DWR for phase-field slit and upper domain goal functional using the fixed-rate strategy with $X = 0.1$.

Dofs	$\lvert J^D(u_h) - J^D(u_{\text{ref}})\rvert$	η	η_{ind}	I_{eff}	I_{ind}
2178	1.02e−02	4.59e−04	1.20e−03	4.50e−02	1.18e−01
3370	7.12e−06	1.01e−03	1.18e−03	1.42e+02	1.66e+02
5442	7.24e−04	1.03e−03	1.98e−03	1.42e+00	2.73e+00
8862	5.14e−04	5.65e−04	1.24e−03	1.10e+00	2.42e+00

10.1.4.2 Example 1b

In this example, we observe in Table 10.2 a much better behavior of the error indicator η, which is however one order of magnitude better than the true error (second column). Consequently, we have an error under-estimation, which is reflected through the effectivity index I_{eff} (fifth column). However, the meshes look much more reasonable: Figure 10.3 (top, right). Here, inside the crack region with the steep gradients ∇u and $\nabla \varphi$ and specifically at the crack tip singularity, we observe mesh refinement.

Table 10.2: Example 1b: PU-DWR for phase-field slit and upper domain goal functional using the averaging strategy with $\alpha = 10$.

Dofs	$\lvert J^D(u_h) - J^D(u_{\text{ref}})\rvert$	η	η_{ind}	I_{eff}	I_{ind}
2178	1.02e−02	4.59e−04	1.20e−03	4.50e−02	1.18e−01
2834	4.84e−04	2.69e−04	4.27e−04	5.56e−01	8.82e−01
4470	1.38e−03	1.53e−04	2.53e−04	1.10e−01	1.83e−01
6938	9.81e−04	9.64e−05	1.86e−04	9.83e−02	1.89e−01
10094	4.62e−04	6.44e−05	1.44e−04	1.39e−01	3.11e−01

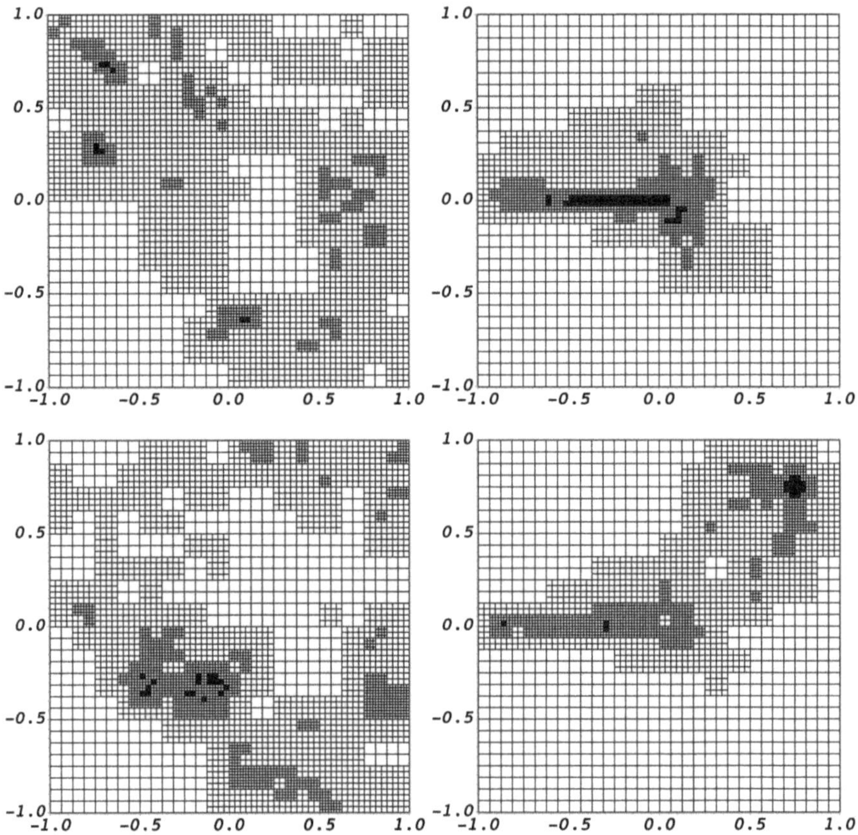

Figure 10.3: Example 1/2: Plots of the adaptively refined meshes. Top row: Example 1. Bottom row: Example 2. Left column: fixed-rate refinement strategy. Right column: averaging refinement strategy.

10.1.5 Example 2a/b: point value as goal functional

We obtain the results shown in Table 10.3 (Example 2a) and 10.4 (Example 2b).

10.1.5.1 Example 2a

The results and their interpretation are similar to Example 1a. The most interesting difference is that the true error is reduced by two orders of magnitude. The reason for this better performance might be that we now consider a goal functional which is further away from the crack zone and, therefore, less influenced by the approximation properties.

Table 10.3: Example 2a: PU-DWR for phase-field slit and point goal functional using the fixed-rate strategy with $X = 0.1$.

Dofs	$\lvert J^P(u_h) - J^P(u_{ref})\rvert$	η	η_{ind}	I_{eff}	I_{ind}
2178	3.87e−03	3.36e−05	6.63e−04	8.66e−03	1.71e−01
3286	2.44e−04	1.69e−04	7.19e−04	6.94e−01	2.95e+00
5470	1.22e−04	1.12e−04	7.51e−04	9.13e−01	6.14e+00
9622	1.28e−04	8.84e−05	5.97e−04	6.90e−01	4.66e+00
16010	5.02e−05	4.30e−05	2.55e−04	8.55e−01	5.08e+00

10.1.5.2 Example 2b

In this last example, the results are definitely improvable. The meshes look fine (Figure 10.3 (bottom, right)) because mesh refinement occurs in the crack region, the crack tip and the point goal functional. However, the behavior of the true error and the error estimator could be better. While the true error decreases, the error estimator is overall unchanged (even slightly worse; while the last row might be too close to the reference value though). Due the behavior of η, the effectivity indices become acceptable, which is in itself satisfying.

Table 10.4: Example 2b: PU-DWR for phase-field slit and point goal functional using the averaging strategy with $\alpha = 10$.

Dofs	$\lvert J^P(u_h) - J^P(u_{ref})\rvert$	η	η_{ind}	I_{eff}	I_{ind}
2178	3.87e−03	3.36e−05	6.63e−04	8.66e−03	1.71e−01
3062	1.89e−04	4.42e−05	2.74e−04	2.34e−01	1.45e+00
4794	2.22e−04	3.16e−05	1.72e−04	1.42e−01	7.74e−01
7046	2.00e−04	3.31e−05	1.04e−04	1.66e−01	5.19e−01
10778	1.07e−04	9.26e−05	2.79e−04	8.62e−01	2.60e+00

10.2 Multigoal-oriented error estimation for the slit without phase-field

To better interpret the previous results, we consider for simplicity a slit modeled inside the geometry. Thus, no phase-field function and no crack irreversibility constraint are necessary. Consequently, we deal with a simple Poisson problem with nonhomogeneous outer boundary conditions. We consider at first a single goal functional and then address multiple goal functionals with the methodology described in Section 9.4. Our raw data is taken from [153, Section 4.4, Example 4a] and reevaluated for the purpose of this monograph.

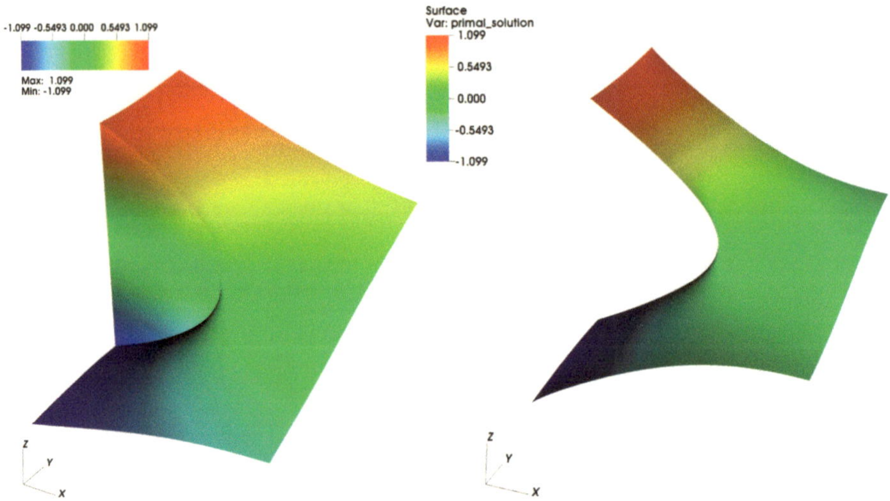

Figure 10.4: Comparison of the solution *u* for the slit modeled with phase-field and a slit built into the geometry.

10.2.1 Problem statement and configuration

Let

$$B = (-1,1) \times (-1,1) \setminus \{(x,0) \mid -1 \leq x \leq 0\},$$

and Γ_D the outer boundary and Γ_i the slit boundary (i. e., interface) inside the domain. The configuration is simply a Poisson problem on a slit domain: Find $u : B \to \mathbb{R}$ such that:

$$\begin{aligned} -\Delta u &= 0 \quad \text{in } B, \\ u &= g \quad \text{on } \Gamma_D, \\ \partial_n u &= 0 \quad \text{on } \Gamma_i. \end{aligned}$$

In Figure 10.4, both solutions, namely with phase-field and a slit inside the geometry, are displayed.

10.2.2 Quantities of interest

We deal with the following four goal functionals (but we notice that not all goal functionals are simultaneously considered in all test cases):

$$\begin{aligned} J_0(u) &:= u(0.75, 0.75), \\ J_1(u) &:= u(-0.5, -0.25), \end{aligned}$$

$$J_2(u) := \int_{\Gamma_1} \nabla u \cdot n \, dx,$$

$$J_3(u) := \int_{\Omega_1} u \, dx,$$

where $\Gamma_1 = \{-1\} \times (-1, -0.25)$ and $\Omega_1 = (0, 1) \times (-1, 0)$. We approximate the exact functionals by values that have been obtained from a very fine mesh:
1. $J_0^{\text{ref}}(u) \approx +0.18949212064$,
2. $J_1^{\text{ref}}(u) \approx -0.66061009755$,
3. $J_2^{\text{ref}}(u) \approx -0.54411579542$,
4. $J_3^{\text{ref}}(u) \approx -0.18268521784$.

10.2.3 Single goal functional

First, we consider some studies for a single goal functional. As discussed earlier, the true slit corresponds to $\varepsilon = 0$ in a phase-field fracture setting. Our findings are summarized in Table 10.5. Therein, we observe that the effectivity indices are not optimal, but they are very stable. Moreover, we observe perfect linear convergence in both the true error and the error estimator.

Table 10.5: Goal functional, errors, and I_{eff} for configuration 1 (single goal functional).

Level	Elements	$J(u_h)$	$\lvert J_0^{\text{ref}}(u) - J(u_h)\rvert$	η	I_{eff}
0	64	1.7935965773e−01	5.65e−02	4.21e−02	7.45e−01
1	208	1.8491217214e−01	2.48e−02	1.84e−02	7.42e−01
2	652	1.8835424968e−01	6.04e−03	4.59e−03	7.60e−01
3	4720	1.8891444682e−01	3.06e−03	2.19e−03	7.16e−01
4	11608	1.8920500524e−01	1.52e−03	1.25e−03	7.05e−01
5	24124	1.8935107801e−01	7.45e−04	5.40e−04	7.25e−01
6	54160	1.8942207581e−01	3.70e−04	2.65e−04	7.17e−01

10.2.4 Multigoal-oriented error estimation

In the multigoal-oriented setting, we are interested in four configurations:
- Config. 1: Evaluating $J_0(u_h)$,
- Config. 2: Evaluating $J_0(u_h)$ and $J_1(u_h)$,
- Config. 3: Evaluating $J_0(u_h)$, $J_1(u_h)$ and $J_2(u_h)$,
- Config. 4: Evaluating $J_0(u_h)$, $J_1(u_h)$, $J_2(u_h)$ and $J_3(u_h)$.

The adjoint combined functional is given by (9.18)

$$J_c(\Phi) = \sum_{i=1}^{4} J_i(\Phi)\frac{\operatorname{sign}(J_i(e_h^{(2)}))}{|J_i(u_h)|}\omega_i,$$

with $\omega_i = 1$. In Table 10.6, the results for configuration 4 are displayed, while for configurations 2 and 3, we refer to [153, Section 4.4, Example 4a]. The error in J_c is underestimated, but we cannot expect to have $I_{\text{eff}} = 1$ for this problem due to the loss of regularity. This confirms the results from the previous Section 10.1 in which we even observed worse I_{eff}. We also notice that only the combined goal functional $J_c(u_h)$ is controlled. Therefore, we have an upper bound, but the single functionals J_i might be more accurate; see [153, 150] for supporting arguments. Adaptively refined meshes are shown in Figure 10.5.

Table 10.6: Errors and I_{eff} for configuration 4 (multiple goal functionals).

Level	Elements	$\lvert J_c^{\text{ref}}(u_h) - J_c(u_h)\rvert$	η	I_{eff}
0	64	1.65e−01	1.16e−01	7.02e−01
1	184	8.16e−02	5.92e−02	7.25e−01
2	580	3.89e−02	2.76e−02	7.09e−01
3	1420	1.92e−02	1.36e−02	7.08e−01
4	3088	9.71e−03	6.84e−03	7.04e−01
5	6784	4.88e−03	3.47e−03	7.13e−01
6	13648	2.44e−03	1.73e−03	7.08e−01
7	27472	1.23e−03	8.81e−04	7.14e−01
8	54712	6.17e−04	4.39e−04	7.12e−01

10.3 Screw tests: error control in time/loading

In this section, our main aim is to study adaptive time/loading step control outlined in Section 9.6. From the mechanical point of view, this test is interesting as well because no initial crack is prescribed. Instead fractures will develop in regions with high stresses. These results were experimentally confirmed in [433] and [432]. Several numerical computations were presented in [444].

10.3.1 Formulation

We work with Proposition 32 with augmented Lagrangian penalization and the quasi-monolithic scheme from Section 7.7.2. Furthermore, we employ the iteration on the extrapolation from Section 7.7.3. This would correspond to Case 6 in Table 8.1 in Chapter 8.

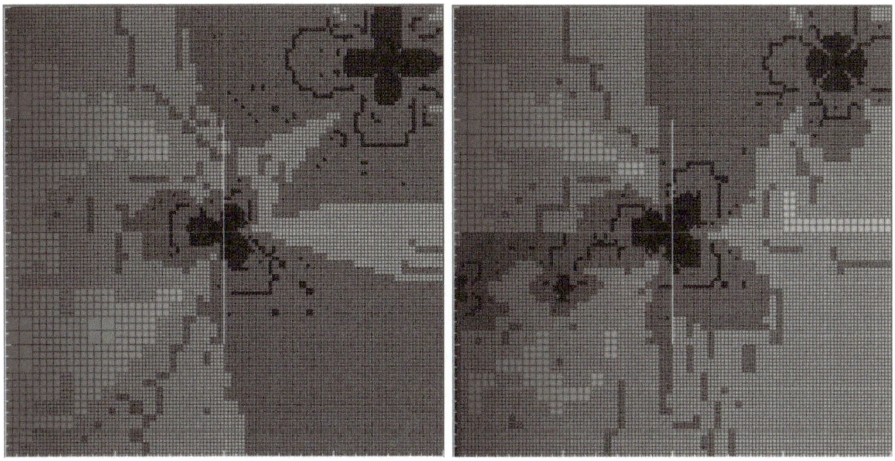

Figure 10.5: Multigoal-oriented error estimation: Meshes for configuration 1 after seven refinement steps with 54 160 elements (left) and configuration 4 after eight refinement steps with 54 712 elements.

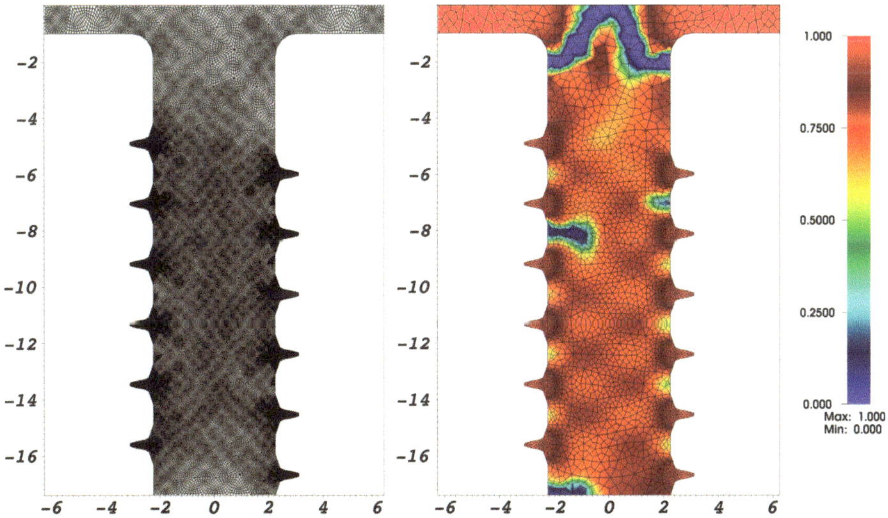

Figure 10.6: Mesh for screw simulations. The screw is fixed at the bottom. At the top, we have nonhomogeneous Dirichlet conditions in y-direction (uniform tension). The units are in mm. At the right, the final crack path in terms of φ is shown.

10.3.2 Configuration

The geometric setting is displayed in Figure 10.6. The total length is 17.20 [mm]. The initial mesh has 3440 elements with 10 800 (7200 + 3600) DoFs. The corresponding minimal cell diameter is $h_{\min} = 0.0689518$. We notice that the mesh is unstructured and not regular since the screw geometry was created using gmsh [195].

10.3.3 Boundary conditions

Crack growth is driven by a nonhomogeneous Dirichlet condition for the displacement field u on Γ_{top}, the head of the screw at $y = 0.0$. We increase the displacement on Γ_{top} at each time step such that the head is pulled, namely

$$u_y = t \times \bar{u}, \quad \bar{u} = 1.0\,\text{mm},$$

where t is the current time. The further displacement boundary conditions are

$$\begin{aligned} u_x &= 0 \quad \text{on } \Gamma_{\text{top}} \times I, \\ u &= 0 \quad \text{on } \Gamma_{\text{bottom}} \times I, \\ \sigma(u)n &= 0 \quad \text{on } \Gamma_{\text{sides}} \times I. \end{aligned}$$

10.3.4 Initial condition

We prescribe no initial data, i. e.,

$$\varphi(0) = 1 \quad \text{in } B \times \{0\}.$$

In particular, there is no initial fracture. Thus, we will observe fracture nucleation.

10.3.5 Parameters

As model parameters, we choose $\gamma = 1$, $\kappa = 10^{-10}h$, $\varepsilon = 2h$ [mm]. We notice that the maximum number of possible augmented Lagrangian iterations is 20 and TOL_{AL} is variable. The Newton tolerance is $\text{TOL}_{\text{New}} = 10^{-8}$. As material parameters, we use $\mu = 80.77\,\text{kN/mm}^2$, $\lambda = 121.15\,\text{kN/mm}^2$ and $G_c = 2.7\,\text{N/mm}$.

The initial time step size is $k = 0.01\,\text{s}$. For adaptive time step control, we set the following parameters (see Section 9.6 for their meaning and final algorithm):

$$\begin{aligned} \text{TOL}_{\text{TS}} &= 10^{-2} \quad \text{and} \quad & \text{TOL}_{\text{TS}} &= 10^{-3}, \\ k_{\max} &= 10^{-2}, & k_{\min} &= 10^{-4}, \\ \gamma &= 1, & K &= 1/15. \end{aligned}$$

10.3.6 Quantity of interests

We study two goal functionals:
- Bulk energy

$$E_b = \int_B g(\varphi)\psi(e)\,dx. \tag{10.1}$$

- Crack energy

$$E_s(\varphi) = \frac{G_c}{2}\int_B \left(\frac{(1-\varphi)^2}{\varepsilon} + \varepsilon|\nabla\varphi|^2\right)dx, \tag{10.2}$$

  with the strain energy functional

$$\psi(e) := \mu\,\mathrm{tr}(e(u)^2) + \frac{1}{2}\lambda\,\mathrm{tr}(e(u))^2,$$

  with $|\nabla\varphi|^2 := \nabla\varphi\cdot\nabla\varphi$.

As goal functional for adaptive time step control, we choose

$$J(\varphi) := E_s(\varphi).$$

10.3.6.1 Numerical results and discussion

In the Figures 10.7 and 10.8 graphical solutions of the stress distribution and displacements are provided. The goal functional, its relative error and the adaptive time step sizes are plotted in Figure 10.9.

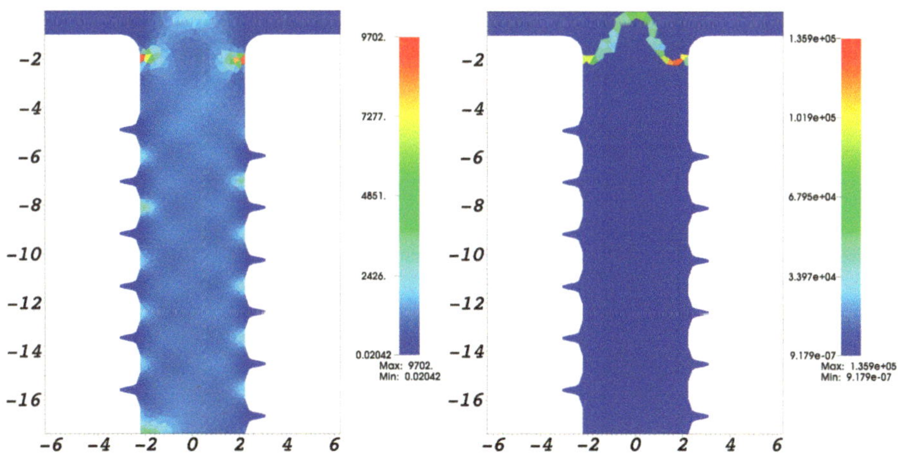

Figure 10.7: Screw simulations: numerical solutions at times $T = 9.92\text{e}{-2}$ s and $T = 1.03\text{e}{-1}$ s. Here, the stresses are shown, which indicate high values in the corners between head and screw piston.

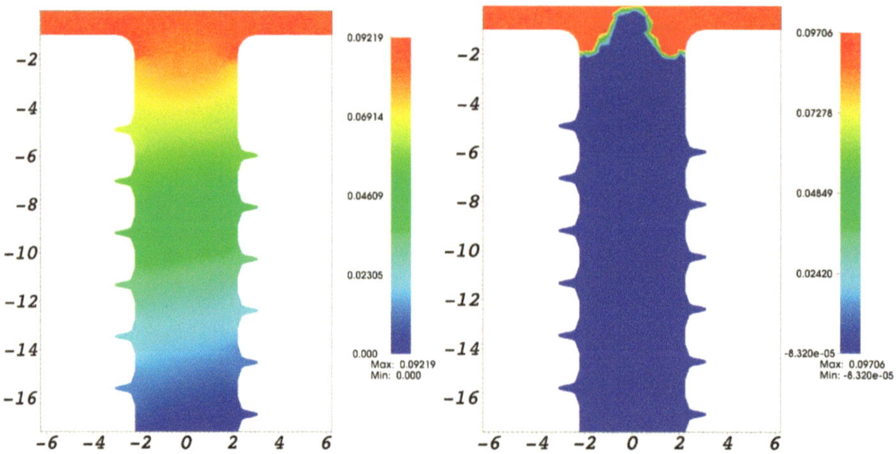

Figure 10.8: Screw simulations: numerical solutions at times $T = 9.92\mathrm{e}{-2}\,\mathrm{s}$ and $T = 1.03\mathrm{e}{-1}\,\mathrm{s}$. The principal displacement direction u_y is displayed.

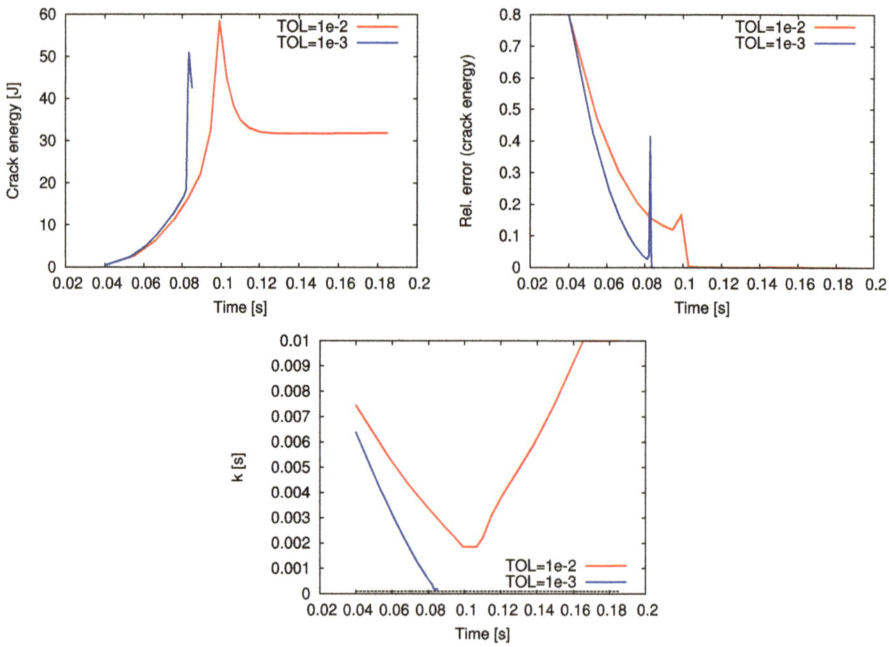

Figure 10.9: Screw simulations: evolution of the crack energy, the relative error and the corresponding time/loading step sizes k in order to satisfy the prescribed tolerances. We observe an increasing crack energy until the screw is broken. Then the crack energy remains constant and the relative error tends to zero as to be expected. Consequently, for the broken screw the time/loading step sizes increase again because the relative error is small anyway.

11 Multiphysics extensions involving fracture

In this chapter, we concentrate on extensions to multiphysics phase-field fracture. From the numerical modeling point of view, the concepts introduced in the Chapters 5, 7, and 9 can be directly employed. To this end, the main topics in this chapter will be on specific modeling aspects such as

- Interface conditions on the crack boundary $\partial\mathcal{C}$;
- Phase-field as level-set and design of indicator functions;
- Crack width;
- Interface conditions in the transition zone: pressurized and nonisothermal fractures;
- Global pressure equation for Darcy flow using generalized parabolic problems;
- Large deformations of solids in Lagrangian and Eulerian coordinates;
- Fluid-structure interaction coupled with phase-field fracture.

Our own published or accepted papers addressing these topics entirely or in parts are [323, 322, 321, 448, 285, 283, 281, 287, 280, 431, 443, 441, 437, 436]. This chapter aims to summarize these developments in a concise way, but new materials are added as well.

11.1 Interface definition and interface conditions

Throughout the entire chapter, the interface (Section 1.3) $\Gamma := \Gamma(t) := \overline{\Omega}_1(t) \cap \overline{\Omega}_2(t)$ plays a crucial role. We also use the other notation $\partial\mathcal{C}$ or Γ_i, specifically when we deal with several interfaces. We recall our definitions made in Section 3.3.2. The usual interface coupling conditions for the solution variable u are (Definition 32):

$$\begin{aligned} u_1 &= u_2 \quad \text{on } \Gamma, \\ \sigma_1 n &= \sigma_2 n \quad \text{on } \Gamma, \end{aligned}$$

with the normal vector n, e. g., $n := n_1$ and the relation $n_1 = -n_2$. We also recall that we assume $\varepsilon > 0$.

11.2 Interpreting phase-field as level-set for capturing the interface

As just motivated, in numerous applications, additional physics inside the fracture are of interest. For $\varepsilon > 0$ (see Figure 2.1 in Chapter 2), the fracture decomposes the domain B into two subdomains. One crucial aspect is now the fixed-mesh feature of phase-field. The interface Γ (i. e., the fracture boundary $\partial\mathcal{C}$) will cut through mesh elements;

https://doi.org/10.1515/9783110497397-011

see back in the Figures 1.7 and 1.8 accompanied with discussions in Section 1.3.4. If we want to describe additional physics in the red subdomain of the left subfigure of Figure 1.8, we deal with the following challenges:

1. The interface is only given in smeared form;
2. We must capture the interface with the help of an indicator function;
3. The indicator function enters as space-time coefficient into the equation to be described inside the fracture.

In the following, we provide some results how the phase-field variable φ can be used as level-set function to decompose the domain into a fracture part Ω_F and a nonbroken subdomain Ω_R.

11.2.1 Constructing and working with indicator functions using φ

Definition 68 (Indicator functions). Given the domain B and the fracture C with the boundary ∂C, we define the following characteristic functions:

$$\begin{aligned} \chi_R &:= \chi_R(x,t) = 1 \quad \text{in } \Omega_R(t), \\ \chi_F &:= \chi_F(x,t) = 1 \quad \text{in } \Omega_F(t), \end{aligned}$$

with the relation

$$\chi_F = 1 - \chi_R.$$

Sometimes, we denote in this context Ω_R simply as Ω; see Figure 2.1 (Section 2).

The next result allows to use χ_R and χ_F to decompose the domain B into two subdomains Ω_R and Ω_F.

Proposition 80 ([285, 431]: Constructing indicator functions using φ). *The indicator functions χ_F and χ_R can be computed as*

$$\chi_F(x,t,\varphi) := \begin{cases} 1, & \text{if } \varphi(x,t) \le x_1 \\ 0, & \text{if } \varphi(x,t) \ge x_2 \\ -\frac{(\varphi - x_2)}{(x_2 - x_1)}, & \text{otherwise} \end{cases}$$

$$\chi_R(x,t,\varphi) := \begin{cases} 1, & \text{if } \varphi(x,t) \ge x_2 \\ 0, & \text{if } \varphi(x,t) \le x_1 \\ \frac{(\varphi - x_1)}{(x_2 - x_1)}, & \text{otherwise} \end{cases}$$

For instance, $x_1 = 0.3$ and $x_2 = 0.7$.

Proof. We know that $0 \le \varphi \le 1$. For the limit cases x_1 and x_2 the above settings are trivial. In between, we use linear interpolations to construct continuous extensions inside the transition zone. □

In the above construction, we notice that φ is itself an unknown variable. In a fully implicit context, this would mean to differentiate χ_R and χ_F with respect to φ when Newton-type methods are employed for solving the overall nonlinear problem. For this reason, often for such problems, methods such as iterative coupling (Section 7.5) are employed.

Algorithm 22 (Iterative coupling with indicator functions). Let $U = (u, \varphi) \in X$ be the solution of the phase-field fracture problem solved by $A_U(U,p)(\Psi) = 0$. Moreover, let for instance $p \in V_p$ be the solution via $A_p(U,p)(\psi_p) = 0$ of some additional physics problem with $p := p_F\chi_F(\varphi) + p_R\chi_R(\varphi)$. For a prototype geometric configuration, we refer the reader to Figure 11.1. Then a basic iterative procedure is: For $j = 1, 2, 3, \ldots$:

1. Given $p^{j-1} \in V_p$. Find $U^j \in X$ such that

$$A_U(U^j, p^{j-1})(\Psi) = 0 \quad \forall \Psi \in X.$$

2. Given $U^j = (u^j, \varphi^j) \in X$. Find $p^j \in V_p$ such that

$$A_p(U^j, p^j)(\psi_p) = 0 \quad \forall \psi_p \in V_p.$$

Specifically, in Step 2, we have $p^j = p_F^j\chi_F(\varphi^j) + p_R^j\chi_R(\varphi^j)$ where φ^j is explicitly known from Step 1.

11.2.2 Constructing a level-set function

With the help of the previous definitions, we have been able to construct indicator functions. But still, the interface Γ has not been yet explicitly defined. At this point, a look into level-set methods is recommended. These were introduced by Osher/Sethian in the late 1980s [346] and the method has been extremely successful since then.

Working with smeared interfaces, we need to define an isosurface Γ (the zero level-set) representing the fracture boundary. This can be conveniently achieved with the help of the phase-field function φ by defining a level-set variable φ_{ls}.

Proposition 81 ([286]: Constructing a level-set variable with phase-field). *Fix a value* $0 < c_{\mathrm{ls}} < 1$ *determining the fracture boundary isosurface. Then a level-set variable is defined as*

$$\varphi_{\mathrm{ls}} := \varphi - c_{\mathrm{ls}}.$$

A possible choice is $c_{\mathrm{ls}} := x_1$ *from before; there* $x_1 = 0.3$.

Definition 69 (Fracture boundary $\partial\mathcal{C}$). With the help of the level-set variable, we define the fracture boundary surface as zero level set:

$$\partial\mathcal{C} := \{x \in B \mid \varphi_{\mathrm{ls}}(x) = 0\}.$$

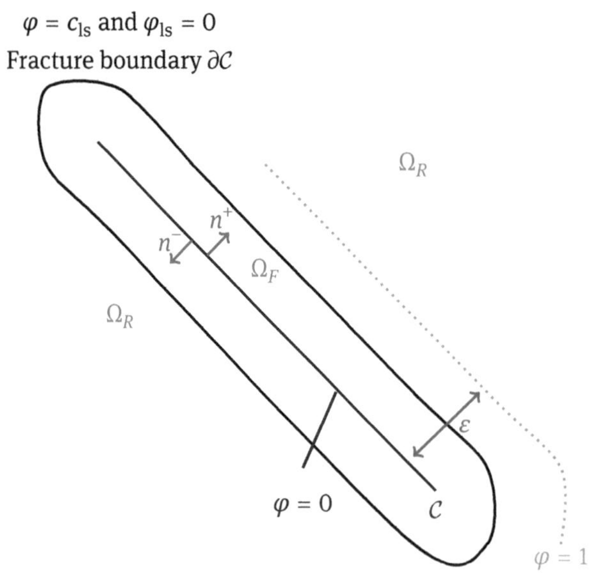

Figure 11.1: Using phase-field as level-set for decomposing the domain B into Ω_F and Ω_R.

11.2.3 Example

Let us revisit Section 3.3.3.5. We aim to solve two Poisson-type problems, but the interface Γ is only implicitly defined with the help of the indicator function χ. Consequently, the domains Ω_1 and Ω_2 are not explicitly known. Formally, we obtain

Proposition 82. *Given the indicator functions χ_1 and χ_2. Find $u = (u_1, u_2) \in X$ with $u_1 := \chi_1 u$ and $u_2 := \chi_2 u$ such that*

$$A(u, \psi) = F(\psi) \quad \forall \psi \in X,$$

where

$$A(u, \psi) := (\chi_1 \nabla u, \nabla \psi) + (\chi_2 \nabla u, \nabla \psi).$$

11.3 Crack width computations and total crack volume

In various applications, the crack width, also known as aperture or crack opening displacement (COD), is of interest.

Definition 70. Formally, the crack width on the lower-dimensional fracture $\mathcal{C}$, $\varepsilon = 0$, is defined as

$$w = [u] \cdot n = u_1 n - u_2 n. \tag{11.1}$$

Here, we use the usual notation $[\cdot]$ for denoting a jump. Furthermore, u_1 and u_2 denote the displacements on each side of $\mathcal{C}$, namely $u_1 := u|_{\mathcal{C}^+}$ and $u_2 := u|_{\mathcal{C}^-}$.

In phase-field fracture, the true fracture middle line C is usually not explicitly known and a fracture boundary ∂C can be constructed using the ideas from the previous section. We basically need to know three things:

1. Defining an interface ∂C (Section 11.2);
2. Defining a normal vector n on ∂C;
3. Interpolating the displacements u to compute the width w.

Proposition 83 (COD computation with phase-field in cartesian coordinates). *Let the domain be defined as $B = (a,b)\times(c,d)$. Let the fracture C be aligned with the x-coordinate axis in a Cartesian system. Thus, the normal direction n equals to the y-coordinate axis. The crack opening displacement can be computed in a point $x_0 \in B$ as*

$$\mathrm{COD}(x_0) := w := w(x_0, y) = \int_c^d u(x_0, y) \cdot \nabla\varphi(x_0, y)\, dy. \tag{11.2}$$

Let $\Gamma := C$ the fracture, $(\varepsilon = 0)$. A sketch of a proof for the Γ limit for the term

$$\int_\Gamma [u] \cdot n\, ds \approx \int_B u \cdot \nabla\varphi\, dx \tag{11.3}$$

is provided in [112, p. 963]. In the following, we derive a COD formula for cracks, which are not aligned with the coordinate axes. Let $x_c \in \Gamma$ be the point in which we want to compute the COD. We extend from x_c to points $x_\varepsilon^+ \in B$ and $x_\varepsilon^- \in B$ each by distance ε into the direction n normal to Γ. Then

$$\mathrm{COD}(x_c) = \int_{x_\varepsilon^+}^{x_\varepsilon^-} u(x, y) \cdot \nabla\varphi(x, y)\, dn. \tag{11.4}$$

In the special case that n is parallel to the y-axis, we obtain (11.2).

To realize (11.4) in practice, we first need to compute the normal vector perpendicular to the fracture surface ∂C. We use simply what is suggested in level-set methods and is finally based on calculus. Our specific ideas were originally inspired by [338] and further developed in [286].

Definition 71 (Normal vector). The normal vector is defined by

$$n := \frac{\nabla\varphi}{\|\nabla\varphi\|}.$$

With the help of this definition, we can realize (11.4). The numerical design is however challenging for mainly two reasons:

- The mesh is fixed and arbitrary normal vectors n require integration rules in that specific direction. If the width must be known on the entire fracture Ω_F, then this procedure appears impracticable.

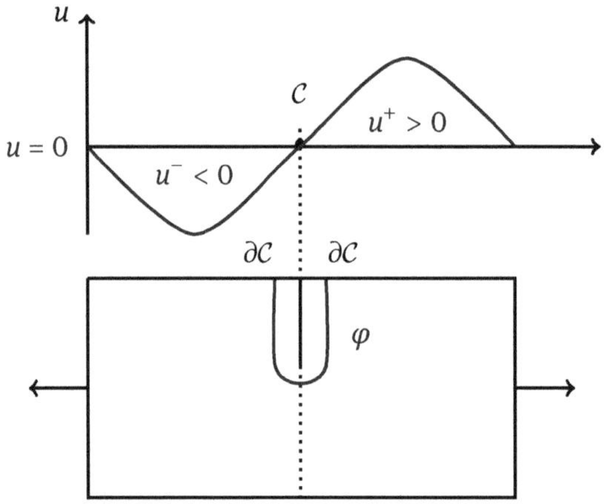

Figure 11.2: Example of a profile of the displacement u and the phase-field variable φ for a crack subject to tensile forces normal to the plane of the crack (Mode I opening).

- The displacement has a sign change and for $\varepsilon > 0$, it is $u = 0$ on the middle isosurface $\mathcal{C}$ (see Figure 11.2). Therefore, if the width should be known pointwise, in particular on $\mathcal{C}$, we have a problem.

For the following, we assume that we mainly deal with Mode I opening (tensile stresses normal to the fracture). We proposed in [286] a five-step procedure:

1. Compute the displacements $u : B \to \mathbb{R}^d$;
2. Take the absolute value $\tilde{w} := \|u\|_{L^\infty}$;
3. Assume symmetry $u^+ = -u^-$ (see again Figure 11.2) and set $w_\varepsilon := 2\tilde{w}$. The justification is

$$w = [u] \cdot n = u^+ n - u^- n = 2u^+ n \approx 2\tilde{w} =: w_\varepsilon.$$

4. Set $w_D := w_\varepsilon$ on the crack boundary $\partial\mathcal{C}$;
5. Interpolate from w_D inside the crack region Ω_F (see Figure 11.3). We notice that there is an approximation error of order ε as illustrated in Figure 11.3.

In the following, we consider Step 5 in more detail. The extension from $\partial\mathcal{C}$ inside Ω_F can be achieved for instance with:

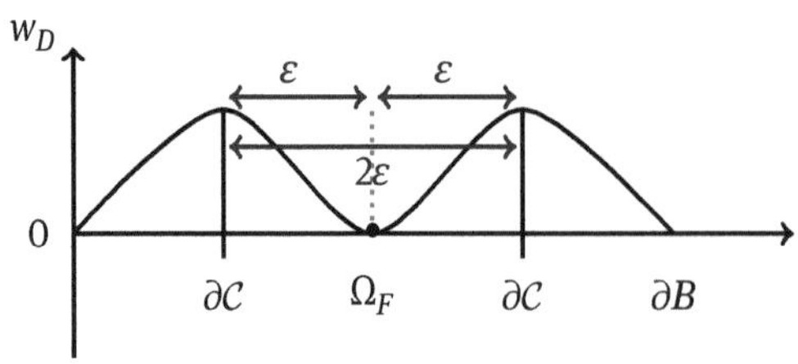

Figure 11.3: Profile of w_D extending from $\partial\mathcal{C}$ inside Ω_F.

1. Constant interpolation if $u^+ = -u^-$:

$$\text{Given } w_D \quad \text{on } \partial\mathcal{C} \quad \rightarrow \quad w_\varepsilon := w_D \quad \text{in } \Omega_F.$$

2. Laplace interpolation by solving a boundary value problem: Find $w_\varepsilon : B \rightarrow \mathbb{R}$ such that

$$\begin{aligned} -\Delta w_\varepsilon &= g \quad \text{in } B, \\ w_\varepsilon &= w_D \quad \text{on } \partial\mathcal{C}, \\ w_\varepsilon &= 0 \quad \text{on } \partial B, \end{aligned}$$

with a forcing function $g := \beta \|w_D\|_{L^\infty}$ and $\beta > 0$. This problem only needs to be solved in the fracture domain Ω_F (see Figure 11.3) but for computational convenience, we solve on the entire domain B.

Finally, (11.3) also yields the total crack volume.

Definition 72 (TCV – total crack volume). Let B be the computational domain. The total crack volume is computed by

$$\text{TCV}_\varepsilon = \int_B u \cdot \nabla\varphi \, dx. \tag{11.5}$$

11.4 Pressurized and nonisothermal phase-field fractures

11.4.1 Pressurized fractures in elasticity

We now add a pressure inside the fracture as a first extension toward multiphysics. This pressure may come from a fluid (Darcy-type, Reynolds' equation, Stokes, Navier–Stokes) in the fracture. Possible applications are fractures in porous media, subsurface modeling, geothermal energy recovery, nuclear waste, groundwater flow, or biomedical applications as aortic dissections. Pressurized and fluid-filled phase-field fracture modeling has been subject to numerous phase-field fracture papers since the year 2012 as listed in Section 1.5.

11.4.1.1 Interface laws for fracture flow coupled to solids/poroelasticity

We first decompose the domain B into Ω_F and Ω_R using the indicator functions defined in Section 11.2. In Ω_F, we model fluid flow with the help of Stokes (e. g., [410, 201, 364]). In Ω_R, we assume elasticity as in the previous chapters of this monograph or later also poroelasticity as already considered in Section 3.4.

We start with an incompressible fluid of Stokes type in the fracture Ω_F.

Formulation 25 (Incompressible Stokes). Find $v : \Omega_F \to \mathbb{R}^d$ and $p : \Omega_F \to \mathbb{R}$ such that

$$-\nabla \cdot \sigma_f(v, p) = 0, \tag{11.6}$$

$$\nabla \cdot v = 0, \tag{11.7}$$

equipped with suitable boundary conditions and with the fluid stress tensor

$$\sigma_f = -pI + v\rho(\nabla v + \nabla v^T),$$

and the fluid density ρ and kinematic viscosity v.

Remark 77 (Notation). With the subscript f, we indicate "fluid" and with the subscript F we denote "fracture."

In the following, we propose interface laws for pressurized and fluid-filled fractures. Here, it is important to distinguish the following.

Definition 73 (Pressurized and fluid-filled fractures). We define the following terminologies:

1. Pressurized fractures: pressure p_F is fixed and given in Ω_F. No fluid equation is actually solved in Ω_F;
2. Fluid-filled fractures: pressure p_F is computed by solving a fluid equation in Ω_F.

Proposition 84 (Interface conditions pressurized fractures: elasticity). *The interface condition for coupling quasi-static elasticity and fracture flow is given by*

$$\sigma_R n = \sigma_F n \quad \text{on } \Gamma \quad \text{(Dynamic condition).} \tag{11.8}$$

Since no equation is actually solved in the fracture, we only need one interface condition (i. e., a boundary condition) in this situation.

Proposition 85 (Interface conditions pressurized fractures: poroelasticity). *The interface coupling conditions of coupling a porous medium with pressure p_R and a fracture pressure p_F are given by*

$$p_R = p_F \quad \text{on } \Gamma \quad \text{(Kinematic condition),} \tag{11.9}$$

$$\sigma_R n = \sigma_F n \quad \text{on } \Gamma \quad \text{(Dynamic condition).} \tag{11.10}$$

The kinematic condition in this case is not a true interface condition, because no pressure equation is actually solved in the fracture Ω_F. For this reason, as before the dynamic condition is sufficient. However, for the sake of modeling, we nonetheless ask for the kinematic condition. The reason becomes clear in Section 11.4.1.5.

Later in Section 11.5, when a pressure flow equation is solved in Ω_F, we will need also the kinematic condition.

Proposition 86 (Interface conditions fluid-filled fractures: poroelasticity). *The interface coupling conditions of coupling a porous medium with pressure p_R and a fracture flow equation yielding p_F are given by*

$$p_R = p_F \quad \text{on } \Gamma \quad \text{(Kinematic condition)}, \tag{11.11}$$

$$\sigma_R n = \sigma_F n \quad \text{on } \Gamma \quad \text{(Dynamic condition)}. \tag{11.12}$$

Remark 78. Interface conditions in porous media are delicate. For further information, we refer to [384, 246, 320, 102, 103, 91, 321].

The constitutive laws depend now on the medium. For Ω_R, we can have for instance:

1. Elasticity: In Ω_R, we deal with
$$\sigma_R := \sigma_s(u) = 2\mu e(u) + \lambda\,\mathrm{tr}(e)I. \tag{11.13}$$
2. Poroelasticity: In Ω_R, we deal with
$$\sigma_R := \sigma_s(u) - \alpha p I,$$
where $\alpha \in [0,1]$ is Biot's coefficient as used in Section 3.4. For general information on effective stresses in porous media, we refer for instance to [119].

For fluid flow, we present as well two ideas:

1. Usually a fracture is much smaller in height than in length. Consequently, Reynolds lubrication theory, e. g., [406], can be assumed. Here, friction forces $\nabla v \approx 0$ do not play a role and the leading order of stress is the hydrostatic pressure:
$$\sigma_F = -pI. \tag{11.14}$$
Applying this idea to phase field fracture was originally proposed in [318] (later fully published as [324]).
2. Stokes flow in the fracture. We assume that viscous forces are present now:
$$\sigma_F = -pI + \rho_f \nu(\nabla v + \nabla v^T).$$

11.4.1.2 Including interface conditions into the energy functional

The previous interface laws must now be included into the PDEs, respectively the energy functional. As starting point Section 4.4.3 (Proposition 15) can be considered.

The dynamic interface condition is a nonhomogeneous Neumann condition, which will enter as boundary integral on the fracture boundary $\Gamma = \partial\mathcal{C}$. In addition, let us consider Neumann conditions on a part ∂B_N of the outer domain boundary. In total, we have

$$\int_{\partial B_N \cup \Gamma} \tau \cdot u\, ds.$$

The modified energy functional reads

$$E_b(u,\varphi) = \frac{1}{2}\int_B g(\varphi)\sigma_R(u) : e(u)\,dx - \int_{\partial B_N \cup \Gamma} \tau \cdot u\,ds + \frac{1}{2}\int_B G_c\left(\frac{1}{\varepsilon}(1-\varphi)^2 + \varepsilon|\nabla\varphi|^2\right)dx, \tag{11.15}$$

where τ is the surface force on the Neumann boundary. As we previously discussed in Section 11.2, the fracture boundary is a priori not known, but can be reconstructed. However, the level-set interface will cut through elements for which either cut-cell approaches or similar techniques must be adopted (see Section 1.3.4). An elegant mathematical alternative is to employ the divergence Theorem 1 (Section 2.2). This idea was originally worked out in [318] later published in [323, 324]. We define the interface traction vector

$$\tau = \sigma_R n \quad \text{on } \Gamma,$$

and assume some other traction forces $\tau : \partial B_N \to \mathbb{R}^d$ on the outer boundary ∂B_N. We also recall

$$\sigma_R n = \sigma_F n \quad \text{on } \Gamma,$$

defined by (11.13) and (11.14), respectively. In the following, we concentrate on the fracture boundary Γ:

$$\begin{aligned} -\int_\Gamma \tau \cdot u\,ds &= -\int_\Gamma \sigma_F n \cdot u\,ds \\ &= \int_\Gamma pn \cdot u\,ds \quad \text{(Interface condition)} \\ &= \int_{\Omega_R} \nabla \cdot (pu)\,dx - \int_{\partial B} pn \cdot u\,ds \quad \text{(Gauss divergence)} \\ &= \int_{\Omega_R} (u\nabla p + p\nabla \cdot u)\,dx - \int_{\partial B} pn \cdot u\,ds. \end{aligned} \tag{11.16}$$

At this point, we assume for the following that we work with homogeneous Dirichlet displacement conditions on ∂B and neglect Neumann parts ∂B_N in the remainder of the current subsection. We come back to them in Section 11.4.1.4. For the moment being, we therefore have

$$-\int_{\partial B} pn \cdot u = 0. \tag{11.17}$$

Second, we only assume slight spatial variations or even constant pressures inside the fracture such that

$$\nabla p \approx 0 \Rightarrow \int_{\Omega_R} u \nabla p \, dx \approx 0.$$

Consequently, (11.16) yields

$$-\int_{\Gamma} \tau \cdot u \, ds = \int_{\Omega_R} p \nabla \cdot u \, dx. \tag{11.18}$$

We finally need to derive a phase-field interpretation and extend the previous integrals from Ω_R to B [318, 323, 324]:

$$\int_{\Omega_R} p \nabla \cdot u \, dx \quad \to \int_{B} g(\varphi) p \nabla \cdot u \, dx. \tag{11.19}$$

With these calculations, we obtain the following result.

Proposition 87. *Let $p \in L^{\infty}(B)$ be given and assume $\nabla p \approx 0$. For $u \in H_0^1(B)$ and $\varphi \in K$ we have the pressurized energy functional:*

$$\begin{aligned} E_T(u, \varphi) = & \int_B g(\varphi) \sigma_R(u) : e(u) + \int_B g(\varphi) p \nabla \cdot u \, dx \\ & + \int_B G_c \left(\frac{1}{2\varepsilon} (1 - \varphi)^2 + \frac{\varepsilon}{2} |\nabla \varphi|^2 \right) dx. \end{aligned} \tag{11.20}$$

In a next step, similar to Section 4.4.3.3, we differentiate this energy functional with respect to u and φ to obtain the Euler–Lagrange system:

Proposition 88 (Pressurized fractures in elasticity). *Set $\sigma := \sigma_R$. Let $p \in L^{\infty}(B)$ and φ^{n-1} be given. Find $(u, \varphi) \in H_0^1(B) \times K$ such that*

$$(g(\varphi)\sigma(u), e(w)) + (g(\varphi)p, \nabla \cdot w) = 0 \quad \forall w \in H_0^1(B), \tag{11.21}$$

and

$$\begin{aligned} & (1 - \kappa)(\varphi \, \sigma(u) : e(u), \psi - \varphi) + 2(1 - \kappa)(\varphi \, p \, \nabla \cdot u, \psi - \varphi) \\ & + \left(-\frac{1}{\varepsilon}(G_c(1 - \varphi), \psi - \varphi) + \varepsilon(G_c \nabla \varphi, \nabla(\psi - \varphi)) \right) \geq 0 \quad \forall \psi \in K. \end{aligned} \tag{11.22}$$

This setting can still be quasi-stationary due to loading-dependent boundary conditions or a time-dependent pressure $p := p(t) : B \times I \to \mathbb{R}$. An example is the form $p(t) = t\bar{p}$ with a given $\bar{p}$ and time $t \in I$. If the given data is time-independent, still a few quasi-stationary steps must be taken in order to satisfy the crack irreversibility constraint.

Remark 79. In [318, 324], we have worked in the pressure terms with φ^2 rather than $g(\varphi)$. For small κ, the result is numerically the same. As long as $g(\varphi)$ is used in the bulk energy term to avoid singular system matrices in the discretization, using φ^2 in the pressure terms is fine. Moreover, we have neglected cut-off terms for φ in order to avoid negative entries, where they may yield wrong results for pressurized fractures [318, 323, 324].

11.4.1.3 The strong form

In order to complete our derivations, we derive the strong form of Proposition 88 in this section. Using integration by parts, we obtain a quasi-stationary elliptic system for the displacements and the phase-field variable.

Proposition 89. *Let $p : B \to \mathbb{R}$ be given. The displacement equation reads: Find $u : B \to \mathbb{R}^d$:*

$$\begin{aligned} -\nabla \cdot (g(\varphi)\sigma(u)) - \nabla \cdot (\varphi^2 p) &= 0 \quad \text{in } B \\ u &= 0 \quad \text{on } \partial B. \end{aligned}$$

The phase-field system consists of three parts: the partial differential equation, the inequality constraint, and the complementarity condition: Find $\varphi : B \to [0,1]$ such that

$$\begin{aligned} (1-\kappa)\sigma(u) : e(u)\,\varphi - G_c\varepsilon\Delta\varphi - \frac{G_c}{\varepsilon}(1-\varphi) + 2\varphi p\nabla\cdot u &\le 0 \quad \text{in } B, \\ \partial_t\varphi &\le 0 \quad \text{in } B, \\ \left[(1-\kappa)\sigma(u) : e(u)\,\varphi - G_c\varepsilon\Delta\varphi - \frac{G_c}{\varepsilon}(1-\varphi) + 2\varphi p\nabla\cdot u\right]\cdot\partial_t\varphi &= 0 \quad \text{in } B, \\ \partial_n\varphi &= 0 \quad \text{on } \partial B. \end{aligned}$$

11.4.1.4 Including the pressure gradient and boundary pressures

We now include the two remaining terms in (11.16) and employ the same ideas as before. In addition, we assume that we deal with Dirichlet and Neumann conditions and split $\partial B = \partial B_D \cup \partial B_N$.

Proposition 90. *Set $\sigma := \sigma_R$. Let traction conditions τ on the Neumann boundary ∂B_N be given. Moreover, let $p \in W^{1,\infty}(B)$ and φ^{n-1} be given. For $u \in V_u^D$ and $\varphi \in K$ we have the pressurized energy functional:*

$$\begin{aligned} E_T(u,\varphi) = \frac{1}{2}\int_B g(\varphi)\sigma(u) : e(u)\,dx + \int_B g(\varphi)(p\nabla\cdot u + \nabla p\cdot u)\,dx - \int_{\partial B_N} (\tau + pn)\cdot u\,ds \\ + \int_B G_c\Big(\frac{1}{2\varepsilon}(1-\varphi)^2 + \frac{\varepsilon}{2}|\nabla\varphi|^2\Big)\,dx. \end{aligned} \tag{11.23}$$

The corresponding Euler–Lagrange system reads as follows:

Proposition 91. *Let $p \in W^{1,\infty}(B)$ and φ^{n-1} be given. Find $(u,\varphi) \in V_u^D \times K$ such that*

$$(g(\varphi)\sigma(u), e(w)) + (g(\varphi)p, \nabla \cdot w) + (g(\varphi)\nabla p, w) \\ - \int_{\partial B_N} (\tau + pn) \cdot w \, ds = 0 \quad \forall w \in V_u^0, \tag{11.24}$$

and

$$(1-\kappa)(\varphi\, \sigma(u) : e(u), \psi - \varphi) + 2(1-\kappa)(\varphi\, (p\, \nabla \cdot u + \nabla p \cdot u), \psi - \varphi) \\ + G_c\Big(-\frac{1}{\varepsilon}(1-\varphi, \psi-\varphi) + \varepsilon(\nabla\varphi, \nabla(\psi - \varphi))\Big) \geq 0 \quad \forall \psi \in K. \tag{11.25}$$

11.4.1.5 Pressurized fractures in poroelasticity

In poroelasticity, we work with

$$\sigma := \sigma_R = \sigma_s - \alpha p I,$$

and use now the interface condition formulated in Proposition 85. We obtain immediately

Proposition 92. *Let traction conditions τ on the Neumann boundary ∂B_N be given. Moreover, let $p \in W^{1,\infty}(B)$ and φ^{n-1} be given. For $u \in V_u^D$ and $\varphi \in K$ we have the pressurized energy functional:*

$$E_T(u,\varphi) = \frac{1}{2}\int_B g(\varphi)\sigma(u) : e(u)\, dx + \int_B g(\varphi)((1-\alpha)p\nabla \cdot u + \nabla p \cdot u)\, dx - \int_{\partial B_N} (\tau + pn) \cdot u \, ds \\ + \int_B G_c\Big(\frac{1}{2\varepsilon}(1-\varphi)^2 + \frac{\varepsilon}{2}|\nabla\varphi|^2\Big)\, dx. \tag{11.26}$$

In the middle integral

$$\int_B g(\varphi)((1-\alpha)p\nabla \cdot u + \nabla p \cdot u)\, dx,$$

we have used the kinematic condition $p := p_R = p_F$ (Proposition 85).

Proposition 93. *Let $p \in W^{1,\infty}(B)$ and φ^{n-1} be given. Find $(u,\varphi) \in V_u^D \times K$ such that*

$$(g(\varphi)\sigma(u), e(w)) + (1-\alpha)(g(\varphi)p, \nabla \cdot w) + (g(\varphi)\nabla p, w) \\ - \int_{\partial B_N} (\tau + pn) \cdot w \, ds = 0 \quad \forall w \in V_u^0, \tag{11.27}$$

and

$$\begin{aligned}&(1-\kappa)(\varphi\,\sigma(u):e(u),\psi-\varphi)+2(1-\kappa)(\varphi\,((1-\alpha)p\,\nabla\cdot u+\nabla p\cdot u),\psi-\varphi)\\&+G_c\Big(-\frac{1}{\varepsilon}(1-\varphi,\psi-\varphi)+\varepsilon(\nabla\varphi,\nabla(\psi-\varphi))\Big)\geq 0\quad\forall\psi\in K.\end{aligned}\tag{11.28}$$

Remark 80. It is clear that incremental or time-dependent processes are readily included in all the previously presented energy functionals by computing $U := U^n$ for $n = 1, 2, 3, \dots, N$ as done in the previous chapters of this monograph. Time-dependencies may be introduced through time-dependent pressures $p := p(t) : B \times I \to \mathbb{R}$ or boundary conditions.

11.4.2 Nonisothermal fractures

In this section, we recapitulate the work from [343] in which we combined nonisothermal derivations from [416] with the pressure interface laws we just considered in Section 11.4.1. To this end, we have to impose several interface conditions on the fracture boundary, namely for the pressure p and the temperature ϑ. In addition, we need to modify the constitutive law for the stress in the thermo-poroelastic medium.

Definition 74 (Constitutive stress tensor). Starting from isotropic linear thermoporoelasticity, we apply the constitutive expression of the stress tensor [119]:

$$\sigma := \sigma_R := \sigma(u, p, \vartheta) = \sigma_0 + \sigma_s(u) - \alpha_B(p - p_0)I - 3\alpha_T K(\vartheta - \vartheta_0)I.$$

Here, σ_0 is the initial stress, α_B the Biot's constant, p the pressure, p_0 the initial pressure, I the identity tensor, $3\alpha_T$ the volumetric skeleton thermal dilation coefficient (or thermal expansion coefficient), the bulk modulus $K := \frac{2}{3}\mu + \lambda$, ϑ the temperature, and ϑ_0 the initial temperature.

Hence, we formulate the energy of the bulk system:

$$\begin{aligned}E_b(u) &= \frac{1}{2}(\sigma_s(u), e(u))_\Omega - \langle\tau, u\rangle_{\partial B_N}\\&= \frac{1}{2}(\sigma_s(u), e(u))_\Omega - \langle\tau, u\rangle_{\partial B_N}\\&\quad - (\alpha_B(p - p_0), \nabla\cdot u) - (3\alpha_T K(\vartheta - \vartheta_0), \nabla\cdot u).\end{aligned}\tag{11.29}$$

In the presence of a fracture C, we obtain as before the regularized formulation.

Proposition 94. *Find $U \in V_u^D \times K$ such that*

$$\begin{aligned}E_\varepsilon(u,\varphi) &= \frac{1}{2}(g(\varphi)\sigma_s(u), e(u))_\Omega - \langle\tau, u\rangle_{\Gamma_n\cup\partial C}\\&\quad - (\alpha_B(p - p_0)\varphi^2, \nabla\cdot u) - (3\alpha_T K(\vartheta - \vartheta_0)\varphi^2, \nabla\cdot u)\end{aligned}$$

$$+\int_B G_c\left(\frac{1}{2\varepsilon}(1-\varphi)^2+\frac{\varepsilon}{2}|\nabla\varphi|^2\right)dx. \tag{11.30}$$

The associated Euler–Lagrange system to the functional (11.30) using an augmented Lagrangian penalization is

Proposition 95. *Given $p, p_0 \in L^\infty(B)$, and $\vartheta, \vartheta_0 \in L^\infty(B)$ and φ^{n-1}. For the loading steps $n = 1, 2, 3, \ldots, N$, find $u \in V_u^D$ such that*

$$\begin{aligned} A_1(U)(w) &:= (g(\varphi)\sigma_s(u), e(w)) - \langle\tau, w\rangle \\ &\quad - (\alpha_B(p-p_0)\varphi^2, \nabla\cdot w) - (3\alpha_T K(\vartheta-\vartheta_0)\varphi^2, \nabla\cdot w) = 0 \quad \forall w \in V_u^0. \end{aligned}$$

Find $\varphi \in V_\varphi$ such that

$$\begin{aligned} A_2(U)(\psi) &:= (1-\kappa)(\varphi\sigma_s(u) : e(u), \psi) \\ &\quad - 2(\alpha_B - 1)((p-p_0)\nabla\cdot u\,\varphi, \psi) - 2(3\alpha_T K)((\vartheta-\vartheta_0)\nabla\cdot u\,\varphi, \psi) \\ &\quad - G_c\left(\frac{1}{\varepsilon}(1-\varphi, \psi) - \varepsilon(\nabla\varphi, \nabla\psi)\right) \\ &\quad + ([\Xi + \gamma(\varphi - \varphi^{n-1})]^+, \psi) = 0 \quad \forall\psi \in V_\varphi. \end{aligned}$$

11.4.2.1 Modeling fracture pressure and temperature as interface law

The energy functional (11.30) and the corresponding Euler–Lagrange system are yet incomplete because only mechanisms in the porous medium are taken into account so far. In the following, we consider pressure and temperature variations in the fracture as well, acting as additional normal stresses on the interface between the porous medium and the fracture. Thus, we get a modification of the normal stress term $\langle\tau, u\rangle_{\partial B_N}$ where $\partial B_N := \Gamma_N \cup \partial\mathcal{C}$.

We follow [318], to derive an expression including temperature effects. For convenience of understanding, we also recapitulate the steps of the pressure and develop the modified terms simultaneously. Assuming that the fracture is a zone of high permeability and its width much smaller than its length, the leading order of the pressure stress in $\mathcal{C}$ is (Section 11.4.1)

$$-(p-p_0)I. \tag{11.31}$$

In addition, we assume (similar to [416]) that the temperature stress in $\mathcal{C}$ is given by

$$C_\vartheta(x,t)(\vartheta_F - \vartheta_0)I. \tag{11.32}$$

The constant $C_\vartheta(x,t)$ is obtained by following [416] to get a relation for the temperature near the fracture. We briefly sketch the basic idea. Here, the heat conduction equation is taken while neglecting convection-dominated terms

$$\partial_t\vartheta - \kappa\Delta\vartheta = 0, \tag{11.33}$$

with the thermal diffusivity $\kappa = \frac{K_r}{\rho_r C_r}$, where ρ_r is the rock density, C_r is the specific heat, K_r thermal conductivity. With the help of the heat conduction equation, a parameter measuring the gradient of the temperature at the fracture interface is derived. This law provides a heuristic argument for the temperature evolution without solving the full temperature equation in the whole domain. Specifically,

$$C_\vartheta = A_\vartheta\left(\frac{\lambda_\vartheta}{2\lambda_\vartheta + 1}\right), \quad A_\vartheta = \frac{E\beta}{1 - \nu_s}$$

where β is a linear thermal expansion coefficient, E Youngs modulus as before, ν_s Poisson's ratio. The parameter λ_ϑ is derived by Hagoort [212] based on the heat conduction equation. Then, [416, equation (42)]

$$\lambda_\vartheta = \sinh^{-1}\left(\frac{\gamma_\vartheta}{0.5 l_0}\sqrt{\pi\kappa t}\right),$$

where $1 \le \gamma_\vartheta \le \frac{4}{\pi}$ and l_0 is the half-length of the crack.

Remark 81. The extension to solving the thermal energy transport equation in the fracture and porous medium coupled to our phase-field approach is subject of ongoing research. Related ideas have been presented by [268, 196]. An example of a transport equation is

$$\rho_r C_r \partial_t \vartheta + \alpha_p \vartheta \partial_t p - \nabla \cdot (K_r \nabla \vartheta) + \nabla \cdot (\alpha_c \vartheta K_p \nabla p) = q_\vartheta,$$

where p is the pressure, α_c is the convective heat capacity, α_p the link between the pore pressure and the temperature, K_p fluid's permeability tensor, and q_ϑ are heat sources or injection. Setting $\alpha_p = 0$ and neglecting the transport term, we obtain equation (11.33).

Remark 82. For recent results in modeling, mathematical analysis and simulations of nonfractured thermo-poroelasticity, we refer the reader to [424, 426].

We pursue in the following by deriving an interface relationship between the porous medium Ω and the fracture boundary ∂C. The continuity of normal stresses on the crack boundary ∂C can be stated as

$$\sigma_R(u, p, \vartheta) n = \sigma_F(p, \vartheta) n.$$

Together with the kinematic conditions we obtain

Proposition 96 (Interface laws for pressurized nonisothermal fracture). *The interface coupling conditions of coupling a nonisothermal porous medium with pressure p_R, temperature ϑ_R and a fracture pressure p_F, fracture temperature ϑ_F are proposed as*

$$p_R = p_F \quad \text{on } \Gamma \qquad \text{(Kinematic condition)},$$

$$\vartheta_R = \vartheta_F \quad \text{on } \Gamma \qquad \textit{(Kinematic condition)},$$
$$\sigma_R(u,p,\vartheta)n = \sigma_F(p,\vartheta)n \quad \text{on } \Gamma \quad \textit{(Dynamic condition)},$$

with

$$\sigma_R = \sigma_0 + \sigma_s - \alpha_B(p - p_0)I - 3\alpha_\vartheta K(\vartheta - \vartheta_0)I,$$

and

$$\sigma_F = -(p - p_0) + C_\vartheta(\vartheta_F - \vartheta_0).$$

Similar to Section 11.4.1.2, we apply Gauss' divergence theorem to rewrite the traction surface integrals as domain integrals in order to combine them with the porous medium terms. For the traction term in (11.30), we obtain

$$\begin{aligned}
\langle \sigma n, u\rangle_{\partial C} &= -\langle (p - p_0)n, u\rangle_{\partial C} + \langle C_\vartheta(\vartheta - \vartheta_0)n, u\rangle_{\partial C} \\
&\quad - (\nabla\cdot(p - p_0)u) + \langle (p - p_0)n, u\rangle_{\Gamma_N} - (C_\vartheta\nabla\cdot(\vartheta - \vartheta_0)u) \\
&\quad + \langle C_\vartheta(\vartheta - \vartheta_0)n, u\rangle_{\Gamma_N} \\
&= -((p - p_0), \nabla\cdot u) - (\nabla(p - p_0), u) + \langle (p - p_0)n, w\rangle_{\Gamma_N} \\
&\quad - (C_\vartheta(\vartheta - \vartheta_0), \nabla\cdot u) - (C_\vartheta\nabla(\vartheta - \vartheta_0), u) + \langle (\vartheta - \vartheta_0)n, u\rangle_{\Gamma_N}.
\end{aligned}$$

These terms are now defined in the domain B and are combined with the existing terms in (11.30) in the porous medium:

$$(\alpha_B(p - p_0), \nabla\cdot u),$$
$$(3\alpha_\vartheta K(\vartheta - \vartheta_0), \nabla\cdot u).$$

Then

$$\begin{aligned}
&\langle \sigma n, u\rangle_{\partial C} + (\alpha_B(p - p_0), \nabla\cdot u) + (3\alpha_\vartheta K(\vartheta - \vartheta_0), \nabla\cdot u) \\
&= (\alpha_B(p - p_0), \nabla\cdot u) \\
&\quad - ((p - p_0), \nabla\cdot u) - (\nabla(p - p_0), u) + \langle (p - p_0)n, u\rangle_{\Gamma_N} \\
&\quad + (3\alpha_\vartheta K(\vartheta - \vartheta_0), \nabla\cdot u) \\
&\quad - (C_\vartheta(\vartheta - \vartheta_0), \nabla\cdot u) - (C_\vartheta\nabla(\vartheta - \vartheta_0), u) + \langle (\vartheta - \vartheta_0)n, u\rangle_{\Gamma_N}.
\end{aligned}$$

In the final step, we combine the previous considerations with a (possibly) given traction on Γ_N. In functional (11.30), we have

$$\begin{aligned}
&- \langle \tau, u\rangle_{\Gamma_N\cup\partial C} - (\alpha_B(p - p_0), \nabla\cdot u) - (3\alpha_\vartheta K(\vartheta - \vartheta_0), \nabla\cdot u) \\
&= -\langle \tau, u\rangle_{\Gamma_N} - \langle \sigma n, u\rangle_C - (\alpha_B(p - p_0), \nabla\cdot u) - (3\alpha_\vartheta K(\vartheta - \vartheta_0), \nabla\cdot u) \\
&= -\langle \tau, u\rangle_{\Gamma_N} - (\alpha_B(p - p_0), \nabla\cdot u) \\
&\quad + ((p - p_0), \nabla\cdot u) + (\nabla(p - p_0), u) - \langle (p - p_0)n, u\rangle_{\Gamma_N}
\end{aligned}$$

$$
\begin{aligned}
&- (3\alpha_\vartheta K(\vartheta - \vartheta_0), \nabla \cdot u) \\
&- (C_\vartheta(\vartheta - \vartheta_0), \nabla \cdot u) - (C_\vartheta \nabla(\vartheta - \vartheta_0), u) + \langle C_\vartheta(\vartheta - \vartheta_0)n, u\rangle_{\Gamma_N}.
\end{aligned}
$$

These expressions are now inserted into the functional (11.30):

$$
\begin{aligned}
E_\varepsilon(u, \varphi) = {} & \frac{1}{2}(((1-\kappa)\varphi^2 + \kappa)\sigma_s(u), e(u))_\Omega - \langle \tilde{\tau}, u\rangle_{\Gamma_N} \\
& - ((\alpha_B - 1)(p - p_0)\varphi^2, \nabla \cdot w) + (\nabla(p - p_0)\varphi^2, w) \\
& - ((3\alpha_\vartheta K + C_\vartheta)(\vartheta_F - \vartheta_0)\varphi^2, \nabla \cdot w) - (C_\vartheta \nabla(\vartheta - \vartheta_0)\varphi^2, w) \\
& + G_c\left(\frac{1}{2\varepsilon}\int_B (1-\varphi)^2\,dx + \frac{\varepsilon}{2}\int_B |\nabla\varphi|^2\,dx\right),
\end{aligned}
$$

where $\tilde{\tau} = \tau + (p - p_0)n - (\vartheta - \vartheta_0)n$, where τ are Neumann traction forces on ∂B_N.

Again, we derive the Euler–Lagrange equations (augmented Lagrangian version) by differentiating the energy functional with respect to u and φ.

Proposition 97. *Given $p \in W^{1,\infty}(B)$, $\vartheta \in W^{1,\infty}(B)$, and φ^{n-1}. For the loading steps $n = 1, 2, 3, \ldots, N$, find $u \in V_u^D$:*

$$
\begin{aligned}
A_1(U)(w) = {} & (((1-\kappa)\varphi^2 + \kappa)\sigma(u), e(w)) - \langle \tilde{\tau}, w\rangle_{\Gamma_N} \\
& + (1 - \alpha_B)((p - p_0)\varphi^2, \nabla \cdot w) + (\nabla(p - p_0)\varphi^2, w) \\
& - (3\alpha_\vartheta K + C_\vartheta)((\vartheta - \vartheta_0)\varphi^2, \nabla \cdot w) - (C_\vartheta \nabla(\vartheta_F - \vartheta_0)\varphi^2, w) \\
= {} & 0 \quad \forall w \in V_u^0,
\end{aligned}
$$

and find $\varphi \in V_\varphi$:

$$
\begin{aligned}
A_2(U)(\psi) = {} & (1-\kappa)(\varphi\sigma(u) : e(u), \psi) \\
& + 2(1 - \alpha_B)((p - p_0)\nabla \cdot u\,\varphi, \psi) + 2(\nabla(p - p_0)u\,\varphi, \psi) \\
& - 2(3\alpha_\vartheta K + C_\vartheta)((\vartheta - \vartheta_0)\nabla \cdot u\,\varphi, \psi) - 2(C_\vartheta \nabla(\vartheta - \vartheta_0)u\,\varphi, \psi) \\
& - G_c\left(\frac{1}{\varepsilon}(1 - \varphi, \psi) - \varepsilon(\nabla\varphi, \nabla\psi)\right) \\
& + ([\Xi + \gamma(\varphi - \varphi^{n-1})]^+, \psi) \\
= {} & 0 \quad \forall \psi \in V_\varphi.
\end{aligned}
$$

Remark 83. We notice that the temperature opens the crack if $\vartheta < \vartheta_0$, i. e., the current (or injected) temperature is lower than the initial temperature. Mathematically, this is due to the different signs in comparison to the pressure terms.

11.4.3 A goal-oriented error estimator

In this section, we formulate a goal-oriented error estimator for quasi-stationary pressurized nonisothermal phase-field fracture using Proposition 97. To this end, we obtain with the help of Section 9.2 an extension of Proposition 68.

Proposition 98 (PU-DWR for pressurized, nonisothermal phase-field fracture). *Let p, $p_0, \vartheta, \vartheta_0$ be given. Let $U = (u, \varphi) \in X^D$ and $U_h = (u_h, \varphi_h) \in X_h^D$ be the continuous and discrete primal solutions, respectively. Let $Z = (z_u, z_\varphi) \in X^0$ and $Z_h \in X_h^0$ be the continuous and discrete adjoint solutions. With the superscript (2) we indicate higher-order solutions. For the finite element approximation Proposition 97, we have the goal-oriented a posteriori error estimate*

$$
|J(U) - J(U_h)| \leq \eta := \left|\sum_{i=1}^{M} \eta_i\right| \leq \sum_{i=1}^{M} |\eta_i|
$$

with

$$
\eta_i = -A(U_h)((Z_h^{(2)} - i_h Z_h^{(2)})\psi_i)
$$

with (neglecting index "h" and superscript (2))

$$
\begin{aligned}
A(U)((Z - i_h Z)\psi_i) &= A_1(U)((z_u - i_h z_u)\psi_i) + A_2(U)((z_\varphi - i_h z_\varphi)\psi_i) \\
&= (g(\varphi)\sigma(u), \nabla(z_u - i_h z_u)\psi_i) - \langle \tilde{\tau}, (z_u - i_h z_u)\psi_i \rangle_{\Gamma_N} \\
&\quad + (1 - \alpha_B)((p - p_0)\varphi^2, \nabla \cdot (z_u - i_h z_u)\psi_i) + (\nabla(p - p_0)\varphi^2, (z_u - i_h z_u)\psi_i) \\
&\quad - (3\alpha_\vartheta K + C_\vartheta)((\vartheta - \vartheta_0)\varphi^2, \nabla \cdot (z_u - i_h z_u)\psi_i) \\
&\quad - (C_\vartheta \nabla(\vartheta_F - \vartheta_0)\varphi^2, (z_u - i_h z_u)\psi_i) + (1 - \kappa)(\varphi\sigma(u) : e(u), (z_\varphi - i_h z_\varphi)\psi_i) \\
&\quad + 2(1 - \alpha_B)((p - p_0)\nabla \cdot u\, \varphi, (z_\varphi - i_h z_\varphi)\psi_i) + 2(\nabla(p - p_0) u\, \varphi, (z_\varphi - i_h z_\varphi)\psi_i) \\
&\quad - 2(3\alpha_\vartheta K + C_\vartheta)((\vartheta - \vartheta_0)\nabla \cdot u\, \varphi, (z_\varphi - i_h z_\varphi)\psi_i) \\
&\quad - 2(C_\vartheta \nabla(\vartheta - \vartheta_0) u\, \varphi, (z_\varphi - i_h z_\varphi)\psi_i) \\
&\quad - G_c \frac{1}{\varepsilon}(1 - \varphi, (z_\varphi - i_h z_\varphi)\psi_i) + \varepsilon G_c(\nabla\varphi, \nabla(z_\varphi - i_h z_\varphi)\psi_i) \\
&\quad + ([\Xi + \gamma(\varphi - \varphi^{n-1})]^+, (z_\varphi - i_h z_\varphi)\psi_i).
\end{aligned}
$$

Proof. Apply Proposition 65 for the primal error estimator part $\rho(U_h)(\cdot)$. Therein, we use the semilinear form $A(U)(\Psi) := A_1(U)(w) + A_2(U)(\psi)$ defined in Proposition 97, set $w := (z_u - i_h z_u)\psi_i$ and $\psi := (z_\varphi - i_h z_\varphi)\psi_i$, where ψ_i is the PU function, and apply the same formal steps as in Proposition 68. □

It follows the question whether it is worth to implement the previous error estimator with so many terms. Supporting arguments are:

1. When small ε are of interest, there are not many alternatives to adaptive mesh refinement in order to achieve accuracy at reasonable computational cost;

2. The semilinear form $A(U)(\Psi)$ in Proposition 97 is at that step already implemented to solve the given problem itself. The presented error estimator "only" requires to replace the test function Ψ by the adjoint solution weighted with the PU function, i. e., $(Z - i_h Z)\psi_i$ such that $A(U)((Z - i_h Z)\psi_i)$ is evaluated.
3. The adjoint problem is linear in which the last Newton system matrix from the primal problem can be simply transposed as suggested in the original work [52].
4. The error estimator yields both: quantification of the error in the goal functional $J(\cdot)$ and local indicators for mesh refinement.

11.5 A pressure diffraction equation in porous media

In the previous sections, we introduced formulations that take given pressures $p : B \to \mathbb{R}$ and given temperatures $\vartheta : B \to \mathbb{R}$. In practical applications, they are however not given, but need to be determined by solving additional equations. Pressurized fractures have been applied often to porous media configurations in which Darcy-type fluids are of interest; see also the overview in Section 1.5.

Since we have now two flow equations, one in Ω_R and another Ω_F, we need two interface conditions (see our discussions around Proposition 86), which shall be given below. Furthermore, we employ the concepts from Section 11.2 and use φ as indicator function to determine Ω_R and Ω_F. To this end, we obtain $\chi_F(\varphi)$ and $\chi_R(\varphi)$, which obviously depend on the solution variable φ. Thus, $\chi_F(\varphi)$ and $\chi_R(\varphi)$ enter as space-time dependent coefficients into a globally defined pressure equation. As discussed in Section 3.4, the pressure equation in porous media is of parabolic type. The idea in [321] was to use the same type of equation in the fracture. This yields a global parabolic pressure equation in which the parameters and constitutive laws are distinguished with the indicator functions χ_F and χ_R. Conceptionally, such problems are known as generalized parabolic equations [289] or also known as diffraction equations [272].

In order to avoid implicit coupling in χ_F and χ_R, we use iterative coupling as proposed in Algorithm 22. Basically, we deal with

Algorithm 23. Computing (u, φ, p) is achieved via the procedure:
1. Given p. Solve for (u, φ).
2. Construct $\chi_F(\varphi)$ and $\chi_R(\varphi)$ (Proposition 80).
3. With the just computed (u, φ), solve a generalized parabolic equation to obtain p with $p = \chi_R p + \chi_F p$.

For obtaining (u, φ) in step 1, we employ a formulation from Section 11.4.1 as for instance Proposition 93. Concerning step 3, a specific example for a pressure diffraction equation is following.

Proposition 99 ([321]: Strong form pressure diffraction equation). *Given $U = (u, \varphi)$. Find a global pressure $p : B \times I \to \mathbb{R}$ with $p = \chi_R p + \chi_F p$ and the initial data $p(0) = p_0$,* *such that*

$$\theta \partial_t p - \nabla \cdot K_{\text{eff}}(\nabla p - \rho^0 g) = \tilde{q} \quad \text{in } B \times I, \tag{11.34}$$

where

$$\theta := \theta(x,t) := \chi_R \theta_R + \chi_F \theta_F = \chi_R\left(\left(\frac{3\alpha^2}{3\lambda + 2\mu} + \frac{1}{M}\right)\right) + \chi_F c_F,$$

$$\tilde{q} := \tilde{q}(x,t) := \chi_R q_R - \chi_{\Omega_R} \partial_t\left(\frac{3\alpha}{3\lambda + 2\mu}\bar{\sigma}\right) + \chi_F q_F,$$

where $M = c_B^{-1}$ (Section 3.4) a compressibility parameter in the porous medium, c_F is the fluid compressibility in the fracture, K_{eff} is an effective permeability ([321]), ρ^0 is gravity, $\tilde{q}$ summarizes source/sink terms, and $\bar{\sigma}$ is a mean stress ([319]). All other parameters are known from previous sections.

The interface conditions are prescribed in a variational-monolithic way (Section 3.3) and are implicitly contained in the weak formulation. Since we work with a global pressure space V_p the kinematic conditions are fulfilled. The integrals of the dynamic coupling condition cancel out in the weak form. Specifically, we have on Γ:

$$p_R = p_F, \tag{11.35}$$

$$[K_{\text{eff}}(\nabla p - \rho^0 g)] \cdot n = 0, \tag{11.36}$$

where $[\cdot]$ denotes the jump. In particular,

$$K_{\text{eff}} = \chi_F \frac{K_F}{\eta_F} + \chi_R \frac{K_R}{\eta_R},$$

where K_F and K_R are the permeabilities in the respective fields and η_F and η_R are kinematic viscosities, respectively. To this end, we arrive at the following weak form.

Proposition 100. *Find $p \in V_p := H^1(B)$ with $p(0) := p_0$ for almost all $t \in I$ such that*

$$(\theta \partial_t p, \phi) + (K_{\text{eff}}(\nabla p - \rho^0 g), \nabla \phi) - (q, \phi) = 0 \quad \forall \phi \in V_p.$$

We notice that this variational-monolithic formulation is an elegant way for this specific problem since any evaluations on cut-elements (interface cuts through single elements on the discrete spatial level) are avoided. This is somewhat similar to two-phase flow computations (some references are provided in Section 1.3.4) in which densities and viscosities change over the interface. To substantiate this section, we briefly recapitulate one result in Section 12.3. For all other findings, we refer to Section 1.5 and “Own related work” in the preface of this monograph.

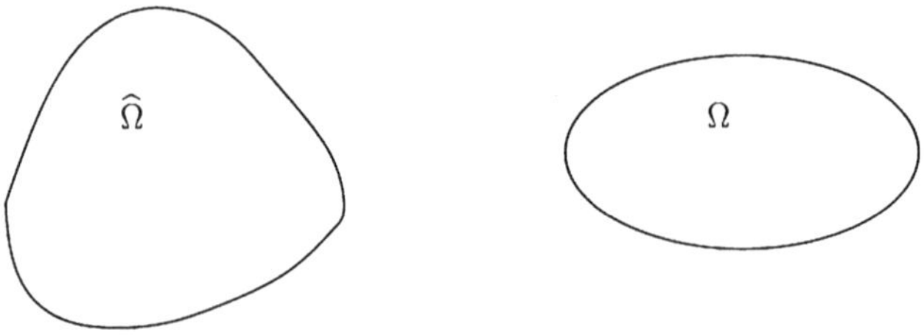

Figure 11.4: Domains: $\widehat{\Omega}$ and Ω.

11.6 A brief introduction to continuum mechanics

In the previous chapters, we worked with infinitesimal solid displacements. For large solid deformations, different coordinate systems come into play: Lagrangian and Eulerian. The Lagrangian (often also called material) system is denoted $\widehat{\Omega}$ or $\widehat{B}$. Eulerian coordinates (the spatial domain) are formulated in the current domain Ω or B.

Definition 75 (Domains). We denote:
- $\widehat{\Omega}$: the reference/undeformed configuration;
- $\Omega(t)$: the current/deformed configuration.

Furthermore, $\Omega(t = 0)$ is the so-called initial configuration. Often, $\widehat{\Omega} := \Omega(t = 0)$. Moreover, we abbreviate as usual $\Omega := \Omega(t)$. We refer the reader to Figure 11.4 for an illustration.

The reason to introduce two coordinate systems is that:
- solids are usually described in the Lagrangian system $\widehat{\Omega}$ because displacements u are observed with respect to some initial or reference configuration. Here, we are interested how specific particles are displaced;
- fluids and transport (i. e., Navier–Stokes, chemical transport, heat) are usually described in Eulerian coordinates Ω because velocities v, concentrations c or temperatures ϑ are observed without paying attention to a specific particle.

Remark 84. In the Lagrangian coordinate system, a specific material point and its deformation (i. e., its evolution in time) are observed. In contrast, using Eulerian coordinates, we observe a fixed point in space and observe what is happening at this spatial point while time is evolving. It might be occupied by different materials while time advances (imagine a river; here you do not identify the single water particle but the general flow pattern). To make it more clear: please think of Kröpcke in Hannover[1] when there is Christmas market or any other day in the year. Imagine you observe people

1 Versailles' place de marché or Linz' Hauptplatz or Munich's Marienplatz or Austin's 6th street or Heidelberg's Hauptstraße.

walking there: do you follow a specific person (that would be Lagrangian modeling) or do you look at the general flow of people (Eulerian point of view)?

The time-evolution of a material point $\hat{x} \in \widehat{\Omega}$ is described by the mapping:

$$t \mapsto x(t, \hat{x}).$$

For the following, let us introduce the upcoming notation to describe (physical) variables:
- Lagrangian coordinates: $\hat{f}(t, \hat{x}) : \widehat{\Omega} \times I \to \mathbb{R}$;
- Eulerian coordinates: $f(t, x) : \Omega \times I \to \mathbb{R}$.

Having two coordinate systems, brings us to the principle question whether one of these two systems might later serve as the computational domain or whether combinations or intermediate domains need to be introduced.

In the following, we deal with points and functions in both types of domains. We introduce the full notation here, but will often abbreviate afterward.

Definition 76 (Points and positions). A point in Lagrangian coordinates is denoted by

$$\hat{x} := \hat{x}(t, x) \in \widehat{\Omega}.$$

A point in Eulerian coordinates is denoted by

$$x := x(t, \hat{x}) \in \Omega.$$

The relation between $\hat{x}$ and x will be just given in the next definition.

Hypothesis 3. *Let us assume:*
- $(t, \hat{x}) \mapsto x(t, \hat{x})$ *is continuously differentiable;*
- *For all* $t \geq t_0$ *the mapping* $\hat{x} \mapsto x(t, \hat{x})$ *is invertible;*
- *Orientation-preserving: the Jacobi determinant is always positive; namely* $\hat{J} = \det(\widehat{\nabla}\widehat{T}) > 0$. *This ensures that volume is not destroyed and that variables under integration can be changed; Section 2.2.3.*

Definition 77 (Deformation field $\widehat{T}$). A deformation of $\widehat{\Omega}$ is a smooth, one-to-one (i. e., injective), orientation-preserving mapping

$$\widehat{T} : \widehat{\Omega} \to \Omega \quad \text{with } (t, \hat{x}) \mapsto (t, x) = (t, \widehat{T}(t, \hat{x})).$$

This mapping associates each point $\hat{x} \in \widehat{\Omega}$ (of a reference domain) to a new position $x \in \Omega$ (of the physical domain). Consequently, with the help of the deformation, $\widehat{T}$, we can represent the deformed configuration as $\Omega = \widehat{T}(\widehat{\Omega})$. In short:

$$\hat{x}(t, x) = \widehat{T}^{-1}(t, x)$$
$$x(t, \hat{x}) = \widehat{T}(t, \hat{x})$$

Definition 78 (Material/Lagrangian description of the displacement field). For $\hat{x} \in \widehat{\Omega}$ and $x \in \Omega$, the displacement is a mapping with

$$\hat{u} : (t, \hat{x}) \to \hat{u}(t, \hat{x}) = x(t, \hat{x}) - \hat{x}.$$

It relates a particle's position in the reference configuration $\hat{x}$ to its corresponding position in the current configuration x at time t.

Definition 79 (Spatial/Eulerian description of the displacement field). It holds

$$u(t, x) = x - \hat{x}(t, x).$$

We recall that also $x = \widehat{T}(t, \hat{x})$.

Definition 80 (Relationship of functions). Clearly a function defined in one system should have the same value in the other system. In detail, let $\hat{u} := \hat{u}(t, \hat{x}) : \widehat{\Omega} \times I \to \mathbb{R}^d$ and $u := u(t, x) : \Omega \times I \to \mathbb{R}^d$. Then we set

$$\hat{u}(t, \hat{x}) = u(t, x).$$

11.6.1 Transformations and deformations

From the previous section, we know how to apply $\widehat{T}$ for solids defined in Ω_s:

$$\widehat{T}(\hat{x}, t) : \widehat{\Omega}_s \times I \to \Omega_s.$$

When solids are coupled to fluids using a so-called ALE (arbitrary Lagrangian–Eulerian) approach [136, 234, 243, 134, 169], then an identical transformation for the fluid domain is constructed:

$$\hat{\mathcal{A}}(\hat{x}, t) : \widehat{\Omega}_f \times I \to \Omega_f.$$

In the following, we therefore use a common notation for the two mappings and we do often not differentiate between the two displacements $\hat{u}_s$ and $\hat{u}_f$; the meaning of a fluid displacement becomes clear when we discuss in more detail ALE fluid-structure interaction. We shall introduce important notations from continuum mechanics, e. g., [236].

Proposition 101. *The deformation, the deformation gradient, and the determinant of the deformation gradient are defined as, respectively,*

$$\hat{\mathcal{A}}(\hat{x}, t) = x = \hat{x} + \hat{u}(\hat{x}, t), \quad \widehat{F} := \widehat{\nabla}\hat{\mathcal{A}} = \hat{I} + \widehat{\nabla}\hat{u}, \quad \hat{J} := \det(\widehat{F}).$$

Here, $\hat{x} \in \widehat{\Omega}$ is a point in the reference domain and $\hat{u}$ is the displacement expressed in Lagrangian coordinates.

Proposition 102 (Deformation gradient $\hat{F}$). *For* $x := x(t,\hat{x}) = \hat{T}(t,\hat{x}) = \hat{x} + \hat{u}$, *it holds for infinitesimal deformations:*

$$dx = \hat{F} \cdot d\hat{x},$$

with $\hat{F} = \hat{\nabla} x$, *i.e,* $\hat{F}_{ij} = \frac{\partial x_i}{\hat{\partial}\hat{x}_j}$.

Proof. Using the relationship $x = \hat{x} + \hat{u}$, we obtain

$$\hat{F} = \frac{\partial x}{\hat{\partial}\hat{x}} = \frac{\partial}{\hat{\partial}\hat{x}}(\hat{x} + \hat{u}) = \frac{\hat{\partial}\hat{x}}{\hat{\partial}\hat{x}} + \frac{\hat{\partial}\hat{u}}{\hat{\partial}\hat{x}} = \hat{I} + \hat{\nabla}\hat{u}.$$

In compact form,

$$\hat{F} = \hat{\nabla}\hat{T} = \hat{I} + \hat{\nabla}\hat{u}.$$

□

The determinant of the deformation gradient (mathematically-speaking we would call it as the functional determinant or Jacobi determinant) is given through

Definition 81. We define

$$\hat{J} := \det(\hat{F}) > 0,$$

which is used to relate volume changes between infinitesimal reference and current domains:

$$d\Omega = \hat{J} d\hat{\Omega}. \tag{11.37}$$

For the following, we need another relation.

Proposition 103 (Eulerian expansion formula). *It holds*

$$\partial_t \hat{J}(t,\hat{x}) = \nabla \cdot v(t,x) \hat{J}(t,\hat{x}).$$

Proof. See for instance [236]. □

Extending Relation (11.37) to the whole domain, we can compute the volume change:

$$|\Omega| = \int_{\Omega} 1 \, dx = \int_{\hat{\Omega}} \hat{J}(t,\hat{x}) \, d\hat{x}, \tag{11.38}$$

which further yields (employing Proposition 103)

$$\frac{d}{dt}|\Omega| = \int_{\Omega} \nabla \cdot v(t,x) \, dx.$$

Consequently, $\hat{J}$ characterizes the volume ratio. Physically, we do not allow for negative volumes $\hat{J} < 0$, which could be mathematically still allowed. Geometrically, $\hat{J} > 0$

preserves the orientation of $\widehat{T}$ and is also assumed in all further considerations. Moreover, we must assume that $\hat{J} > 0$; see Hypothesis 3. If a body does not move, we have $\widehat{F} = \hat{I}$ and it follows $\hat{J} = 1$. On the other hand, a motion is called isochoric when $\hat{J} = 1$ with possibly $\widehat{F} \neq \hat{I}$.

Definition 82. The inverse transformation is defined as

$$\mathcal{A}(x,t) : \Omega \times I \to \widehat{\Omega},$$

with

$$\hat{x} = \mathcal{A}(x,t) = x - u(x,t).$$

For illustrations, we refer to the Figures 11.5 and 11.6.

It is trivial to see that the following identity holds.

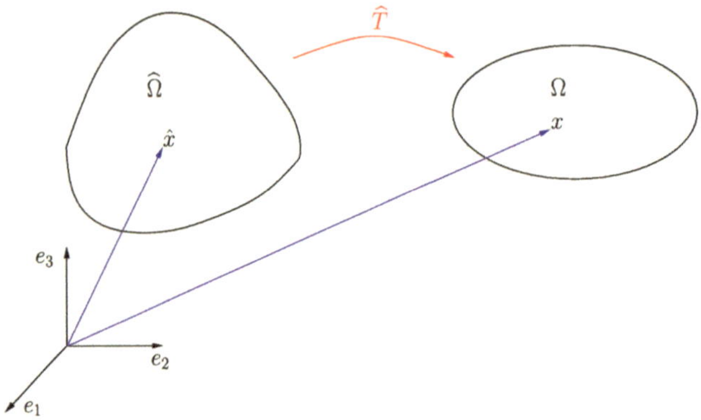

Figure 11.5: Definition of the deformation $\widehat{T} : \widehat{\Omega} \to \Omega$.

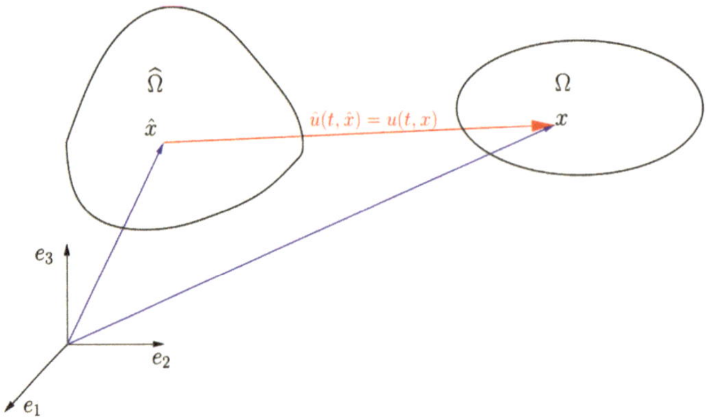

Figure 11.6: Descriptions of displacements: Going from the origin via $\widehat{\Omega}$ means $\hat{x} + \hat{u} = x$. Going from the origin via Ω leads to $x - u = \hat{x}$.

Proposition 104. *For the composition of both transformations, we have*

$$\mathcal{A}(\hat{\mathcal{A}}(\hat{x}, t), t) = \hat{x}. \tag{11.39}$$

Proof. We have immediately

$$\mathcal{A}(\hat{\mathcal{A}}(\hat{x}, t), t) = \mathcal{A}(x, t) = \hat{x}.$$ □

Spatial differentiation of (11.39) yields

Proposition 105. *It holds for the Eulerian and Lagrangian deformation gradients, respectively,*

$$\underbrace{(I - \nabla u)}_{=F}\underbrace{(\hat{I} + \widehat{\nabla}\hat{u})}_{=\widehat{F}} = \hat{I},$$

where I and $\hat{I}$ denote the identity matrices.

Proof. Using the previous proposition and the definitions of $\mathcal{A}$ and $\hat{\mathcal{A}}$, it holds

$$\begin{aligned} \nabla\mathcal{A}(\hat{\mathcal{A}}, t) &= \nabla\hat{x} \\ \Leftrightarrow \nabla\mathcal{A}\cdot\widehat{\nabla}\hat{\mathcal{A}} &= \hat{I} \\ \Leftrightarrow \nabla(x - u)\,\widehat{\nabla}(\hat{x} + \hat{u}) &= \hat{I} \\ \Leftrightarrow (I - \nabla u)\,(\hat{I} + \hat{u}) &= \hat{I}. \end{aligned}$$

Consequently, we have in Eulerian coordinates

$$\begin{aligned} F &= (I - \nabla u), \\ J &= \det(F). \end{aligned}$$ □

It can be inferred that

Proposition 106 ($\widehat{F}$ vs. F and $\hat{J}$ vs. J). *It holds:*

$$\begin{aligned} \widehat{F} &= (\hat{I} + \widehat{\nabla}\hat{u}) = (I - \nabla u)^{-1} =: F^{-1}, \\ \hat{J} &= \det(\widehat{F}) = \det(F^{-1}) =: J^{-1}. \end{aligned}$$

Specifically, if $\hat{J} > 0$, it follows that $J > 0$.

With the help of these relations, we now state versions for strain tensors.

Proposition 107 (Green–Lagrange strain tensors). *It holds for the Eulerian and Lagrangian versions, respectively:*

$$E := \frac{1}{2}(F^{-T}F^{-1} - I), \quad \text{and} \quad \widehat{E} := \frac{1}{2}(\widehat{F}^T\widehat{F} - \hat{I}).$$

11.6.2 Identification of Eulerian and Lagrangian coordinates

We now come to an essential result that justifies why we have not distinguished Eulerian and Lagrangian coordinates so far in this monograph.

Proposition 108. *For infinitesimally small deformations, we have*

$$E \approx \hat{E} \approx e(u) = \frac{1}{2}(\nabla u + \nabla u^T).$$

Furthermore, we have

$$\hat{J} \approx J \approx 1 + \nabla \cdot u.$$

Proof. The key idea is to work with first-order approximations under the condition that we deal with a infinitesimal transformation, i. e.,

$$\|\nabla u\| \ll 1.$$

In this limit, we can work with
- $\Omega \approx \hat{\Omega}$;
- $\hat{\nabla} \approx \nabla$.

For the strain $\hat{E}$ we obtain $e(u)$ while neglecting the second-order term $\nabla u^T \cdot \nabla u$. Moreover, $e(u)$ and ∇u have the same order, i. e., $\|\nabla u\| \ll 1$ yields $\|E\| \ll 1$. But the inverse direction does not hold true. In fact, a rigid body motion has $\hat{E} = e(u) = 0$ but the transformation can be arbitrarily large. Furthermore, for $\hat{F}$ we obtain no specific simplification (except $\hat{F} \to I$ for $\|\nabla u\| \to 0$) but in $\hat{J}$ some terms cancel:

$$\hat{J} \approx J \approx 1 + \partial_z u_z + \partial_y u_y + \partial_x u_x = 1 + \nabla \cdot u = 1 + (e(u))_{ii=1,2,3}. \tag{11.40}$$

□

11.6.3 Transformation of derivatives

We recall from the previous sections that for $x \in \Omega$ and $\hat{x} \in \hat{\Omega}$ we have $\hat{f}(t, \hat{x}) = f(t, x)$ with $\hat{f} : \hat{\Omega} \times I \to \mathbb{R}$ and $f : \Omega \times I \to \mathbb{R}$. Let us now compute derivatives in both coordinate systems.

Definition 83 (Total/Material time derivative of a Lagrangian field). It holds:

$$\frac{d}{dt}\hat{f}(t, \hat{x}) = \hat{\partial}_t \hat{f}(t, \hat{x}).$$

The material time derivative measures the rate at which $\hat{f}$ changes in time, following the path-line of this particle. This means that we measure the rate-change in time of exactly the same particle at all times.

Definition 84 (Local/spatial time derivative of an Eulerian field). The local time derivative of an Eulerian field is denoted by

$$\partial_t f(t, x).$$

The current position x is held fixed while measuring the rate at which f changes in time at this fixed point. This means, at each time, f represents a new particle at x.

Proposition 109 (Total/Material time derivative of an Eulerian field). *Let $f(t,x) : \Omega \times I \to \mathbb{R}$ with $t \in I \subset \mathbb{R}$ and $x := (x_1, \ldots, x_d) \in B \subset \mathbb{R}^d$. Then*

$$\begin{aligned}
\frac{d}{dt} f(t,x) &= \frac{d}{dt} f(t, \hat{A}(t,\hat{x})) \\
&= \partial_t f(t,x) + \nabla f(t,x) \cdot \partial_t \hat{A}(t,\hat{x}) \\
&= \partial_t f(t,x) + \nabla f(t,x) \cdot \partial_t x(t,\hat{x}) \\
&= \partial_t f(t,x) + \nabla f(t,x) \cdot v(t,x).
\end{aligned}$$

The material derivative at a fixed material point $\hat{x}$ describes the change of an Eulerian variable f that is at time t at point x and travels with the velocity $v(t,x)$. Here, the first term prescribes the local change whereas the second term, the convective part, is due to the movement of the particle through the domain.

Proof. Follows immediately from the chain rule from Section 2.2.5. □

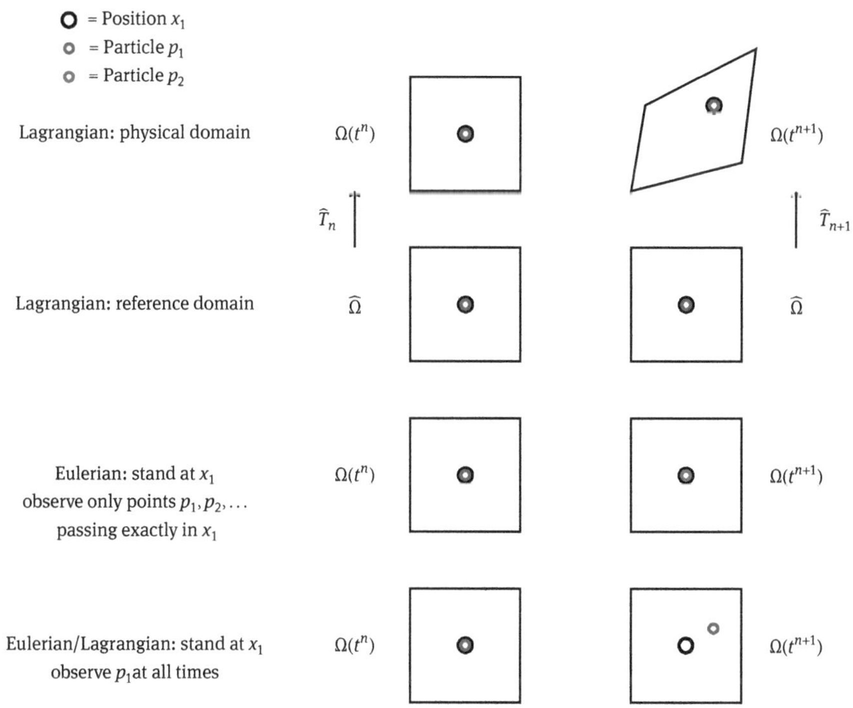

Proposition 110. *Let* $\hat{u} : \widehat{\Omega} \times I \to R^d$ *and* $u : \Omega \times I \to \mathbb{R}^d$. *We obtain for the temporal derivative of the displacements in Lagrangian coordinates*

$$d_t\hat{u} = \partial_t\hat{u}$$

and in Eulerian coordinates as total time derivative

$$d_t u = \partial_t u + (v \cdot \nabla)u,$$

with the vector-valued velocity $v : \Omega \times I \to \mathbb{R}^d$. *We notice that*

$$(v \cdot \nabla)u = \nabla u \cdot v,$$

and furthermore that the Eulerian total time derivative is a transport equation.

Proof. Follows immediately from Definition 83 and Proposition 109. □

Proposition 111 (Spatial differentiation). *Let* $f(t,x) : \Omega \times I \to \mathbb{R}^d$ *be an Eulerian function and let* $\hat{f}(t,\hat{x}) : \widehat{\Omega} \times I \to \mathbb{R}^d$ *a corresponding Lagrangian function. To compute the Eulerian spatial derivative, we have the relation:*

$$\nabla f = \hat{\nabla}\hat{f}\,\widehat{F}^{-1}.$$

The inverse $\widehat{F}^{-1}$ *exists due to the assumptions on* $\widehat{T}$, *respectively,* $\hat{\mathcal{A}}$.

Proof. Chain rule:

$$\hat{\nabla}\hat{f} = \hat{\nabla}\hat{f}(t,\hat{x}) = \hat{\nabla}f(t,\widehat{T}(t,\hat{x})) = \sum_{j=1}^{d} \partial_j f(t,\widehat{T}(t,\hat{x})) \cdot \frac{\hat{\partial}\widehat{T}(t,\hat{x})}{\hat{\partial}\hat{x}} = \nabla f\,\widehat{F}.$$

Since $\widehat{F}$ is invertible due to the assumptions on $\widehat{T}$ (Hypothesis 3) the assertion follows. □

11.6.4 Nanson's formula and the Piola transformation

We now introduce one of the most important transformations in continuum mechanics. Let $\widehat{T}$ be injective on $\widehat{\Omega}$ such that $\widehat{F} = \hat{\nabla}\widehat{T}$ is invertible at all points in $\widehat{\Omega}$. Let $\sigma := (\sigma_{ij})_{ij=1}^{3}$ be a tensor in Ω defined at the point $x = \widehat{T}(t,\hat{x}) \in \Omega$. Then we are able to define a corresponding tensor $\hat{\sigma}$ in $\widehat{\Omega}$ with the help of the following.

Theorem 9 (Piola transformation). *The Piola transformation is defined as*

$$\hat{\sigma}(\hat{x}) := \hat{J}(\hat{x})\sigma(x)\widehat{F}^{-T}(\hat{x}) \quad \text{for } x = \widehat{T}(t,\hat{x}).$$

In short,

$$\hat{\sigma} := \hat{J}\sigma\widehat{F}^{-T} \quad \text{for } x = \widehat{T}(t,\hat{x}).$$

Proof. To show the result, we first need a relationship between normal vectors in both coordinate systems, namely n and $\hat{n}$. For infinitesimal elements, we have

$$ds \cdot dx = \hat{J} d\hat{s} \cdot d\hat{x}$$

where we define with abuse of notation $ds := ds \cdot n$ and $d\hat{s} := d\hat{s} \cdot \hat{n}$ (for details see [236, p. 75]). We then encounter a previous result $dx = \hat{F} d\hat{x}$. Then it follows:

$$(\hat{F}^T ds - \hat{J} d\hat{s}) d\hat{x} = 0,$$

which yields (similar to a complementarity condition)

$$(\hat{F}^T ds - \hat{J} d\hat{s}) = 0.$$

Thus

$$\hat{F}^T ds = \hat{J} d\hat{s}$$

from which we obtain Nanson's formula

$$ds = \hat{J} \hat{F}^{-T} d\hat{s}.$$

We now employ Cauchy's stress theorem (see e. g., [236, 79, 113]):

$$\sigma n \, ds = \hat{\sigma} \hat{n} \, d\hat{s}.$$

Thus, with Nanson's formula we have immediately

$$\sigma n \, ds = \sigma \hat{n} \hat{J} \hat{F}^{-T} d\hat{s} = \hat{\sigma} \hat{n} d\hat{s}$$

from which we obtain the result

$$\hat{J} \sigma \hat{F}^{-T} = \hat{\sigma}. \qquad \square$$

Often, we use $\hat{\Sigma}(\hat{x}) := \hat{\sigma}_s(\hat{x})$. The Piola transformation relates tensor fields between the deformed Ω and the reference configuration $\hat{\Omega}$. One important application is to transform the Cauchy stress tensor from Eulerian coordinates to the Piola–Kirchhoff stress tensor into the Lagrangian framework.

11.6.5 Reynolds transport theorem and conservation of momentum

In order to interchange integration and differentiation, we apply parameter integral differentiation from Section 2.2.3.

Proposition 112 (Reynolds' transport theorem). *As before, let $\Omega := \Omega(t)$ be a time-dependent domain and $(t,\hat{x}) \mapsto x(t,\hat{x})$ be given and assume Hypothesis 3. Furthermore, let the functions $(t,x) \mapsto \partial_t x(t,\hat{x})$ and $(t,x) \mapsto f(t,x)$ be continuously differentiable. Then it holds*

$$\frac{d}{dt}\int_\Omega f(t,x)\,dx = \int_\Omega \left[\frac{d}{dt}f(t,x) + f(t,x)\nabla\cdot v(t,x)\right]dx$$
$$= \int_\Omega [\partial_t f(t,x) + \nabla\cdot(f(t,x)\cdot v(t,x))]\,dx.$$

The second term on the right-hand side represents the rate of transport (also known as the outward normal flux) of the quantity fv across the boundary surface $\partial\Omega$. This contribution originates from the moving domain Ω. The first contribution is the local time rate of change of the spatial scalar field f. If the domain Ω does not move, the second term on the right-hand side will of course vanish.

Proof. We calculate

$$\frac{d}{dt}\int_\Omega f(x,t)\,dx = \frac{d}{dt}\int_{\hat{\Omega}} f(\hat{\mathcal{A}}(\hat{x},t),t)\hat{J}(\hat{x},t)d\hat{x}.$$

We eliminated time dependence on the right-hand side integral, and thus differentiation and integration commute now:

$$\int_{\hat{\Omega}} \frac{d}{dt}(f(\hat{\mathcal{A}}(\hat{x},t),t)\hat{J}(\hat{x},t))d\hat{x} = \int_{\hat{\Omega}}\left(\frac{d}{dt}f(\hat{\mathcal{A}}(\hat{x},t),t)\cdot\hat{J}(\hat{x},t) + f(\hat{\mathcal{A}}(\hat{x},t),t)\frac{d}{dt}\hat{J}(\hat{x},t)\right)d\hat{x}.$$

Here, $\frac{d}{dt}f(\hat{\mathcal{A}}(\hat{x},t),t)$ is the material time derivative of a spatial field. In the last step, we need the Eulerian expansion formula

$$\partial_t\hat{J} = \nabla\cdot v\hat{J}.$$

Then

$$\int_{\hat{\Omega}}\left(\frac{d}{dt}f(\hat{\mathcal{A}}(\hat{x},t),t)\cdot\hat{J}(\hat{x},t) + f(\hat{\mathcal{A}}(\hat{x},t),t)\frac{d}{dt}\hat{J}(\hat{x},t)\right)d\hat{x}$$
$$= \int_{\hat{\Omega}}\left(\frac{d}{dt}f(\hat{\mathcal{A}}(\hat{x},t),t)\cdot\hat{J}(\hat{x},t) + f(\hat{\mathcal{A}}(\hat{x},t),t)\nabla\cdot v(x,t)\hat{J}\right)d\hat{x}$$
$$= \int_{\hat{\Omega}}\left(\frac{d}{dt}f(\hat{\mathcal{A}}(\hat{x},t),t) + f(\hat{\mathcal{A}}(\hat{x},t),t)\nabla\cdot v(x,t)\right)\hat{J}(\hat{x},t)d\hat{x}$$
$$= \int_\Omega\left(\frac{d}{dt}f(x,t) + f(x,t)\nabla\cdot v\right)dx$$

$$= \int_\Omega \big(\partial_t f(x,t) + \nabla f(x,t)\, v(x,t) + f(x,t)\nabla \cdot v(x,t)\big)dx$$

$$= \int_\Omega \big(\partial_t f + \nabla \cdot (fv)\big)dx. \qquad \square$$

Reynolds' transport theorem is frequently applied in the derivation of balance equations. We refer to classical textbooks, e. g., [236], and provide one example in the following, namely conservation of momentum, that is useful in view of this monograph.

11.6.6 Conservation of momentum

Let $f : \Omega \times I \to \mathbb{R}^d$ be a given force density acting on the volume (e. g., the gravity g); and let $\tau := \tau(x,t,n) : \partial\Omega \to \mathbb{R}^d$ a surface-related force density (traction force). For the velocity v, the mass m and the force F, Newton's second law states:

$$\frac{d}{dt}(mv) = F,$$

where F is the force that acts on the mass. In continuum mechanics, Newton's second law reads

$$\frac{d}{dt}\int_\Omega \rho v\, dx = \int_\Omega \rho f\, dx + \int_{\partial\Omega} \tau\, ds.$$

Employing Reynolds' transport theorem to the left-hand side yields:

$$\int_\Omega [\partial_t(\rho v) + \nabla \cdot (\rho v v)]\, dx = \int_\Omega \rho f\, dx + \int_{\partial\Omega} \tau\, ds.$$

Applying Cauchy's existence (from 1827) [79, 113] of the stress tensor (also used before in the Piola transformation),

$$\tau = \sigma \cdot n,$$

further allows to write and using Gauss' divergence theorem 1 (Section 2.2)

$$\int_{\partial\Omega} \tau\, ds = \int_{\partial\Omega} \sigma \cdot n\, ds = \int_\Omega \nabla \cdot \sigma\, dx.$$

We notice that the divergence operator applied to a matrix is calculated component wise and leads to a vector of the form:

$$\nabla \cdot \sigma = \begin{pmatrix} \sum_{j=1}^d \partial_{x_j}\sigma_{1j} \\ \sum_{j=1}^d \partial_{x_j}\sigma_{2j} \\ \vdots \\ \sum_{j=1}^d \partial_{x_j}\sigma_{dj} \end{pmatrix}.$$

We further obtain

$$\int_{\Omega} (\partial_t(\rho v_j) + \nabla \cdot (\rho v_j v) - \rho f_j - (\nabla \cdot \sigma)_j)\, dx = 0. \tag{11.41}$$

At this point we now use the product rule and the continuity equation (conservation of mass), namely

$$\partial_t \rho = -\nabla \cdot (\rho v_j)$$

for the first two terms in (11.41) and calculate

$$\begin{aligned}\partial_t(\rho v_j) + \nabla \cdot (\rho v_j v) &= \rho \partial_t v_j + v_j \partial_t \rho + \nabla \cdot (\rho v_j v)\\ &= \rho \partial_t v_j - v_j \nabla \cdot (\rho v_j) + \nabla \cdot (\rho v_j v)\\ &= \rho \partial_t v_j + \rho v \cdot \nabla v_j.\end{aligned}$$

Therefore,

$$\int_{\Omega} (\rho \partial_t v_j + \rho v \cdot \nabla v_j - (\nabla \cdot \sigma)_j - \rho f_j)\, dx = 0, \quad j = 1, \ldots, d$$

and written in compact form, we obtain

Proposition 113 (Conservation of momentum in Eulerian coordinates). *The conservation of momentum of a medium in Ω is given by*

$$\int_{\Omega} (\rho \partial_t v + \rho v \cdot \nabla v - \nabla \cdot \sigma - \rho f)\, dx = 0.$$

We further obtain the following differential form:

Proposition 114 (Conservation of momentum in Eulerian coordinates). *For smooth functions ρ, v, σ, and arbitrary Ω, it holds the differential form:*

$$\rho \partial_t v + \rho (v \cdot \nabla) v - \nabla \cdot \sigma = \rho f.$$

In Lagrangian coordinates, the convection term $\rho v \cdot \nabla v$ will not arise and we have in the reference configuration:

Proposition 115 (Conservation of momentum in Lagrangian coordinates). *The conservation of momentum of a medium in $\widehat{\Omega}$ is given by*

$$\int_{\widehat{\Omega}} (\hat{\rho} \partial_t^2 \hat{u} - \widehat{\nabla} \cdot \widehat{\Sigma} - \hat{\rho} \hat{f})\, d\hat{x} = 0.$$

We further obtain as differential form:

Proposition 116 (Conservation of momentum in Lagrangian coordinates). *For smooth functions $\hat{\rho}$, $\hat{u}$, $\hat{\sigma}$, and arbitrary $\widehat{\Omega}$, it holds the differential form:*

$$\hat{\rho}\partial_t^2\hat{u} - \widehat{\nabla}\cdot\widehat{\Sigma} = \hat{\rho}\hat{f}.$$

These last derivations close the loop to the displacement equations that we already introduced in Chapter 4, respectively, in Section 2.3 and Section 2.5.

11.6.7 Constitutive laws

With the previously definitions, we recall the constitutive stress tensors in the respective frameworks:

$$\sigma_f := \sigma_f(v_f, p_f) = -p_f I + 2\rho_f \nu_f(\nabla v_f + \nabla v_f^T), \tag{11.42}$$

$$\widehat{\sigma}_f := \widehat{\sigma}_f(\hat{v}_f, \hat{p}_f) = -\hat{p}_f\hat{I} + 2\hat{\rho}_f\nu_f(\widehat{\nabla}\hat{v}_f\widehat{F}^{-1} + \widehat{F}^{-T}\widehat{\nabla}\hat{v}_f^T), \tag{11.43}$$

with the velocity v_f, the pressure p_f, the identity matrix I, the density ρ_f, and the (kinematic) viscosity ν_f and their respective "hat" coordinates. The latter is required for the previously mentioned ALE approach for one of our fluid-structure interaction models that we shall study in the remainder of this chapter. The ALE formulation of the stress tensor is simply obtained by applying the spatial derivatives from Proposition 111, namely

$$\nabla v_f \quad \to \quad \widehat{\nabla}\hat{v}_f\widehat{F}^{-1}.$$

For the structure, we use the laws based on the Saint Venant–Kirchhoff (STVK) material, e. g., [114]

$$\sigma_s := \sigma_s(u_s) = JF^{-1}(\lambda_s(\mathrm{tr}\, E)I + 2\mu_s E)F^{-T}, \tag{11.44}$$

$$\widehat{\sigma}_s := \widehat{\sigma}_s(\hat{u}_s) = \hat{J}^{-1}\widehat{F}(\lambda_s(\mathrm{tr}\,\widehat{E})\hat{I} + 2\mu_s\widehat{E})\widehat{F}^T, \tag{11.45}$$

in which the material is characterized by the Lamé coefficients λ_s and μ_s. We recall the relationship between the first (non-symmetric) $\widehat{\Pi} := \hat{J}\widehat{\sigma}_s\widehat{F}^{-T}$ and second (symmetric) $\widehat{\Sigma}$ Piola-Kirchhoff stress tensors:

$$\widehat{\Sigma} = \widehat{F}^{-1}\widehat{\Pi} = \widehat{F}^{-1}\hat{J}\widehat{\sigma}_s\widehat{F}^{-T}$$

11.7 Fractures inside large solid deformations

11.7.1 Lagrangian perspective for dynamic fracture

We now apply the previous concepts to re-write the formulations presented in Section 5.6 to obtain a representation of phase-field fracture in a fixed reference domain $\widehat{B}$.

Since displacements are naturally written in Lagrangian coordinates, it remains for the phase-field system to be transformed to $\widehat{B}$. To this end, we first obtain the following CVIS.

Proposition 117. *Given the initial data* $\hat{u}(0) = \hat{u}^0$, $\partial_t\hat{u}(0) = \hat{v}^0$ *and* $\hat{\varphi}(0) = \hat{\varphi}^0$. *Find* $(\hat{u}, \hat{\varphi}) \in \hat{V}_u^D \times \hat{K}$ *for almost all times* $t \in I$ *such that*

$$
\begin{aligned}
&(\hat{\rho}_s \partial_t^2 \hat{u}, \hat{\psi}^u) + (g(\hat{\varphi})\, \widehat{\Sigma}(\hat{u}), \widehat{\nabla}\hat{\psi}^u) = 0 \quad \forall \hat{\psi}^u \in \hat{V}_u^0,\\
&(1-\kappa)(\hat{J}\hat{\varphi}\ (\widehat{\Sigma}(\hat{u})) : e(\hat{u}), \hat{\psi}^{\hat{\varphi}} - \hat{\varphi})\\
&\quad + G_c\Big(-\frac{1}{\varepsilon}(\hat{J}(1-\hat{\varphi}), \hat{\psi}^{\hat{\varphi}} - \hat{\varphi}) + \varepsilon(\hat{J}(\widehat{\nabla}\hat{\varphi}\widehat{F}^{-1})\widehat{F}^{-T}, \widehat{\nabla}(\hat{\psi}^{\hat{\varphi}} - \hat{\varphi}))\Big) \geq 0 \quad \forall \hat{\psi}^{\hat{\varphi}} \in \hat{K}.
\end{aligned}
$$

Proof. We only comment on the transformations. The two basic rules, using the previous sections, are

$$(\varphi, \psi)_\Omega \quad \to \quad (\hat{J}\hat{\varphi}, \hat{\psi})_{\widehat{\Omega}}$$

following directly from Section 2.2.3, and again change of variables together with Proposition 111, it holds

$$(\nabla\varphi, \nabla\psi)_\Omega \quad \to \quad (\hat{J}\widehat{\nabla}\hat{\varphi}\widehat{F}^{-1}, \widehat{\nabla}\hat{\psi}\widehat{F}^{-1})_{\widehat{\Omega}} \quad \Leftrightarrow \quad (\hat{J}(\widehat{\nabla}\hat{\varphi}\widehat{F}^{-1})\widehat{F}^{-T}, \widehat{\nabla}\hat{\psi})_{\widehat{\Omega}}. \qquad \square$$

Next, we split the second order in time elastodynamics equation into a first-order system. This step later allows us to employ a One-step-θ time-stepping scheme for more flexibility in the choice of temporal discretization.

Proposition 118. *Given the initial data* $\hat{u}(0) = \hat{u}^0$, $\hat{v}(0) = \hat{v}^0$ *and* $\hat{\varphi}(0) = \hat{\varphi}^0$. *Find* $(\hat{v}, \hat{u}, \hat{\varphi}) \in L^2(\widehat{B}) \times \hat{V}_u^D \times \hat{K}$ *for almost all times* $t \in I$ *such that*

$$
\begin{aligned}
&(\hat{\rho}_s \partial_t \hat{v}, \hat{\psi}^v) + (g(\hat{\varphi})\, \widehat{\Sigma}(\hat{u}), \widehat{\nabla}\hat{\psi}^v) = 0 \quad \forall \hat{\psi}^v \in \hat{V}_u^0,\\
&\hat{\rho}_s(\partial_t \hat{u} - \hat{v}, \hat{\psi}^u) = 0 \quad \forall \hat{\psi}^u \in L^2(\widehat{B}),\\
&(1-\kappa)(\hat{J}\hat{\varphi}\ \widehat{\Sigma}(\hat{u}) : e(\hat{u}), \hat{\psi}^{\varphi} - \hat{\varphi})\\
&\quad + G_c\Big(-\frac{1}{\varepsilon}(\hat{J}(1-\hat{\varphi}), \hat{\psi}^{\varphi} - \hat{\varphi}) + \varepsilon(\hat{J}(\widehat{\nabla}\hat{\varphi}\widehat{F}^{-1})\widehat{F}^{-T}, \widehat{\nabla}\hat{\psi}^{\varphi} - \widehat{\nabla}\hat{\varphi})\Big) \geq 0 \quad \forall \hat{\psi}^{\varphi} \in \hat{K}.
\end{aligned}
$$

11.7.2 Lagrangian perspective for pressurized dynamic fractures

We extend now Proposition 88 (Section 11.4.1.2) to the reference configuration $\widehat{B}$. The transformation of the pressure term from Eulerian coordinates to the Lagrangian system reads

$$(\varphi^2 p, \nabla \cdot w)_\Omega \quad \to \quad (\hat{\varphi}^2 \hat{p}, \nabla \cdot (\hat{J}\widehat{F}^{-1}\hat{w}))_{\widehat{\Omega}}.$$

We then obtain

Proposition 119. *Given $\hat{\rho}_F \in L^\infty(\widehat{B})$ and the initial data $\hat{u}(0) = \hat{u}^0$, $\partial_t\hat{u}(0) = \hat{v}^0$ and $\hat{\varphi}(0) = \hat{\varphi}^0$. Find $(\hat{u}, \hat{\varphi}) \in \hat{V}_u^D \times \hat{K}$ for almost all times $t \in I$ such that*

$$
\begin{aligned}
&(\hat{\rho}_s\partial_t^2\hat{u}, \hat{\psi}^u) + (g(\hat{\varphi})\,\widehat{\Sigma}(\hat{u}), \widehat{\nabla}\hat{\psi}^u) + (\hat{\varphi}^2\hat{p}_F, \widehat{\nabla}\cdot(\hat{J}\widehat{F}^{-1}\hat{\psi}^v)) = 0 \quad \forall\hat{\psi}^u \in \hat{V}_u^0,\\
&(1-\kappa)(\hat{J}\hat{\varphi}\;(\widehat{\Sigma}(\hat{u})) : e(\hat{u}), \hat{\psi}^{\hat{\varphi}} - \hat{\varphi}) + 2(\hat{\varphi}\;\hat{p}_F\;\widehat{\nabla}\cdot(\hat{J}\widehat{F}^{-1}\hat{u}), \hat{\psi}^{\hat{\varphi}} - \hat{\varphi})\\
&\quad + G_c\Big(-\frac{1}{\varepsilon}(\hat{J}(1-\hat{\varphi}), \hat{\psi}^{\hat{\varphi}} - \hat{\varphi}) + \varepsilon(\hat{J}(\widehat{\nabla}\hat{\varphi}\widehat{F}^{-1})\widehat{F}^{-T}, \widehat{\nabla}(\hat{\psi}^{\hat{\varphi}} - \hat{\varphi}))\Big) \geq 0 \quad \forall\hat{\psi}^{\hat{\varphi}} \in \hat{K}.
\end{aligned}
$$

As before, we next split the second-order in time elastodynamics equation into a first-order system.

Proposition 120. *Given $\hat{\rho}_F \in L^\infty(\widehat{B})$ and the initial data $\hat{u}(0) = \hat{u}^0$, $\hat{v}(0) = \hat{v}^0$ and $\hat{\varphi}(0) = \hat{\varphi}^0$. Find $(\hat{v}, \hat{u}, \hat{\varphi}) \in L^2(\widehat{B}) \times \hat{V}_u^D \times K$ for almost all times $t \in I$ such that*

$$
\begin{aligned}
&(\hat{\rho}_s\partial_t\hat{v}, \hat{\psi}^v) + (g(\hat{\varphi})\,\widehat{\Sigma}(\hat{u}), \widehat{\nabla}\hat{\psi}^v) + (\hat{\varphi}^2\hat{p}_F, \widehat{\nabla}\cdot(\hat{J}\widehat{F}^{-1}\hat{\psi}^v)) = 0 \quad \forall\hat{\psi}^v \in \hat{V}_u^0,\\
&\hat{\rho}_s(\partial_t\hat{u} - \hat{v}, \hat{\psi}^u) = 0 \quad \forall\hat{\psi}^u \in L^2(\widehat{B}),\\
&(1-\kappa)(\hat{J}\hat{\varphi}\;\widehat{\Sigma}(\hat{u}) : e(\hat{u}), \hat{\psi}^{\varphi} - \hat{\varphi}) + 2(\hat{\varphi}\hat{p}_F\;\widehat{\nabla}\cdot(\hat{J}\widehat{F}^{-1}\hat{u}), \hat{\psi}^{\varphi} - \hat{\varphi})\\
&\quad + G_c\Big(-\frac{1}{\varepsilon}(\hat{J}(1-\hat{\varphi}), \hat{\psi}^{\varphi} - \hat{\varphi}) + \varepsilon(\hat{J}(\widehat{\nabla}\hat{\varphi}\widehat{F}^{-1})\widehat{F}^{-T}, \widehat{\nabla}\hat{\psi}^{\varphi} - \widehat{\nabla}\hat{\varphi})\Big) \geq 0 \quad \forall\hat{\psi}^{\varphi} \in K.
\end{aligned}
$$

11.7.3 Fixed-mesh fully Eulerian framework

11.7.3.1 Background

The fully Eulerian framework for solids is a fixed-mesh approach and naturally corresponds to the original definitions of the phase-field fracture approach. We recall that the purpose for using phase-field as an indicator function to locate fracture is the fact that we do not need to adapt the mesh according to the fracture surface. The previous section jeopardizes this objective. Now, we go back to a fixed mesh. However, this means that we now need to transform the displacement equations from the reference domain $\widehat{B}$ to the Eulerian domain B. Since the natural framework for structural displacements is however the Lagrangian framework, the equations will change. Most severely, convection terms of first-order hyperbolic type enter as we have previously seen (see in particular Proposition 110). This follows immediately from Reynolds' transport theorem from Section 11.6.5.

We and others have studied in great detail fully Eulerian approaches for solids in combination with fluid flow in the past [138, 377, 373, 437]. A breakthrough was the PhD thesis from Stefan Frei [177] (see also Chapter 3 in [178] and [376, Chapters 6 and 12]) in which an accurate treatment of the interface cut was achieved by proposing a locally modified finite element method [179]. I was again involved in a carefully designed comparison for a challenging solid growth-contact/fluid-structure interaction problem [182] and later an open-source implementation (Chapter 13 [Software 4]) to demonstrate the methodology of the locally modified finite element method [183].

11.7.3.2 Equations

We use these concepts as motivation to derive a fixed-mesh approach for large displacements involving the phase-field framework. With the preliminaries from before, we obtain

Proposition 121 (Solid in Eulerian coordinates). *Find* $(v_s, u_s) \in L_s \times \{u_s^D + V_s^0\}$, *such that* $v_s(0) = v_s^0$ *and* $u_s(0) = u_s^0$ *are satisfied, and for almost all* $t \in I$, *it holds:*

$$\begin{aligned}(\rho_s J \partial_t v_s, \psi^v)_{\Omega_s} + (\rho_s J (v_s \cdot \nabla) v_s, \psi^v)_{\Omega_s} \\ + (\sigma_s, \nabla \psi^v)_{\Omega_s} - (\rho_s J f_s, \psi^v)_{\Omega_s} &= 0 \quad \forall \psi^v \in V_s^0, \\ \rho_s (\partial_t u_s + (v_s \cdot \nabla) u_s - v_s, \psi^u)_{\Omega_s} &= 0 \quad \forall \psi^u \in L_s.\end{aligned}$$

Since we work now in a fixed-mesh approach, the solid particles are not anymore attached to their position (Section 11.6). This means that deforming boundaries must be captured with the help of an indicator function. We omit details in this section in order to keep the focus on the equations. All information is provided later in Section 11.8.2. Therein, the indicator function is applied to the interface, but the same rules apply for boundary movements.

Now, we combine this solid representation with Proposition 19 and obtain as CVIS:

Proposition 122 (Phase-field for solid deformations in Eulerian coordinates). *Find* $(v_s, u_s, \varphi) \in L_s \times \{u_s^D + V_s^0\} \times K$, *such that* $v_s(0) = v_s^0$ *and* $u_s(0) = u_s^0$ *and* $\varphi(0) = \varphi_0$ *are satisfied, and for almost all* $t \in I$, *it holds:*

$$\begin{aligned}(\rho_s J \partial_t v_s, \psi^v)_{\Omega_s} + (\rho_s J (v_s \cdot \nabla) v_s, \psi^v)_{\Omega_s} \\ + (g(\varphi)\sigma_s, \nabla \psi^v)_{\Omega_s} - (\rho_s J f_s, \psi^v)_{\Omega_s} &= 0 \quad \forall \psi^v \in V_s^0, \\ \rho_s (\partial_t u_s + (v_s \cdot \nabla) u_s - v_s, \psi^u)_{\Omega_s} &= 0 \quad \forall \psi^u \in L_s, \\ ((1-\kappa)\varphi \sigma(u) : e(u), \psi - \varphi)_{\Omega_s} \\ + \left(-\frac{G_c}{\varepsilon}(1-\varphi), \psi - \varphi\right)_{\Omega_s} + (G_c \varepsilon \nabla \varphi, \nabla(\psi - \varphi))_{\Omega_s} &\geq 0 \quad \forall \psi \in K.\end{aligned}$$

Remark 85. As discussed in the motivation to this subsection, the fixed-mesh fully Eulerian approach is appealing because it allows for large displacements, topology changes and fits naturally with the phase-field idea, but has some challenging technical issues such low-regularity convection terms and nonaligned boundaries or interfaces. Nonetheless, with the efforts specifically in [177], the method is a true alternative because it benefits from all the advantages we have listed in Chapter 1.

Remark 86. The formulation of a fully Eulerian phase-field approach allowing for large fractured-solid displacements has been carefully derived in this section. All further investigations such as the implementation, numerical tests, numerical analysis are open and topics for future research.

11.8 Coupling fluid-structure interaction with phase-field fracture

In this section, we now couple fractured solids (subject to an internal pressure) from the previous sections with flow described by the incompressible Navier–Stokes equations. Mathematical modeling and numerical simulations were presented in [441, 443]. We first recall typical challenges of such fluid-structure interaction problems:

1. Dealing and coupling of different classes of partial differential equations (PDEs): elliptic, parabolic, hyperbolic that require different tools for both the mathematical analysis and numerical modeling;
2. Combining different coordinate systems: Eulerian and Lagrangian;
3. Nonlinearities in various equations and nonlinear coupling terms;
4. Multidomain character with interface coupling conditions;
5. Moving interface $\widehat{\Gamma}_i$ requires accurate discretizations, nonstandard preconditioners, and sufficient mathematical regularity;
6. Preconditioners require careful design with respect to the interface conditions, e. g., [220, 275, 274, 188, 376, 375, 376, 248, 251].

11.8.1 An arbitrary Lagrangian–Eulerian framework

11.8.1.1 Notation and function spaces

For the following, we need some additional notation and function spaces. We denote by $\Omega := \Omega(t) \subset \mathbb{R}^d$, the domain, which consists of three time-dependent subdomains $\Omega_f(t)$, $\Omega_s(t)$ and $\mathcal{C}(t)$. We assume $\mathcal{C}(t) \subset \Omega_s(t)$. The FSI-interface between $\Omega_f(t)$ and $\Omega_s(t)$ is denoted by $\Gamma_i(t) = \overline{\partial\Omega_f(t)} \cap \overline{\partial\Omega_s(t)}$. The initial (or later reference) domains are denoted by $\widehat{\Omega}$, $\widehat{\Omega}_f$, and $\widehat{\Omega}_s$, respectively, with the interface $\widehat{\Gamma}_i = \overline{\partial\widehat{\Omega}_f} \cap \overline{\partial\widehat{\Omega}_s}$. Furthermore, we denote the outer boundary by $\partial\widehat{\Omega} = \widehat{\Gamma} = \widehat{\Gamma}_{\text{in}} \cup \widehat{\Gamma}_D \cup \widehat{\Gamma}_{\text{out}}$ where $\widehat{\Gamma}_D$ and $\widehat{\Gamma}_{\text{in}}$ are Dirichlet boundaries (for the velocities and displacements) and $\widehat{\Gamma}_{\text{out}}$ denotes a fluid outflow Neumann boundary, respectively. The displacements are set to zero on $\widehat{\Gamma}_{\text{out}}$. For a prototype setting, we refer the reader to Figure 11.7.

For the function spaces in the (fixed) reference domains $\widehat{\Omega}$, $\widehat{\Omega}_f$, $\widehat{\Omega}_s$, $\widehat{C}$, we define spaces for spatial discretization. In the fluid domain, we work with

$$
\begin{aligned}
\hat{L}_f &:= L^2(\widehat{\Omega}_f),\\
\hat{L}_f^0 &:= L^2(\widehat{\Omega}_f)/\mathbb{R},\\
\hat{V}_f^0 &:= \{\hat{v}_f \in H^1(\widehat{\Omega}_f)^d : \hat{v}_f = 0 \text{ on } \widehat{\Gamma}_{\text{in}} \cup \widehat{\Gamma}_D\},\\
\hat{V}_{f,\hat{u}}^0 &:= \{\hat{u}_f \in H^1(\widehat{\Omega}_f)^d : \hat{u}_f = \hat{u}_s \text{ on } \widehat{\Gamma}_i,\ \hat{u}_f = 0 \text{ on } \widehat{\Gamma}_{\text{in}} \cup \widehat{\Gamma}_D \cup \widehat{\Gamma}_{\text{out}}\},\\
\hat{V}_{f,\hat{u},\widehat{\Gamma}_i}^0 &:= \{\hat{\psi}_f \in H^1(\widehat{\Omega}_f)^d : \hat{\psi}_f = 0 \text{ on } \widehat{\Gamma}_i \cup \widehat{\Gamma}_{\text{in}} \cup \widehat{\Gamma}_D \cup \widehat{\Gamma}_{\text{out}}\}.
\end{aligned}
$$

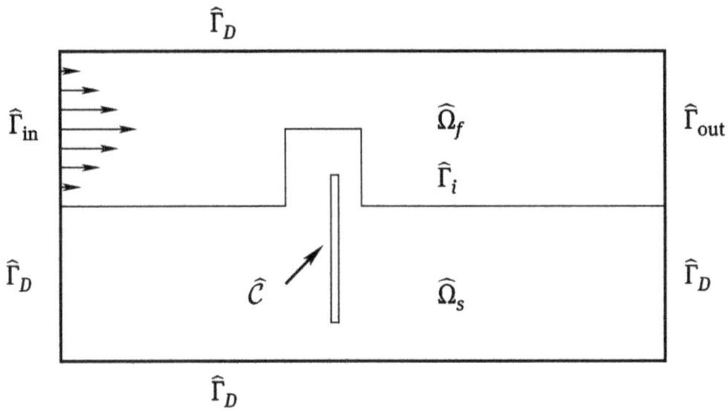

Figure 11.7: Prototype setting with the notation of domains and boundaries. Reprinted from [441] with permission from Elsevier.

In the solid domain, we work with

$$\begin{aligned}\hat{L}_s &:= L^2(\widehat{\Omega}_s)^d,\\ \hat{V}_s^0 &:= \{\hat{u}_s \in H^1(\widehat{\Omega}_s)^d : \hat{u}_s = 0 \text{ on } \widehat{\Gamma}_D\},\\ \hat{W} &:= \{\hat{\varphi} \in H^1(\widehat{\Omega}_s \cup \widehat{C}) : \hat{\varphi} \leq \hat{\varphi}^{n-1} \text{ a. e. on } \widehat{\Omega}_s \cup \widehat{C}\}.\end{aligned}$$

As explained in Section 3.3.3.4, we adopt variational-monolithic interface coupling. The velocity spaces are extended from $\widehat{\Omega}_f$ and $\widehat{\Omega}_s$ to the entire domain $\widehat{\Omega}$ such that we can work with global H^1 functions; see also Section 3.3.3.4. For FSI references, we refer to [239, 139, 140, 435, 376]. Thus, we define

$$\hat{V}^0 := \{\hat{v} \in H^1(\widehat{\Omega})^d : \hat{v} = 0 \text{ on } \widehat{\Gamma}_{\text{in}} \cup \widehat{\Gamma}_D\}. \tag{11.46}$$

By this choice, the kinematic coupling conditions are automatically satisfied in a variational sense (Section 3.3.3.4). Alternative definitions with a similar aim can be found, for example, in [165, 166, 90].

As before, we notice that the spaces on the current domains Ω, Ω_f, Ω_s, C are defined correspondingly without "hat" notation.

11.8.1.2 Governing systems

For the solid, we adopt Proposition 120. For the fluid, the incompressible Navier–Stokes equations are employed.

Formulation 26 (Fluid flows in Eulerian coordinates). Find $(v_f, p_f) \in \{v_f^D + V_f^0\} \times L_f^0$ such that the initial data satisfy $v_f(0) = v_f^0$, and for almost all $t \in I$ holds:

$$\begin{aligned}&\rho_f(\partial_t v_f, \psi^v)_{\Omega_f} + \rho_f(v_f \cdot \nabla v_f, \psi^v)_{\Omega_f}\\ &+(\sigma_f, \nabla\psi^v)_{\Omega_f} - \langle \sigma_f n_f, \nabla\psi^v\rangle_{\Gamma_{f,N}\cup\Gamma_i} + \rho_f(f, \psi^v)_{\Omega_f} = 0 \quad \forall \psi^v \in V_f^0,\\ &\qquad (\operatorname{div} v_f, \psi^p)_{\Omega_f} = 0 \quad \forall \psi^p \in L_f^0,\end{aligned} \tag{11.47}$$

with the Cauchy stress tensor

$$\sigma_f = -p_f I + \rho_f \nu_f (\nabla v_f + \nabla v_f^T), \tag{11.48}$$

and a correction term on the outflow boundary [229]:

$$g := -\rho_f \nu_f \nabla v_f^T \quad \text{on } \Gamma_{f,N} = \Gamma_{\text{out}}.$$

11.8.1.3 Arbitrary Lagrangian–Eulerian

We now need to bring different coordinate systems together. The coordinate system of the fractured solid (Lagrangian) and the Navier–Stokes flow (Eulerian) must be coupled. Here, the previously mentioned ALE (arbitrary Lagrangian–Eulerian) approach is adopted [136, 234, 243, 134, 169]. This is an intermediate approach that is Lagrangian at the FSI interface and Eulerian far away from the interface. Between, there is a transition from Lagrangian to Eulerian. The ALE mapping is realized by solving an additional partial-differential equation, the so-called mesh motion problem. Various possibilities are presented in [223, 397, 435, 135]. One option is to solve a second-order problem of Poisson-type:

$$-\widehat{\nabla} \cdot (\hat{\sigma}_{\text{mesh}}) = 0, \quad \hat{u}_f = \hat{u}_s \text{ on } \hat{\Gamma}_i, \quad \hat{u}_f = 0 \text{ on } \partial\hat{\Omega}_f \setminus \hat{\Gamma}_i, \tag{11.49}$$

with $\hat{\sigma}_{\text{mesh}} = \hat{\alpha}_u \widehat{\nabla} \hat{u}_f$.

We also introduce the ALE time-derivative, which is the total derivative of an Eulerian field (see also Proposition 109 in Section 11.6.3) and is important when working on moving domains $\Omega(t)$:

$$\partial_t|_{\hat{\mathcal{A}}} v_f(x,t) = w \cdot \nabla v_f + \partial_t v_f(x,t), \tag{11.50}$$

where the transport term appears due to the motion of the computational domain. Specifically, we have:

- In a Lagrangian description, we have $w = v_f$, i. e., the domain $\Omega(t)$ is moving with the fluid velocity v_f and the ALE time derivative corresponds to the total time derivative in the Lagrangian sense (Definition 83): $\partial_t|_{\hat{\mathcal{A}}} v_f(x,t) = \frac{d}{dt} v_f(t,\hat{x})$;
- In a fixed Eulerian setting, it holds $w = 0$, i. e., the domain $\Omega(t) =: \Omega$ is fixed. Here, we deal with classical Eulerian descriptions and the ALE time derivative corresponds to the local time derivative in the Eulerian sense (Definition 84): $\partial_t|_{\hat{\mathcal{A}}} v_f(x,t) = \partial_t v_f(x,t)$;
- In ALE, we have $0 \le w \le v_f$. It is $v_f = w$ at the FSI-interface $\hat{\Gamma}_i$ and a bit away, we have $0 < w < v_f$, while far away $w = 0$ (the mesh is not moving anymore).

11.8.1.4 Interface coupling conditions

We deal with two different types of interfaces:

1. Interface $\widehat{\Gamma}_i$ between fluid and solids (FSI interface)
2. Interface $\partial\widehat{C}$ between solid and fracture

The first condition is of type 2 (Section 1.3) and will be realized as an interface-tracking method in this section. The second condition takes care of the fracture boundary of type 3 and will be realized with the phase-field method via interface-capturing.

11.8.1.4.1 FSI interface conditions

We first describe the FSI interface conditions in more detail. Coupling fluids with solids requires two physical conditions; namely continuity of velocities and balance of normal stresses. A third condition of geometric nature is necessary when working with the ALE framework: continuity of displacements, which couples the physical solid $\hat{u}_s$ and the fluid mesh motion $\hat{u}_f$.

As before, in variational-monolithic coupling, these Dirichlet conditions are built as shown in Section 3.3.2 into the corresponding function space by employing a globalized Sobolev space $\hat{V}^0$ (see (11.46)). Neumann-type conditions are weakly incorporated through interface integrals, which cancel again out because of their variational continuity thanks to working with the space $\hat{V}^0$; see Section 3.3.3.4.

For the fluid problem, continuity of velocities is required (i. e., a kinematic coupling condition):

$$\hat{v}_f = \hat{v}_s \quad \text{on } \widehat{\Gamma}_i. \tag{11.51}$$

To complete the solid problem, we must enforce the balance of the normal stresses on the interface (i. e., a dynamic coupling condition):

$$\hat{J}\hat{\sigma}_f\hat{F}^{-T}\hat{n}_f + \hat{F}\hat{\Sigma}\hat{n}_s = 0 \quad \text{on } \widehat{\Gamma}_i. \tag{11.52}$$

For the geometric problem due to the ALE framework, we have

$$\hat{u}_f = \hat{u}_s \quad \text{on } \widehat{\Gamma}_i, \tag{11.53}$$

from which we obtain immediately $\partial_t\hat{u}_s = \hat{v}_s = \hat{w} = \hat{v}_f$ on $\widehat{\Gamma}_i$ by temporal differentiation. For the geometric condition, we notice that we only need a one-way transport, namely $\hat{u}_s \to \hat{u}_f$. Back coupling from the (artificial) fluid displacements to the physical solid displacements must be avoided. Mathematically, this is achieved by choosing the test space $\hat{V}^0_{f,\hat{u},\widehat{\Gamma}_i}$. Numerically, either we work with a small scaling parameter $\alpha > 0$ [139, 377, 435, 438] or by eliminating in the system matrix those entries that represent the back-coupling nodes.

11.8.1.4.2 Phase-field fracture interface

The interface condition between the solid and the fracture is conceptionally the same as in Section 11.4. They only need to be transformed into the reference system $\widehat{\Omega}$. To this end, we have

$$\widehat{F}\widehat{\Sigma}\hat{n}_s = \widehat{\sigma}_F \hat{n}_s \quad \text{on } \partial\widehat{C},$$

where $\widehat{\sigma}_F := -\hat{p}_F \hat{I}$, where $\hat{p}_F$ is a fracture pressure, which is not linked to the fluid pressure $\hat{p}_f$ obtained by solving the Navier–Stokes equations.

11.8.1.5 Final model using an augmented Lagrangian penalization

We collect the previous pieces and formulate a computational framework [441].

Proposition 123 (Variational-monolithic ALE-FSI phase-field fracture). *Let $\hat{p}_F \in H^1(\widehat{\Omega}_s \cup \widehat{C})$ be given. Find a global vector-valued velocity, vector-valued displacements, a scalar-valued fluid pressure, and a scalar-valued phase-field function, that is to say that $(\hat{v}, \hat{u}_f, \hat{u}_s, \hat{p}_f, \hat{\varphi}) \in \{\hat{v}^D + \hat{V}^0\} \times \{\hat{u}_f^D + \hat{V}^0_{f,\hat{u}}\} \times \hat{V}^0_s \times \hat{L}^0_f \times \hat{V}_\varphi$, such that $\hat{v}(0) = \hat{v}^0$, $\hat{u}_f(0) = \hat{u}_f^0$, $\hat{u}_s(0) = \hat{u}_s^0$ and $\hat{\varphi}(0) = \hat{\varphi}^0$ are satisfied, and for almost all times $t \in I$ holds:*

$$\textit{Fluid/solid momentum} \quad \begin{cases} (\hat{J}\hat{\rho}_f \partial_t \hat{v}, \hat{\psi}^v)_{\widehat{\Omega}_f} + (\hat{\rho}_f \hat{J}(\widehat{F}^{-1}(\hat{v} - \hat{w}) \cdot \widehat{\nabla}\hat{v}), \hat{\psi}^v)_{\widehat{\Omega}_f} \\ \quad + (\hat{J}\widehat{\sigma}_f \widehat{F}^{-T}, \widehat{\nabla}\hat{\psi}^v)_{\widehat{\Omega}_f} \\ \quad + \langle \rho_f \nu_f \hat{J}(\widehat{F}^{-T}\widehat{\nabla}\hat{v}^T \hat{n}_f)\widehat{F}^{-T}, \hat{\psi}^v \rangle_{\widehat{\Gamma}_{\text{out}}} \\ \quad + (\hat{\rho}_s \partial_t \hat{v}, \hat{\psi}^v)_{\widehat{\Omega}_s} + (g(\hat{\varphi})\widehat{F}\widehat{\Sigma}, \widehat{\nabla}\hat{\psi}^v)_{\widehat{\Omega}_s} \\ \quad + (\hat{\varphi}_s^2 \hat{p}_F, \widehat{\nabla} \cdot (\hat{J}\widehat{F}^{-1}\hat{\psi}^v)) = 0 \quad \forall \hat{\psi}^v \in \hat{V}^0, \end{cases}$$

$$\textit{Fluid mesh motion} \quad \left\{ (\widehat{\sigma}_{mesh}, \widehat{\nabla}\hat{\psi}_f^u)_{\widehat{\Omega}_f} = 0 \quad \forall \hat{\psi}_f^u \in \hat{V}^0_{f,\hat{u},\widehat{\Gamma}_i}, \right.$$

$$\textit{Solid momentum, 2nd equation} \quad \left\{ \hat{\rho}_s(\partial_t \hat{u}_s - \hat{v}_s, \hat{\psi}_s^u)_{\widehat{\Omega}_s} = 0 \quad \forall \hat{\psi}_s^u \in \hat{L}_s, \right.$$

$$\textit{Fluid mass conservation} \quad \left\{ (\widehat{\operatorname{div}}\,(\hat{J}\widehat{F}^{-1}\hat{v}), \hat{\psi}_f^p)_{\widehat{\Omega}_f} = 0 \quad \forall \hat{\psi}_f^p \in \hat{L}_f^0, \right.$$

$$\textit{Phase-field} \quad \begin{cases} (1-\kappa)(\hat{J}\hat{\varphi}\,\widehat{\Sigma} : \widehat{E}, \hat{\psi}_s^\varphi)_{\widehat{\Omega}_s} + 2(\hat{\varphi}\hat{p}_F\,\widehat{\nabla} \cdot (\hat{J}\widehat{F}^{-1}\hat{u}_s), \hat{\psi}_s^\varphi)_{\widehat{\Omega}_s} \\ \quad - \langle 2\hat{J}\hat{\varphi}\hat{p}_F \widehat{F}^{-T}\hat{n}_s \hat{u}_s, \hat{\psi}_s^\varphi \rangle_{\widehat{\Gamma}_i} \\ \quad + G_c(-\frac{1}{\varepsilon}(\hat{J}(1-\hat{\varphi}), \hat{\psi}_s^\varphi) + G_c\varepsilon(\hat{J}(\widehat{\nabla}\hat{\varphi}\widehat{F}^{-1})\widehat{F}^{-T}, \widehat{\nabla}\hat{\psi}_s^\varphi))_{\widehat{\Omega}_s} \\ \quad + (\hat{J}[\widehat{\Xi} + \gamma(\hat{\varphi} - \hat{\varphi}^{n-1})]^+, \hat{\psi}_s^\varphi)_{\widehat{\Omega}_s} = 0 \quad \forall \hat{\psi}_s^\varphi \in \hat{V}_\varphi. \end{cases}$$

Remark 87. As discussed in Section 3.3, the Neumann coupling conditions on $\widehat{\Gamma}_i$ are fulfilled in a monolithic-variational way and appear only implicitly. Thus, the condition

$$\langle \hat{J}\widehat{\sigma}_f \widehat{F}^{-T}\hat{n}_f, \hat{\psi}^v \rangle_{\widehat{\Gamma}_i} + \langle g(\hat{\varphi})[\widehat{F}\widehat{\Sigma} - \hat{J}\hat{p}_F\widehat{F}^{-T}]\hat{n}_s, \hat{\psi}^v \rangle_{\widehat{\Gamma}_i} = 0 \quad \forall \hat{\psi}^v \in \hat{V}^0 \tag{11.54}$$

is implicitly contained in the above system because of $\hat{\psi}_f^v = \hat{\psi}_s^v$ on $\widehat{\Gamma}_i$ in $\hat{V}^0$.

Remark 88. In [441], we extensively discussed the influence of solid damping terms, which are neglected in this monograph.

The discretization and solution algorithm follows the steps explained in:
- Chapter 5: regularization (already done with augmented Lagrangian), temporal and spatial discretization
- Chapter 7: nonlinear and linear solutions

11.8.1.6 Temporal discretization

We first group the above problem according to Section 5.3. Afterward, we present both One-Step-θ and the Fractional-Step-θ time-stepping discretizations.

Formulation 27 (The time-discretized abstract problem). Given $\hat{U}^{n-1}$, find $\hat{U}^n = \{\hat{v}_f^n, \hat{v}_s^n, \hat{u}_f^n, \hat{u}_s^n, \hat{p}_f^n, \hat{\varphi}_s^n\} \in \widehat{X}_D^0$, where $\widehat{X}_D^0 := \{\hat{v}_f^D + \hat{V}_{f,\hat{v}}^0\} \times \hat{L}_s \times \{\hat{u}_f^D + \hat{V}_{f,\hat{u}}^0\} \times \{\hat{u}_s^D + \hat{V}_s^0\} \times \hat{L}_f^0 \times H^1(\widehat{\Omega}_s \cup \widehat{C})$ and $\widehat{X} = \hat{V}_{f,\hat{v}}^0 \times \hat{L}_s \times \hat{V}_{f,\hat{u},\hat{\Gamma}_i}^0 \times \hat{V}_s^0 \times \hat{L}_f^0 \times H^1(\widehat{\Omega}_s \cup \widehat{C})$, for $n = 1, 2, \ldots, N$ such that

$$\hat{A}(\hat{U}^n)(\hat{\Psi}) = 0 \quad \forall \hat{\Psi} \in \widehat{X}, \tag{11.55}$$

where the semilinear form $\hat{A}(\cdot)(\cdot)$ is split into

$$\hat{A}(\hat{U}^n)(\hat{\Psi}) := \hat{A}_T(\hat{U}^{n,k})(\hat{\Psi}) + \hat{A}_I(\hat{U}^n)(\hat{\Psi}) + \hat{A}_E(\hat{U}^n)(\hat{\Psi}) + \hat{A}_P(\hat{U}^n)(\hat{\Psi}).$$

Details of this decomposition are provided in Definition 85.

Definition 85 (Arranging the semilinear form $\hat{A}(\hat{U}^n)(\hat{\Psi})$ into groups). We formally split the semilinear form into four categories: time equation terms (including the time derivatives); implicit terms (such as the fluid incompressibility and also the phase-field equation); pressure terms; and finally all 'standard' terms (stress terms and fluid convection). In detail, we have the decomposition:

$$\begin{aligned}
\hat{A}_T(\hat{U})(\hat{\Psi}) &= (\hat{J}\hat{\rho}_f \partial_t \hat{v}_f, \hat{\psi}_f^v)_{\widehat{\Omega}_f} - (\hat{\rho}_f \hat{J}(\hat{F}^{-1}\hat{w} \cdot \widehat{\nabla}\hat{v}_f), \hat{\psi}_f^v)_{\widehat{\Omega}_f} \\
&\quad + (\hat{\rho}_s \partial_t \hat{v}_s, \hat{\psi}_s^v)_{\widehat{\Omega}_s} + (\hat{\rho}_s \partial_t \hat{u}_s, \hat{\psi}_s^u)_{\widehat{\Omega}_s}, \\
\hat{A}_I(\hat{U})(\hat{\Psi}) &= (\hat{\sigma}_{\text{mesh}}, \widehat{\nabla}\hat{\psi}_f^u)_{\widehat{\Omega}_f} + (\widehat{\operatorname{div}}\,(\hat{J}\hat{F}^{-1}\hat{v}_f), \hat{\psi}_f^p)_{\widehat{\Omega}_f} \\
&\quad + (1-\kappa)(\hat{J}\hat{\varphi}_s\, \hat{\Sigma} : \hat{E}, \hat{\psi}^\varphi)_{\widehat{\Omega}_s} + 2(\hat{J}\hat{\varphi}_s\, \hat{p}_F\, \widehat{\nabla} \cdot (\hat{J}\hat{F}^{-1}\hat{u}_s), \hat{\psi}^\varphi)_{\widehat{\Omega}_s} \\
&\quad + G_c\left(-\frac{1}{\varepsilon}(\hat{J}(1-\hat{\varphi}_s), \hat{\psi}^\varphi) + \varepsilon(\hat{J}(\widehat{\nabla}\hat{\varphi}_s \hat{F}^{-1})\hat{F}^{-T}, \widehat{\nabla}\hat{\psi}^\varphi)\right)_{\widehat{\Omega}_s} \\
&\quad + (\hat{J}[\hat{\Xi} + \gamma(\hat{\varphi}_s - \hat{\varphi}_s^{n-1})]^+, \hat{\psi}^\varphi)_{\widehat{\Omega}_s}, \\
\hat{A}_P(\hat{U})(\hat{\Psi}) &= (\hat{J}\hat{\sigma}_{f,p}\hat{F}^{-T}, \widehat{\nabla}\hat{\psi}_f^v)_{\widehat{\Omega}_f}, \\
\hat{A}_E(\hat{U})(\hat{\Psi}) &= (\hat{\rho}_f \hat{J}(\hat{F}^{-1}\hat{v}_f \cdot \widehat{\nabla}\hat{v}_f), \hat{\psi}_f^v)_{\widehat{\Omega}_f} + (\hat{J}\hat{\sigma}_{f,vu}\hat{F}^{-T}, \widehat{\nabla}\hat{\psi}_f^v)_{\widehat{\Omega}_f} \\
&\quad + \langle \rho_f \nu \hat{J}(\hat{F}^{-T}\widehat{\nabla}\hat{v}_f^T)\hat{F}^{-T}\hat{n}, \hat{\psi}_f^v \rangle_{\hat{\Gamma}_{\text{out}}}
\end{aligned}$$

$$+ (g(\hat\varphi_s)\hat F\hat\Sigma, \hat\nabla\hat\psi_s^v)_{\widehat\Omega_s}$$
$$+ (\hat\varphi_s^2\hat p_F, \hat\nabla\cdot(\hat J\hat F^{-1}\hat\psi^v))_{\widehat\Omega_s} - (\hat\rho_s\hat v_s, \hat\psi_s^u)_{\widehat\Omega_s},$$

where the fluid stress tensor $\hat\sigma_f$ is further split into $\hat\sigma_{f,vu}$, $\hat\sigma_{f,p}$:

$$\hat\sigma_{f,p} = -\hat p_f\hat I, \quad \hat\sigma_{f,vu} = \rho_f\nu_f(\hat\nabla\hat v_f\hat F^{-1} + \hat F^{-T}\hat\nabla\hat v_f^T).$$

The (nonlinear) time derivative in $\hat A_T(\hat U)(\hat\Psi)$ is approximated by a backward difference quotient. For $n = 1, 2, \ldots, N$, we compute $\hat v_i := \hat v_i^n$, $\hat u_i := \hat u_i^n$ $(i = f, s)$ via

$$\begin{aligned}\hat A_T(\hat U^{n,k})(\hat\Psi) := &\frac{1}{k}(\hat\rho_f\hat J^{n,\theta}(\hat v_f - \hat v_f^{n-1}), \hat\psi^v)_{\widehat\Omega_f} - \frac{1}{k}(\hat\rho_f(\hat J\hat F^{-1}(\hat u_f - \hat u_f^{n-1})\cdot\hat\nabla)\hat v_f, \hat\psi^v)_{\widehat\Omega_f}\\ &+ \frac{1}{k}(\hat\rho_s(\hat v_s - \hat v_s^{n-1}), \hat\psi^v)_{\widehat\Omega_s} + (\hat u_s - \hat u_s^{n-1}, \hat\psi^u)_{\widehat\Omega_s},\end{aligned} \tag{11.56}$$

where we employ the parameter $\theta \in [0, 1]$ (see Section 5.3). Furthermore, with the help of Section 5.3.3, we use

$$\hat J^{n,\theta} = \theta\hat J^n + (1-\theta)\hat J^{n-1},$$

and $\hat u_i^n := \hat u_i(t_n)$, $\hat v_i^n := \hat v_i(t_n)$, and $\hat J := \hat J^n := \hat J(t_n)$. The former time step is given by $\hat v_i^{n-1}$, etc. for $i = f, s$.

The following two time-stepping schemes are defined in a bit more general way while assuming a right-hand side functional $\hat F(\hat\Psi)$ as well; but we notice that in the previous systems $\hat F(\hat\Psi) = 0$.

Formulation 28 (One-Step-θ scheme). Let the previous time step solution $\hat U^{n-1} = \{\hat v_f^{n-1}, \hat v_s^{n-1}, \hat u_f^{n-1}, \hat u_s^{n-1}, \hat p_f^{n-1}, \hat\varphi_s^{n-1}\}$ and the time step $k := k_n = t_n - t_{n-1}$ be given. In order to solve (11.55), we seek $\hat U^n = \{\hat v_f^n, \hat v_s^n, \hat u_f^n, \hat u_s^n, \hat p_f^n, \hat\varphi_s^n\} \in \hat X^0$ by employing One-step-θ splitting:

$$\begin{aligned}&\hat A_T(\hat U^{n,k})(\hat\Psi) + \theta\hat A_E(\hat U^n)(\hat\Psi) + \hat A_P(\hat U^n)(\hat\Psi) + \hat A_I(\hat U^n)(\hat\Psi)\\ &= -(1-\theta)\hat A_E(\hat U^{n-1})(\hat\Psi) + \theta\hat F^n(\hat\Psi) + (1-\theta)\hat F^{n-1}(\hat\Psi).\end{aligned} \tag{11.57}$$

In the following, we now state the version of the Fractional-Step-θ scheme (see back in Section 5.3) that includes also pressure terms:

Formulation 29 (Fractional-Step-θ scheme). We choose $\theta = 1 - \frac{\sqrt{2}}{2}$, $\theta' = 1 - 2\theta$, and $\alpha = \frac{1-2\theta}{1-\theta}$, $\beta = 1 - \alpha$. The time step is split into three consecutive subtime steps. Let $\hat U^{n-1} = \{\hat v_f^{n-1}, \hat v_s^{n-1}, \hat u_f^{n-1}, \hat u_s^{n-1}, \hat p_f^{n-1}, \hat p_s^{n-1}\}$ and the time step $k := k_n = t_n - t_{n-1}$ be given. Find $\hat U^n = \{\hat v_f^n, \hat v_s^n, \hat u_f^n, \hat u_s^n, \hat p_f^n, \hat p_s^n\}$ such that

$$\begin{aligned}&\hat A_T(\hat U^{n-1+\theta,k})(\hat\Psi) + \alpha\theta\hat A_E(\hat U^{n-1+\theta})(\hat\Psi)\\ &+ \theta\hat A_P(\hat U^{n-1+\theta})(\hat\Psi) + \hat A_I(\hat U^{n-1+\theta})(\hat\Psi) = -\beta\theta\hat A_E(\hat U^{n-1})(\hat\Psi) + \theta\hat F^{n-1}(\hat\Psi),\end{aligned}$$

$$\begin{aligned}
&\hat{A}_T(\hat{U}^{n-\theta,k})(\hat{\Psi}) + \alpha\theta'\hat{A}_E(\hat{U}^{n-\theta})(\hat{\Psi}) \\
&\quad + \theta'\hat{A}_P(\hat{U}^{n-\theta})(\hat{\Psi}) + \hat{A}_I(\hat{U}^{n-\theta})(\hat{\Psi}) = -\alpha\theta'\hat{A}_E(\hat{U}^{n-1+\theta})(\hat{\Psi}) + \theta'\hat{F}^{n-\theta}(\hat{\Psi}), \\
&\quad \hat{A}_T(\hat{U}^{n,k})(\hat{\Psi}) + \alpha\theta\hat{A}_E(\hat{U}^{n})(\hat{\Psi}) \\
&\qquad + \theta\hat{A}_P(\hat{U}^{n})(\hat{\Psi}) + \hat{A}_I(\hat{U}^{n})(\hat{\Psi}) = -\beta\theta\hat{A}_E(\hat{U}^{n-\theta})(\hat{\Psi}) + \theta\hat{F}^{n-\theta}(\hat{\Psi}).
\end{aligned}$$

11.8.1.7 Spatial discretization

The time discretized forms from the previous subsection are the starting points for a Galerkin discretization in space. Spatial discretization was formally introduced in Section 5.4. To this end, we construct finite dimensional subspaces $\widehat{X}_h^0 \subset \widehat{X}^0$ to find an approximate solution to the continuous problem. As previously explained, in the context of our variational-monolithic formulation, the computations are done on the reference configuration $\widehat{\Omega}$. We use shape-regular meshes. A mesh consists of cells $\hat{K}$ (Section 5.4). The discretization parameter in the reference configuration is denoted by $\hat{h}$.

On $\hat{\mathcal{T}}_h$, the conforming finite element space for $\{\hat{v}_{f,h}, \hat{v}_{s,h}, \hat{u}_{f,h}, \hat{u}_{s,h}, \hat{p}_{f,h}, \hat{\varphi}_{s,h}\}$ is denoted by the space $\widehat{X}_h \subset \widehat{X}$. For Navier–Stokes flow, we prefer the biquadratic, discontinuous-linear Q_2^c/P_1^{dc} element [201]. The definitions of the spaces for the velocities and pressure unknowns read

$$\begin{aligned}
\hat{V}_{f,s,h} &:= \{\hat{v}_h \in [C(\hat{\Omega}_h)]^d,\ \hat{v}_h|_{\hat{K}} \in [Q_2^c(\hat{K})]^d\ \forall \hat{K} \in \hat{\mathcal{T}}_h, \hat{v}_h|_{\hat{\Gamma}\setminus\hat{\Gamma}_i} = 0\}, \\
\hat{P}_{f,h} &:= \{\hat{p}_h \in [\hat{L}^2(\hat{\Omega}_h)],\ \hat{p}_h|_{\hat{K}} \in [P_1^{dc}(\hat{K})]\ \forall \hat{K} \in \hat{\mathcal{T}}_h\}.
\end{aligned}$$

For the solid and the phase-field variable, we use

$$\begin{aligned}
\hat{U}_{f,s,h} &:= \{\hat{u}_h \in [C(\hat{\Omega}_h)]^d, \hat{u}_h|_{\hat{K}} \in [Q_2^c(\hat{K})]^d\ \forall \hat{K} \in \hat{\mathcal{T}}_h, \hat{u}_h|_{\hat{\Gamma}\setminus\hat{\Gamma}_i} = 0\}, \\
\hat{\Phi}_{s,h} &:= \{\hat{\varphi}_h \in [C(\hat{\Omega}_h)]^d, \hat{\varphi}_h|_{\hat{K}} \in [Q_1^c(\hat{K})]^d\ \forall \hat{K} \in \hat{\mathcal{T}}_h, \hat{\varphi}_h|_{\hat{\Gamma}\setminus\hat{\Gamma}_i} = 0\}.
\end{aligned}$$

The property of the Q_2^c/P_1^{dc} element is continuity of the velocity values across different mesh cells. However, the pressure is defined by discontinuous test functions. In addition, this element preserves local mass conservation, is of low order, satisfies the inf-sup stability (see, e. g., [201]) and, therefore, is an optimal choice for both fluid problems and fluid-structure interaction problems.

Formulation 30. Let the previous, spatially discretized, time step solution $\hat{U}_h^{n-1} = \{\hat{v}_{f,h}^{n-1}, \hat{v}_{s,h}^{n-1}, \hat{u}_{f,h}^{n-1}, \hat{u}_{s,h}^{n-1}, \hat{p}_{f,h}^{n-1}, \hat{\varphi}_{s,h}^{n-1}\}$ be given. Using a One-Step-θ discretization as in Formulation 28, we seek $\hat{U}_h^n = \{\hat{v}_{f,h}^n, \hat{v}_{s,h}^n, \hat{u}_{f,h}^n, \hat{u}_{s,h}^n, \hat{p}_{f,h}^n, \hat{\varphi}_{s,h}^n\} \in \widehat{X}_h^0$ via

$$\begin{aligned}
&\hat{A}_T(\hat{U}_h^{n,k})(\hat{\Psi}_h) + \theta\hat{A}_E(\hat{U}_h^n)(\hat{\Psi}_h) + \hat{A}_P(\hat{U}_h^n)(\hat{\Psi}_h) + \hat{A}_I(\hat{U}_h^n)(\hat{\Psi}_h) \\
&= -(1-\theta)\hat{A}_E(\hat{U}_h^{n-1})(\hat{\Psi}_h) \quad \forall \hat{\Psi} \in \widehat{X}_h.
\end{aligned}$$

11.8.1.8 Nonlinear solution

We finally show in a formal way, how the defect-correction step of Newton's method reads. Both variants the residual-based version from Section 7.9.1 and the error-oriented version from Section 7.9.2 have been implemented and tested. In [441], we obtained better results with the error-oriented technique since the residual might increase at the beginning. We start from Formulation 30, which is fully discretized but still nonlinear.

Formally, the Jacobian at a time point t_n reads: Find $\delta\hat{U}^j \in \widehat{X}_h^D$ such that

$$\begin{aligned} \hat{A}'(\hat{U}_h^j)(\delta\hat{U}_h^j, \hat{\Psi}_h) &= -\hat{A}(\hat{U}_h^j)(\hat{\Psi}_h) \quad \forall \hat{\Psi}_h \in \widehat{X}_h^0, \\ \hat{U}_h^{j+1} &= \hat{U}_h^j + \lambda_j \delta\hat{U}_h^j. \end{aligned}$$

Specifically, using the previous groupings (Definition 85), we have for the right-hand side residual using a One-Step-θ scheme (Formulation 28):

$$\begin{aligned} \hat{A}(\hat{U}_h^j)(\hat{\Psi}_h) := {} & \hat{A}_T(\hat{U}_h^{n,k})(\hat{\Psi}_h) + \theta\hat{A}_E(\hat{U}_h^n)(\hat{\Psi}_h) \\ & + \hat{A}_P(\hat{U}_h^n)(\hat{\Psi}_h) + \hat{A}_I(\hat{U}_h^n)(\hat{\Psi}_h) \\ & + (1-\theta)\hat{A}_E(\hat{U}_h^{n-1})(\hat{\Psi}_h). \end{aligned}$$

For the Jacobian, we need to compute the directional derivatives and obtain formally:

$$\begin{aligned} \hat{A}'(\hat{U}_h^j)(\delta\hat{U}_h^j, \hat{\Psi}_h) := {} & \hat{A}_T'(\hat{U}_h^{n,k})(\delta\hat{U}_h^j, \hat{\Psi}_h) + \theta\hat{A}_E'(\hat{U}_h^n)(\delta\hat{U}_h^j, \hat{\Psi}_h) \\ & + \hat{A}_P'(\hat{U}_h^n)(\delta\hat{U}_h^j, \hat{\Psi}_h) + \hat{A}_I'(\hat{U}_h^n)(\delta\hat{U}_h^j, \hat{\Psi}_h). \end{aligned}$$

Inside each bilinear form of $\hat{A}'(\hat{U}_h^j)(\delta\hat{U}_h^j, \hat{\Psi}_h)$, the directional derivatives need to be evaluated now. For classical fluid-structure interaction (without phase-field) details are provided in [434, Section 5.6] with C++ implementations provided in [438],

https://media.archnumsoft.org/10305/

and DOpElib [207, 413],

http://www.dopelib.net

in Example `dopelib/Examples/PDE/InstatPDE/Example2`. Furthermore, we notice that we employ in our implementation global H^1 spaces for $\hat{v}_h^n := (\hat{v}_{f,h}^n, \hat{v}_{s,h}^n)$ and $\hat{u}_h^n := (\hat{u}_{f,h}^n, \hat{u}_{s,h}^n)$. Therefore, we finally obtain at each t_n the solution $\hat{U}_h^n = (\hat{v}_h^n, \hat{u}_h^n, \hat{p}_h^n, \hat{\varphi}_h^n)$.

11.8.2 A fixed-mesh fully Eulerian model

Similar to Section 11.7.3, we propose now the promised alternative approach, which is more natural in terms of the phase-field coordinate system and fluid flow. In extension to Section 11.7.3, we first need to define an indicator function (level-set) that determines the solid and fluid subdomains. This task is more critical than in ALE fluid-structure interaction because the interface Γ is not anymore aligned with mesh faces and must be captured. This is conceptionally the same difficulty as described in Section 11.2.

Using the reverse ALE transformation (Definition 82), the domains are implicitly defined by the characteristic function χ_f (same way as in Section 11.2).

Proposition 124. *The indicator functions for determining Ω_f and Ω_s are obtained via*

$$\chi_f := \begin{cases} 1, & x - u \in \widehat{\Omega}_f, \\ 0, & x - u \in \widehat{\Omega}_s \cup \widehat{\Gamma}_i, \end{cases}$$

$$\chi_s := 1 - \chi_f.$$

Here, $\widehat{\Omega}_f$ and $\widehat{\Omega}_s$ denote the reference configurations of the Eulerian system: $\widehat{\Omega}_f = \Omega_f(t = 0)$. The key ingredient in this framework is the initial position set (IPS), which is used to determine the position of the structural points at the initial time step $t = 0$ [138, 377, 373, 437] and [376, Section 6.2]. For this reason, the inverse mapping $\mathcal{A}_s(x,t)$ (see also Section 11.6.1) is strictly-speaking only required for the solid displacements u_s.

Remark 89. While in Section 11.2, the indicator functions depend on φ, they depend here on the displacement solution u.

Definition 86. The zero level set of the (time-evolving) interface Γ_i (or outer boundary) is given by

$$\Gamma_i = \{x \in \Omega \mid x - u \in \widehat{\Gamma}_i\},$$

in which $u := \chi_f u_f + \chi_s u_s$. As in phase-field fracture itself, the interface is allowed to cut through mesh elements. An illustration is provided in Figure 11.9.

The previous results assume that u is computable. The question remains how this can be done. For instance in level-set methods [346, 401], the value of u is transported inside domain with a certain velocity w [329]. Consequently, at a given material point at the position $x \in \Omega$ and time $t \in I$, the value $u(x,t)$ determines the initial position of this point at the initial time $t = 0$. In Eulerian solid formulations, we need to construct such a transport equation. Luckily, one candidate appears naturally within a fully Eulerian setting.

Formulation 31 (Reduced initial position set). Find $u \in u_0^D + V^0$ and for almost all $t \in I$, it holds

$$(\partial_t u - w + (w \cdot \nabla)u, \psi) = 0 \quad \forall \psi \in V^0,$$

in which u_0^D is a suitable extension of the Dirichlet data along the boundaries. The initial and boundary conditions are given by

$$u(x,0) = 0, \quad x \in \Omega,$$
$$u(x,t) = 0, \quad x \in \partial\Omega,\ t \in I.$$

Then the value of u will be determined by the choice of the convection velocity w. Here, one choice is to simply take $w := v$.

11.8.2.1 Governing systems

For the solid, we adopt Proposition 122 for dynamic phase-field fracture in Eulerian coordinates. This system is augmented with terms for the pressure similar to Section 11.4.1.2. For the fluid, the incompressible Navier–Stokes equations from Formulation 26 are taken.

11.8.2.2 Interface conditions

Analogously to Section 11.8.1.4, we deal with two types of interfaces, but now in Eulerian coordinates:

1. Interface Γ_i between fluid and solids (FSI interface)
2. Interface ∂C between solid and fracture

Again, the first interface is of type 2 and the second one of type 3. But now both need to be captured with indicator functions, which we formulated in Proposition 124 and Proposition 80, respectively. The FSI coupling conditions on the interface Γ_i read

$$\begin{aligned} v_f &= v_s, \\ \sigma_f n &= \sigma_s n, \end{aligned}$$

where σ_f is defined in (11.48) and σ_s defined in (11.13). For pressurized fractures, according to Proposition 84, we have at the fracture boundary:

$$\sigma_s n = \sigma_F n \quad \text{on } \partial C,$$

where $\sigma_F = -p_F I$ with p_F being the fracture pressure.

We notice that fully Eulerian fluid-structure interaction is delicate because the FSI coupling conditions require a highly accurate discretization for the correct balance of contact forces and smeared transition zones result in an important information loss. This is mainly due to that the whole constitutive stress tensor σ (i. e., the material coefficients ν_f and μ_s, λ_s, and the densities ρ_f, ρ_s) vary; see also Section 11.7.3.1 for further background literature. For instance, this is less problematic in two-phase flows in which the density and the viscosity change across Γ_i, but the principle stress tensor is composed by the same state variables; we refer the reader also to Section 11.5. In pressurized phase-field fracture, we could even avoid computing quantities inside the transition zone by using Gauss' divergence theorem (Section 11.4). However, in fluid-filled phase-field fracture (Section 11.5), dependencies on ε were again observed and further research is necessary.

11.8.2.3 The final system

We have now all ingredients to pose the final problem in Eulerian coordinates:
- Navier–Stokes equations;
- Solid displacements;
- Phase-field variational inequality.

We obtain the following CVIS.

Proposition 125 (Coupling phase-field with FSI in Eulerian coordinates). *Given the indicator functions $\chi_f(u_s)$ and $\chi_s(u_s)$ through Proposition 124 and a fixed fracture pressure $p_F \in L^\infty$. Find $(v_f, v_s, u_f, u_s, p_f, \varphi) \in \{v_f^D + V_f^0\} \times L_s \times \{u_f^D + V_s^0\} \times \{u_s^D + V_s^0\} \times L_f^0 \times K$, such that $v_f(0) = v_f^0$, $u_f(0) = u_f^0$, $v_s(0) = v_s^0$, $u_s(0) = u_s^0$, $\varphi(0) = \varphi^0$ are satisfied, and for almost all $t \in I$ holds:*

$$
\begin{aligned}
(\chi_f\rho_f\partial_t v_f, \psi_f^v) + (\chi_f\rho_f(v_f\cdot\nabla)v_f, \psi_f^v)&\\
+ (\chi_f\sigma_f, \nabla\psi_f^v) - \langle g, \psi_f^v\rangle - (\chi_f\rho_f f_f, \psi_f^v) &= 0 \quad \forall\psi_f^v \in V_f^0,\\
(\chi_s J\rho_s\partial_t v_s, \psi_s^v) + (\chi_s J\rho_s(v_s\cdot\nabla)v_s, \psi_s^v)&\\
+ (\chi_s g(\varphi)\sigma_s, \nabla\psi_s^v) - (\chi_s J\rho_s f_s, \psi_s^v)&\\
+ (\chi_s g(\varphi)p_F, \nabla\cdot\psi_s^v) &= 0 \quad \forall\psi_s^v \in V_s^0,\\
(\chi_f\nabla u_f, \nabla\psi^u) &= 0 \quad \forall\psi_f^u \in V_f^0,\\
(\chi_s(\rho_s\partial_t u_s + (v_s\cdot\nabla)u_s - v_s, \psi_s^u)) &= 0 \quad \forall\psi^u \in L_s,\\
(\chi_f\nabla\cdot v_f, \psi_f^p) &= 0 \quad \forall\psi_f^p \in L_f,\\
(\chi_s(1-\kappa)\varphi\sigma(u) : e(u), \psi-\varphi)&\\
+ 2(1-\kappa)(\chi_s\varphi\,(p_F\,\nabla\cdot u), \psi-\varphi) - 2\langle\chi_s\varphi p_F n\cdot u, \psi-\varphi\rangle_{\Gamma_i}&\\
+ \left(-\chi_s\frac{G_c}{\varepsilon}(1-\varphi), \psi-\varphi\right) + (\chi_s G_c\varepsilon\nabla\varphi, \nabla(\psi-\varphi)) &\ge 0 \quad \forall\psi \in K.
\end{aligned}
$$

Remark 90 (Outlook). This model of fully Eulerian phase-field fluid-structure-interaction framework is the outcome of recent research and first-time outlined in this monograph. The implementation has to be done yet. The algorithmic realization of time-stepping schemes and spatial discretization follows the concepts outlined in this monograph. Here, the recent efforts in temporal and spatial discretization [182, 177, 180, 181, 183] of locally modified finite elements and fully Eulerian FSI should be taken into account.

11.8.3 Fluid-structure interaction with fracture Navier–Stokes flow

In this section, we augment the previous fixed-mesh fully Eulerian frameworks with Navier–Stokes flow inside the fracture. Previously, the pressure p_F inside the fracture

Ω_F was given. We shall now solve the incompressible Navier–Stokes equations in Ω_F to obtain p_F. Since the pressure is not anymore necessarily constant, we also need to consider the pressure gradient ∇p from (11.16) (see also Section 11.4.1.4). Conceptionally, all ingredients have been introduced in the previous section. We propose two further frameworks in the following:

1. Phase-field fracture in elasticity with fracture Navier–Stokes flow:
   Starting from Proposition 125, we have Ω_s as before, and $\Omega_F := \Omega_f$ as fracture fluid domain. With this, the fracture boundary $\partial\mathcal{C}$ becomes now the FSI interface Γ_i. The interface coupling conditions are

$$v_F = v_S,$$
$$\sigma_F n = \sigma_S n,$$

   with $\sigma_F := \sigma_f$ defined in (11.48) and σ_s defined in (11.13). Most importantly, but conceptionally simple, the indicator functions are now defined with the help of the phase-field function: $\chi_F(\varphi)$ and $\chi_S(\varphi)$ (Proposition 80). Consequently, the arising situation is related to Section 11.5.
2. Fluid-structure interaction coupled to phase-field with fracture Navier–Stokes flow:
   In this setting, we again start from Proposition 125, but now deal with three subdomains Ω_s (solid), Ω_f (some fluid domain a priori independent of the fracture), Ω_F (fracture fluid domain). In the two flow domains, the fluid equations might differ by their material parameters and constitutive laws. We deal with a FSI interface Γ_i (determined by $\chi_s(u)$ and $\chi_f(u)$) and a FSI-fracture boundary $\partial\mathcal{C}$ (determined by $\chi_S(\varphi)$ and $\chi_F(\varphi)$). The interface coupling conditions are

$$\begin{aligned} v_f &= v_s, && \text{on } \Gamma_i,\\ \sigma_f n &= \sigma_s n, && \text{on } \Gamma_i,\\ v_F &= v_S, && \text{on } \partial\mathcal{C},\\ \sigma_F n &= \sigma_S n, && \text{on } \partial\mathcal{C}. \end{aligned}$$

Based on these two descriptions, we formulate the following CVISs:

Proposition 126 (Phase-field fracture in elasticity with fracture Navier–Stokes flow). *Given the indicator function $\chi_F(\varphi)$ through Proposition 80 and set $p_F := p_f$. Find $(v_F, v_S, u_F, u_S, p_F, \varphi) \in \{v_F^D + V_F^0\} \times L_s \times \{u_F^D + V_S^0\} \times \{u_S^D + V_S^0\} \times L_F^0 \times K$, such that $v_F(0) = v_F^0$, $u_F(0) = u_F^0$, $v_s(0) = v_S^0$, $u_s(0) = u_S^0$, $\varphi(0) = \varphi^0$ are satisfied, and for almost all $t \in I$ holds:*

$$\begin{aligned} &(\chi_F\rho_F\partial_t v_F, \psi_F^v) + (\chi_F\rho_F(v_F\cdot\nabla)v_F, \psi_F^v)\\ &+ (\chi_F\sigma_F, \nabla\psi_F^v) - \langle g, \psi_F^v\rangle - (\chi_F\rho_F f_F, \psi_F^v) = 0 \quad \forall \psi_F^v \in V_F^0, \end{aligned}$$

$$
\begin{aligned}
(J\rho_s\partial_t v_s, \psi_s^v) + (J\rho_s(v_s\cdot\nabla)v_s, \psi_s^v) + (g(\varphi)\sigma_s, \nabla\psi_s^v) - (J\rho_s f_s, \psi_s^v)&\\
+ (g(\varphi)p_F, \nabla\cdot\psi_s^v) + (g(\varphi)\nabla p_F, \psi_s^v) &= 0 \quad \forall\psi_s^v \in V_s^0,\\
(\chi_F\nabla u_F, \nabla\psi^u) &= 0 \quad \forall\psi_F^u \in V_F^0,\\
(\rho_s\partial_t u_s + (v_s\cdot\nabla)u_s - v_s, \psi_s^u) &= 0 \quad \forall\psi^u \in L_s,\\
(\chi_F\nabla\cdot v_F, \psi_F^p) &= 0 \quad \forall\psi_F^p \in L_F,\\
((1-\kappa)\varphi\sigma(u) : e(u), \psi-\varphi)&\\
+ 2(1-\kappa)(\varphi\,(p_F\,\nabla\cdot u + \nabla p_F\cdot u), \psi-\varphi)&\\
+\left(-\frac{G_c}{\varepsilon}(1-\varphi), \psi-\varphi\right) + (G_c\varepsilon\nabla\varphi, \nabla(\psi-\varphi)) &\geq 0 \quad \forall\psi\in K.
\end{aligned}
$$

Next, we present the three domain problem.

Proposition 127 (FSI coupled to phase-field with fracture Navier–Stokes flow). *Given the indicator function $\chi_F(\varphi)$ through Proposition 80 to determine Ω_F. Given $\chi_f(u_s)$ and $\chi_s(u_s)$ through Proposition 124 to determine Ω_f and Ω_s. Find $(v_f, v_F, v_s, u_f, u_s, p_f, p_F, \varphi) \in \{v_f^D + V_f^0\}\times\{v_F^D + V_F^0\}\times L_s\times\{u_f^D + V_s^0\}\times\{u_s^D + V_s^0\}\times L_f^0\times L_F^0\times K$, such that $v_f(0) = v_f^0$, $v_F(0) = v_F^0$, $u_f(0) = u_f^0$, $v_s(0) = v_s^0$, $u_s(0) = u_s^0$, $\varphi(0) = \varphi^0$ are satisfied, and for almost all $t\in I$ holds:*

$$
\begin{aligned}
(\chi_f\rho_f\partial_t v_f, \psi_f^v) + (\chi_f\rho_f(v_f\cdot\nabla)v_f, \psi_f^v)&\\
+ (\chi_f\sigma_f, \nabla\psi_f^v) - \langle g, \psi_f^v\rangle - (\chi_f\rho_f f_f, \psi_f^v) &= 0 \quad \forall\psi_f^v\in V_f^0,\\
(\chi_F\rho_F\partial_t v_F, \psi_F^v) + (\chi_F\rho_F(v_F\cdot\nabla)v_F, \psi_F^v)&\\
+ (\chi_F\sigma_F, \nabla\psi_F^v) - \langle g, \psi_F^v\rangle - (\chi_F\rho_F f_F, \psi_F^v) &= 0 \quad \forall\psi_F^v\in V_F^0,\\
(\chi_s J\rho_s\partial_t v_s, \psi_s^v) + (\chi_s J\rho_s(v_s\cdot\nabla)v_s, \psi_s^v)&\\
+ (\chi_s g(\varphi)\sigma_s, \nabla\psi_s^v) - (\chi_s J\rho_s f_s, \psi_s^v)&\\
+ (\chi_s g(\varphi)p_F, \nabla\cdot\psi_s^v) + (\chi_s g(\varphi)\nabla p_F, \psi_s^v) &= 0 \quad \forall\psi_s^v\in V_s^0,\\
(\chi_f\nabla u_f, \nabla\psi^u) &= 0 \quad \forall\psi_f^u\in V_f^0,\\
(\chi_s(\rho_s\partial_t u_s + (v_s\cdot\nabla)u_s - v_s), \psi_s^u) &= 0 \quad \forall\psi^u\in L_s,\\
(\chi_f\nabla\cdot v_f, \psi_f^p) &= 0 \quad \forall\psi_f^p\in L_f,\\
(\chi_F\nabla\cdot v_F, \psi_F^p) &= 0 \quad \forall\psi_F^p\in L_F,\\
(\chi_s(1-\kappa)\varphi\sigma(u) : e(u), \psi-\varphi)&\\
+ 2(1-\kappa)(\chi_s\varphi\,(p_F\,\nabla\cdot u + \nabla p_F\cdot u), \psi-\varphi)&\\
- 2\langle\chi_s\varphi p_F n\cdot u, \psi-\varphi\rangle_{\Gamma_i}&\\
+\left(-\chi_s\frac{G_c}{\varepsilon}(1-\varphi), \psi-\varphi\right) + (\chi_s G_c\varepsilon\nabla\varphi, \nabla(\psi-\varphi)) &\geq 0 \quad \forall\psi\in K.
\end{aligned}
$$

11.8.4 Illustrating comparisons of ALE with fixed-mesh fully Eulerian

In the Figures 11.8 and 11.9, newly evaluated results from Section 12.4.2 (using raw data from [443]) are displayed. We however notice that the fully Eulerian result is obtained from postprocessing the ALE solution with the inverse transformation $\mathcal{A}(x,t)$ (Definition 82, Section 11.6.1), i. e., the fully Eulerian figures are not actual computations, but only for illustration purposes—the implementation has yet to be done. Using the ALE approach (Section 11.8.1.3), all computations (Figure 11.8) are done in $\widehat{\Omega}$ (top), and in the physical domain (middle), specifically the mesh is deformed. Also, the FSI interface Γ_i is aligned with the mesh edges (Figure 11.9), while the fracture boundary $\partial\mathcal{C}$ is smeared. In the fully Eulerian approach, the mesh is fixed and both the FSI interface Γ_i and the (smeared) fracture boundary $\partial\mathcal{C}$ cut through mesh elements (Figure 11.9). The advantage becomes obvious because solid deformations and fracture propagation are independent of the mesh.

11.9 Specific challenges in multiphysics phase-field fracture

In this section, we highlight some challenges that we encountered over the last years in multiphysics phase-field fracture. Never ending need is required for better numerical solvers, parallel computing, more numerical analysis, and better mathematical models. Rather, we focus on very specific points from which some are a priori not obvious:

1. Too coarse meshes inside the fracture: Solving additional equations inside the fracture also require their own accurate discretization. However, phase-field with for instance $\varepsilon = 2h$ only has approximately 4 elements in perpendicular direction; see, for instance, Figure 5.4. Such a discretization is far from being accurate in terms of the discretization error.
2. Multiple interacting interfaces of different types (Section 1.3) and different numerical modeling (Section 1.3.4).
3. Computing leak-off (in porous media) and other physics at the fracture tip. Physics suggest negative pressures around the fracture tip in a porous medium, which we could reproduce in some of our studies [322].
4. Coupling fracture flow equations with other flow equations when they interact on interfaces; see e. g., Section 11.8.3.
5. Dependencies on ε and the level-set threshold c_{ls}. The numerical results depend on the fracture boundary location $\partial\mathcal{C}$.
6. Usually porous media equations using Biot's theory are upscaled equations. Here, the interface conditions on $\partial\mathcal{C}$ become more challenging [321].
7. Relation of ε and heterogeneous materials with spatially varying material parameters.
8. Validity of stress-splitting laws in multiphysics applications.

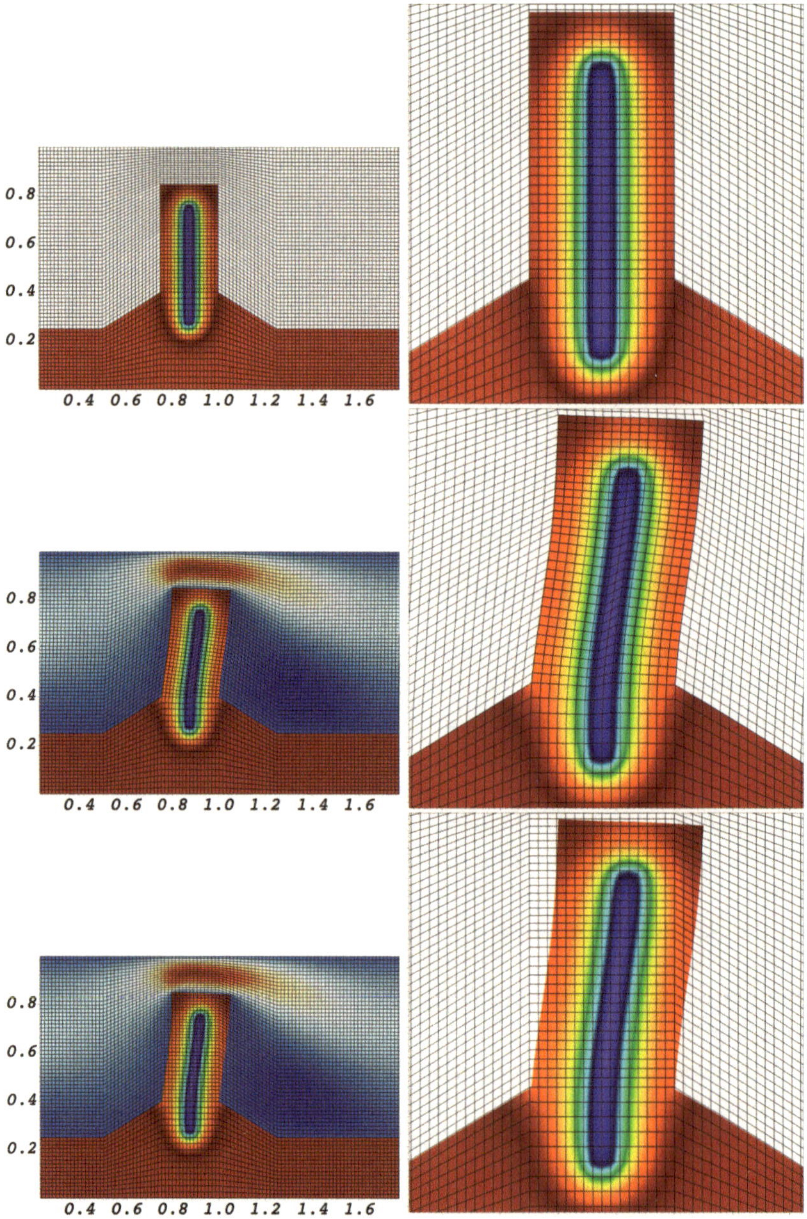

Figure 11.8: Comparison of IPS (initial point set) $\Omega(0)$, i. e., the reference domain $\widehat{\Omega}$ (top), the ALE configuration (middle), and a fully Eulerian perspective (bottom). The fully Eulerian subfigure is not a result of a computation (see also Remark 90), but taking the inverse transformation $\mathcal{A}(x,t)$ (Definition 82, Section 11.6.1) from the ALE computation. Consequently, the last subfigure is only an illustration. The figures are newly created from the raw data obtained in [443].

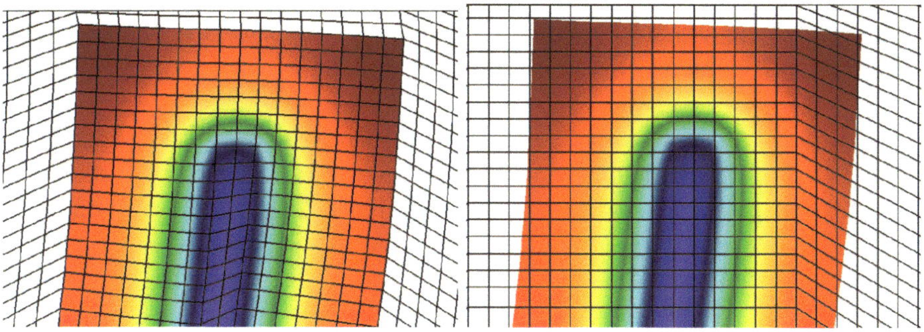

Figure 11.9: Zoom-in of Figure 11.8. At left, the ALE approach is shown, where the mesh fits with the FSI interface Γ_i. At right in the fully Eulerian approach, where the mesh is fixed and the FSI interface cuts through mesh elements.

9. Influence of boundary conditions on simulation results and goal functional evaluations. We have found quantitative influences in [222]. Furthermore, we have found influences on the fracture path near the boundary. It is not always clear, whether in these cases, the fracture path is physics or due to boundary conditions.
10. Initiatives and willingness to further compare models, algorithms, discretizations, and programming code. As an example, nearly every group has its own model for fluid-filled phase-field fractures: please see the list at the beginning of Section 1.5.

11.9.1 A good practice example from Frei/Richter/Wick

In view of the upcoming simulations and also the software chapter, let us hint on the last bullet point of the previous list. It is useful to have benchmarks! It is useful to compare different numerical models, different discretizations, and different solvers. An excellent example to me is this multiphysics work (without fractures though): Frei/Richter/Wick [182], but which is in good continuation of Section 11.8, we just have had. Therein, we partially developed a new mathematical model for solid growth with contact interacting with an ODE and incompressible Navier-Stokes flow. The resulting model is a complex fluid-structure interaction system.

The entire setting was developed with two different numerical models, namely ALE fluid-structure interaction and fully Eulerian fluid-structure interaction. Furthermore, two different programming codes were employed: in deal.II [27] and in Gascoigne [48]. The solid growth models in Eulerian coordinates were newly developed and continuous comparisons with the ALE version allowed to detect:

- mathematical model errors in the Eulerian constitutive laws; specifically in the development of multiplicative splitting of the deformation tensors $\widehat{F}$ and F;
- errors in the algorithmic realization in both codes: ALE and fully Eulerian;

- programming bugs in both codes: equations, goal functional evaluations, wrong boundary conditions, errors in the derivatives of the Jacobian matrices.

With continuous feedback discussions on the models and programming codes, reliable monitoring of our developments was achieved. The final results were furthermore tested for various spatial meshes (uniform mesh refinement) and different time step sizes in order to verify the numerical robustness and computational convergence.

12 Simulations IV

In this chapter, we substantiate some concepts from the previous chapter with representative numerical examples. For more studies, we refer the reader back to the preface "Own related work" and specifically therein to "Multiphysics modeling and applications" and "Benchmarking." Specifically, we notice that in our recent review article [431], several numerical simulations of multiphysics porous media phase-field fracture applications are shown. These results are complementary to the simulations presented in this chapter.

12.1 Quasi-static pressurized fractures

12.1.1 Sneddon's test

The first example is motivated by [71, 430] and is based on Sneddon's theoretical calculations [396, 395]. In the book [396], we refer to Chapter 2.4 for 2D calculations and to Chapter 3.3 for 3D manufactured solutions. Specifically, we consider the 2D case where a (constant) fracture pressure $p := p_F$ is used acting on the fracture boundary $\partial\mathcal{C}$; see specifically Proposition 84 and then the subsequent Section 11.4.1.2. To this end, we work with Proposition 88 with simple penalization (Section 5.2.2). We assume a dimensionless form of the equations in this section.

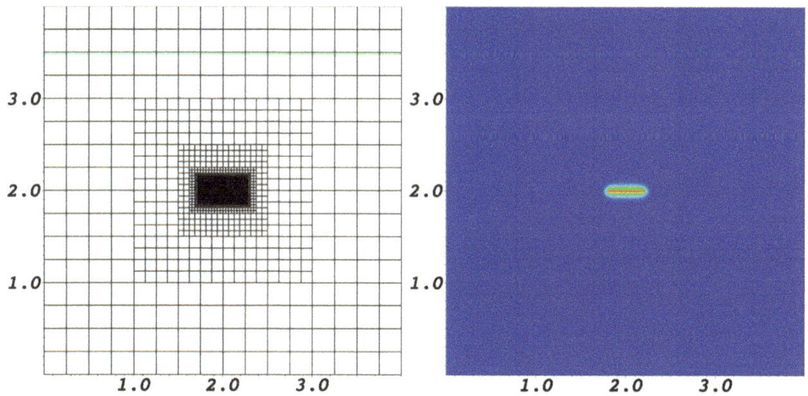

Figure 12.1: Sneddon's test: Configuration (left) and crack pattern (right). Figures reprinted by permission from Springer Nature: Springer, GEM – International Journal on Geomathematics, [324], 2019.

https://doi.org/10.1515/9783110497397-012

12.1.1.1 Benchmarking and further results

The Sneddon test case has been serving often as benchmark in our previous work. This configuration provides several appealing features (similar to Chapter 6) to test numerical discretizations, error analyses, and numerical solver behavior. Benchmark results for various goal functionals are provided in two works:

1. [222]: Results in 2D and 3D measuring the influence of ε under mesh refinement, influence of the domain size, and parallel scalability studies. Moreover, an open-source code is provided that allows to reproduce our findings.
2. [449]: Within the DFG priority program[1] SPP 1748: A selection of benchmark problems in solid mechanics and applied mathematics; benchmark No. 6. Therein, uniform mesh refinement was adopted and various goal functionals were computed.

Own further results are provided in:

1. [42]: pressurized fractures in incompressible solids and residual-based mesh adaptivity;
2. [324] and Section 12.1.2 below: influence of Neumann boundary conditions and numerical verification of the boundary integrals on ∂B_N in the formulation presented in Section 11.4.1.4;
3. [444]: development of monolithic nonlinear solvers;
4. [442]: goal-oriented error estimation and local mesh adaptivity;
5. [285]: step-wise increasing pressure and computational mesh convergence studies.
6. [430]: our first results in 2D and 3D: first comparisons and verification of the augmented Lagrangian penalization. Therein, the coupled problem is solved with iterative coupling (Section 7.5).

12.1.1.2 Configuration

The configuration is displayed in Figure 12.1. The domain is $B = (0,4)^2$. The test is stationary because there are neither time dependencies in boundary conditions nor in the given pressure. However, we perform two (pseudo) time steps, i. e., $N = 2$, in order to account for the crack irreversibility condition.

12.1.1.3 Boundary conditions

We use

$$\begin{aligned} u &= 0 \quad \text{on } \partial B \times I, \\ \varepsilon \partial_n \varphi &= 0 \quad \text{on } \partial B \times I. \end{aligned}$$

1 https://www.uni-due.de/spp1748/

12.1.1.4 Initial conditions

We prescribe the initial crack implicitly with the help of the phase-field function as recently carefully described in [449]; see also [430]. The initial crack length is $L = 2l_0 = 0.4$ with

$$\Omega_F = (1.8, 2.2) \times (2 - h, 2 + h) \subset B,$$

where h is the local mesh size. Therefore,

$$\varphi^0 = 0 \quad \text{in } \Omega_F \times \{0\}$$

and $\varphi^0 = 1$ in $B \setminus \Omega_F$. Thus, we deal with a *2D* crack with a length much larger than its width.

12.1.1.5 Parameters

The mechanical parameters Young's modulus and Poisson's ratio are set to be $E = 1.0$ and $\nu_s = 0.2$. The applied fracture pressure is $p = 10^{-3}$. The critical energy release rate is chosen $G_c = 1.0$. We work with simple penalization (Section 5.2.2) and choose $\gamma = 100 \times h^{-2}$. Furthermore, the phase-field regularization parameters are chosen as $\kappa = 10^{-10}$ (see Section 5.5.3.6) and $\varepsilon = 0.5\sqrt{h}$.

12.1.1.6 Quantities of interest

The goal is to measure the crack opening displacement (COD) and the volume of the crack (TCV) under spatial mesh refinement. For their definitions, we refer the reader to Section 11.3. We note that the integration is perpendicular to the crack direction. Here, the crack is aligned with the x-axis and, therefore, integration into the normal direction coincides with the y-direction.

Second, following [128], page 710, the volume of the fracture is $V - \pi w l_0$. The analytical expression for the fracture width (to which we compare) [128] is $w = 4\frac{(1-\nu_s^2) l_0 p}{E}$. Then the analytical expression for the volume becomes

$$V = 2\pi \frac{(1 - \nu_s^2) l_0^2 p}{E}, \tag{12.1}$$

where we recall from before $l_0 = 0.2$.

12.1.1.7 Discussion of our findings

The crack pattern and the corresponding mesh are displayed in Figure 12.1. Our findings for different spatial mesh parameters h are summarized in Figure 12.2. Specifically, we observe overall convergence to Sneddon's analytical solution [396] as well as a much better approximation of the crack tips under local mesh refinement. The obtained crack volumes are displayed in Table 12.1 in which the exact value is computed by Formula (12.1).

Table 12.1: Example 1: Fracture volume. The exact formula is given in (12.1).

h	8.8×10^{-2}	4.4×10^{-2}	2.2×10^{-2}	1.1×10^{-2}	exact
V	3.02×10^{-4}	2.77×10^{-4}	2.57×10^{-4}	2.49×10^{-4}	2.41×10^{-4}

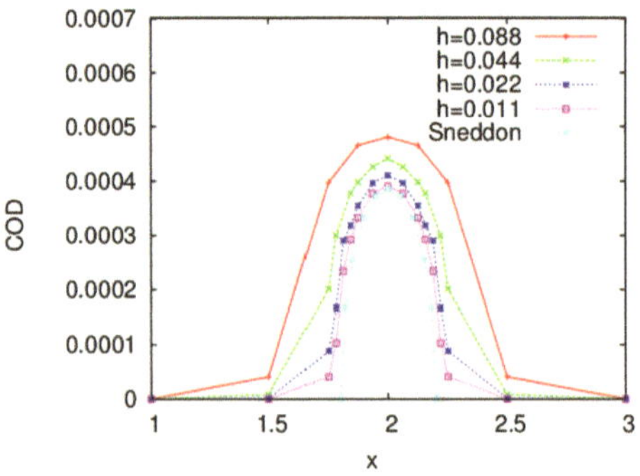

Figure 12.2: Sneddon's test: COD for different h. Sneddon's turquoise line with squares corresponds to his analytical solution. It is well observed that the crack tips must be resolved correctly as they are not approximated well on coarse meshes. Figures reprinted by permission from Springer Nature: Springer, GEM – International Journal on Geomathematics, [324], 2019.

12.1.2 Mixed boundary conditions

In this second test, we address mixed boundary conditions from [324]. We highlight that the developments in (11.16) from Section 11.4.1.2 are correct. Specifically, prescribing conditions other than Dirichlet on boundary parts ∂B_N, the boundary term

$$\int_{\partial B_N} (\tau + pn) \cdot u \, ds$$

must be considered according to Section 11.4.1.4. Therein, $\tau : \partial B_N \to \mathbb{R}^d$ are traction forces.

12.1.2.1 Configuration and governing system

The domain and parameters are the same as in the previous example. As governing CVIS, we employ Proposition 91 with an augmented Lagrangian treatment of the crack irreversibility constraint. Specifically, we have $p = 10^{-3}$ and $\nabla p = 0$.

12.1.2.2 Boundary conditions

The only (major) change in comparison to Example 1 concerns the boundary conditions. The top Γ_{top} and bottom Γ_{bottom} boundaries now form a Neumann boundary $\partial B_N = \Gamma_{\text{top}} \cup \Gamma_{\text{bottom}}$. Here, we prescribe Neumann conditions τ of homogeneous and nonhomogeneous type, respectively. Then we compare these results to the previous setting. In total, we design the following tests:

– Case 1 (as in Example 1 before): let $\partial B_N = \emptyset$,

$$u = 0 \quad \text{on } \partial_D B;$$

- Case 2:

$$\tau_{t,b} = (0,0)^T \quad \text{on } \partial B_N,$$
$$u = 0 \quad \text{on } \partial_D B;$$

- Case 3:

$$\tau_t = (0, 0.001)^T \quad \text{on } \Gamma_{\text{top}},$$
$$\tau_b = (0, -0.001)^T \quad \text{on } \Gamma_{\text{bottom}},$$
$$u = 0 \quad \text{on } \partial_D B;$$

- Case 4:

$$\tau_t = (0, 0.1)^T \quad \text{on } \Gamma_{\text{top}},$$
$$\tau_b = (0, -0.1)^T \quad \text{on } \Gamma_{\text{bottom}},$$
$$u = 0 \quad \text{on } \partial_D B.$$

These computations are performed on a 7 times uniformly refined mesh with $h = 0.044$. The phase-field function is similar to Example 1, displayed in Figure 12.1 (right subfigure).

12.1.2.3 Discussion of our findings

The maximal crack width openings $w_{\max}$ at $x = 2$ computed with the help of Proposition 83 in the middle of the fracture are:

- Case 1: $w_{\max}(x = 2;\ 0 \le y \le 4) = 5.25244 \times 10^{-4}$;
- Case 2: $w_{\max}(x = 2;\ 0 \le y \le 4) = 5.52572 \times 10^{-4}$;
- Case 3: $w_{\max}(x = 2;\ 0 \le y \le 4) = 1.31092 \times 10^{-3}$;
- Case 4: $w_{\max}(x = 2;\ 0 \le y \le 4) = 7.67588 \times 10^{-2}$.

Let us interprete these findings in the following. In Case 2, zero traction forces ($\tau = 0$) are applied on the top and the bottom boundaries and the fracture pressure keeps the fracture open. In addition, the maximal crack opening displacement is very similar (as expected) to Case 1. We further remark that the pressure boundary term $\int_{\partial B_N} p w_n \, dS$ (see Section 11.4.1) is important when working with Neumann conditions. For instance, when $\int_{\partial B_N} p w_n \, dS$ is not used in Case 2, we obtain a negative width, which is of course nonphysical for this setting. The influence is significant for all cases when $\|\tau\| < \|p\mathbf{n}\|$ (Case 2) and $\|\tau\| \approx \|p\mathbf{n}\|$ (Case 3).

In the Cases 3 and 4, the domain is under tension, since now the traction forces are strictly positive/negative, respectively. Consequently, the fracture opens more than in the first two cases. Specifically, when the traction force is increased by a magnitude of order 2, the fracture width is also higher about a magnitude of order 2. The y-displacement fields (here these correspond to the crack opening displacements since the fracture is aligned with the x-axis) are displayed in Figure 12.3.

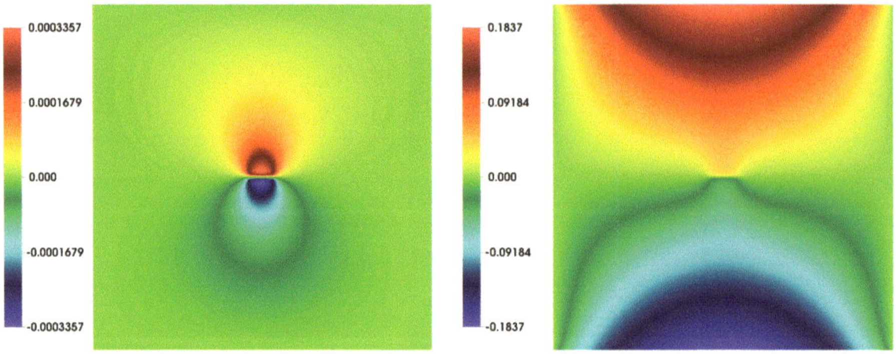

Figure 12.3: Pressurized fracture with mixed boundary conditions: the y-displacements for the Cases 1 and 4. Figures reprinted by permission from Springer Nature: Springer, GEM – International Journal on Geomathematics, [324], 2019.

12.2 Nonisothermal phase-field fracture

In this second representative numerical test, we consider fractures subject to pressure and temperature variations as modeled in Section 11.4.2. Specifically, we work with Proposition 97. The material parameters are taken from [416]. We emphasize that based on Sneddon's calculations for pressurized fracture, the authors in [416] developed manufactured solutions for pressurized nonisothermal fractures. Our other results are published in [343].

Configuration

The half-crack length is 10 m in a porous medium of size $B := (0\,\mathrm{m}, 200\,\mathrm{m})^2$. The two-dimensional test cases are based on the configuration shown in Figure 12.4.

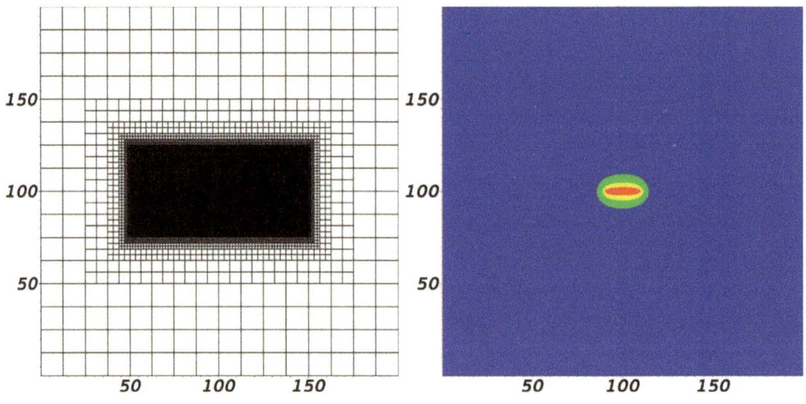

Figure 12.4: Nonisothermal tests: Mesh and geometry and initial crack location (in red) for all three examples. The domain is $B := (0, 200)^2$ [m].

Parameters and problem data

For the flow and temperature, we have the thermal diffusivity $\kappa_T = 10^{-6}\,\mathrm{m}^2/\mathrm{s}$, the initial temperature $T_0 = 100\,°\mathrm{C}$, (the apparent temperature is $T_F = 30\,°\mathrm{C}$), the total initial stress $p_0 = 12130\,\mathrm{kPa}$. In the stationary-in-length test cases (only evolution of the aperture), we use the constant pressure $p_F = 15834\,\mathrm{kPa}$.

For the crack propagation test cases, we use a higher and time-increasing pressure:

$$p = t \times 10 \times p_F.$$

The geomechanical parameters are Young's modulus $E = 1.5 \times 10^{10}\,\mathrm{Pa}$, Poisson ratio $\nu = 0.15$, the linear thermal-expansion coefficient $\beta = 10^{-5} 1/C$, and finally the critical energy release rate $G_c = 10^{10}\,\mathrm{Pa/m}$. Furthermore, the phase-field parameters are chosen as $\kappa = 10^{-12}$ and $\varepsilon = 10 \times \sqrt{h}$.

12.2.1 Example 1: Sneddon's 2d setting with pressure

In the first example, we consider the same stationary setting with a pressurized fracture as in the previous numerical test. The only extension is that we employ more realistic material and geometry data. These are motivated from [416]. The loading step size is $k = 1s$ and five steps are computed in order to satisfy the crack irreversibility constraint. We observe results on three different refinement levels using local mesh adaptivity. The formula to compute the half-width is

$$w(x,t) = \frac{2(1-\nu_s^2)l_0}{E}\sqrt{1-\rho^2}(p - p_0 - C_T(T_F - T_0)),$$

where l_0 is the half-length of the crack and $0 < \rho < 1$, i. e., $\rho = x/l_0$ with x being the distance to the origin of the crack. Our findings are displayed in Figure 12.5 [left subfigure].

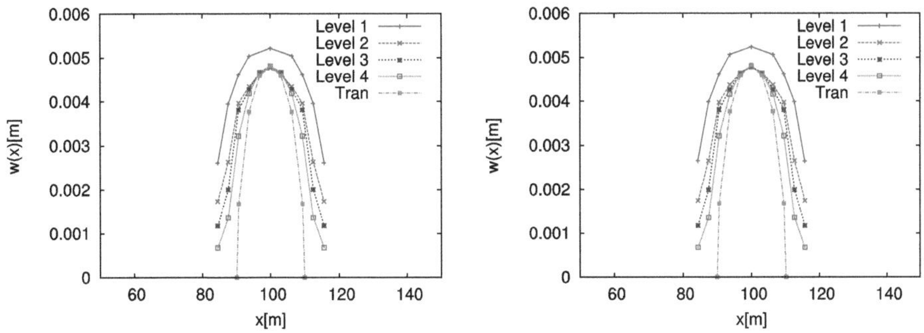

Figure 12.5: Examples 1 and 2: Maximum width in the middle of the fracture. "Tran" denotes the reference solution from [416]. "Level" denotes the mesh level obtained by refining locally around the fracture region.

12.2.2 Example 2: Sneddon's 2d setting with pressure and constant temperature

We set $\lambda_T = 10^{-4}$ constant and we work still in a stationary setting where the time step loop terminates after one step. As in the previous test case, we use $k = 1\,\mathrm{s}$. The crack opening discplacement is shown in Figure 12.5 [right subfigure].

12.2.3 Example 3: Setting with pressure, temperature, and decline constant

In this third example, we use Hagoort's decline constants (Section 11.4.2.1) to heuristically account for the temperature variation in time. Since this test case is now time-dependent, we set the time step size to one day $k = 86400\,\mathrm{s}$ and observe $T = 365$ days (one year). Our results are displayed in Figure 12.6.

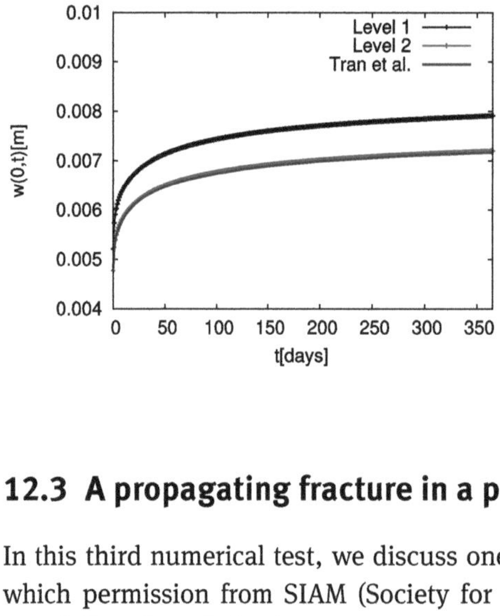

Figure 12.6: Example 3: Maximum width evolution over time in the nonisothermal case. Injection of cold water $T_F < T_0$ leads to an increasing aperture over large time scales. Tran's et al. reference solution can be computed with the formulae provided in [416]. "Level" denotes the mesh level obtained by refining locally around the fracture region.

12.3 A propagating fracture in a porous medium

In this third numerical test, we discuss one result from [321, Section 4.1, $\alpha = 1$] for which permission from SIAM (Society for Industrial and Applied Mathematics) is gratefully acknowledged.

12.3.1 Outline of the main concepts

The configuration, boundary conditions, material coefficients, model, and discretization parameters can be found in [321]. In short, we consider a porous medium with Biot's coefficient $\alpha = 1$ and inject a fluid via a source q into the middle of a given fracture. For the displacements and the phase-field variable, we define as usual $U = (u, \varphi)$. Here, Proposition 93 from Section 11.4.1.5 is employed. The numerical solution uses iterative coupling as presented in Section 7.6. Once solved, we then obtain (u, φ) at a given time point t_n. This solution is then coupled to the pressure diffraction system as outlined in Section 11.5. Employing Algorithm 23, we obtain a global pressure p. To distinguish fracture pressure and porous media pressure, we adopt the phase-field indicator function (Section 11.2), i. e., it holds $p = \chi_R p + \chi_F p$.

12.3.2 Discussions of our findings

We discuss in the following aspects that are characteristic in terms of this monograph. The crack pattern and pressure distribution are displayed in Figure 12.8. Quantitative results are shown in Figure 12.7. According to Section 5.5.3, we took the following design decisions:

1. Proof of concept: numerical model of a propagating fracture inside a porous medium, detecting negative pressures at the fracture tip;
2. Robustness of numerical discretization for fixed $\varepsilon > 0$ and only varying the spatial discretization parameter h.

These two aspects are shown in Figure 12.8 in terms of the phase-field function (top row) and the pressure distribution (bottom row). Measuring the half length of the crack and the maximum pressure in terms of the $L^\infty(B)$ norm, we obtain the quantitative results as show in Figure 12.7. For fixed $\varepsilon > 0$, here taken as $\varepsilon = h_{\text{coarse}} = 0.088$, we observe convergence of both quantities of interest under spatial mesh refinement.

12.4 Fractured solid coupled with Navier–Stokes flow

In this fourth numerical test, we consider a fixed (nonpropagating) fracture in the solid part coupled to Navier–Stokes flow. This configuration belongs to Section 11.8 and is a phase-field fluid-structure interaction problem modeled in ALE coordinates.

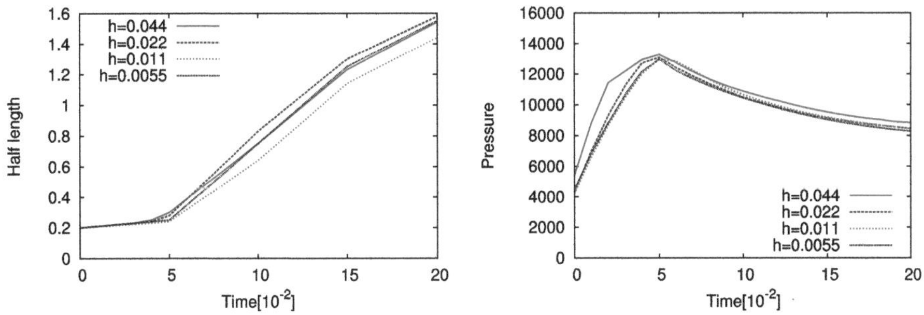

Figure 12.7: Fluid-filled fracture in a porous medium: Evolution of the half-length (left) and of the pressure p (right). [321]: Copyright @ 2015 Society for Industrial and Applied Mathematics. Reprinted with permission. All rights reserved.

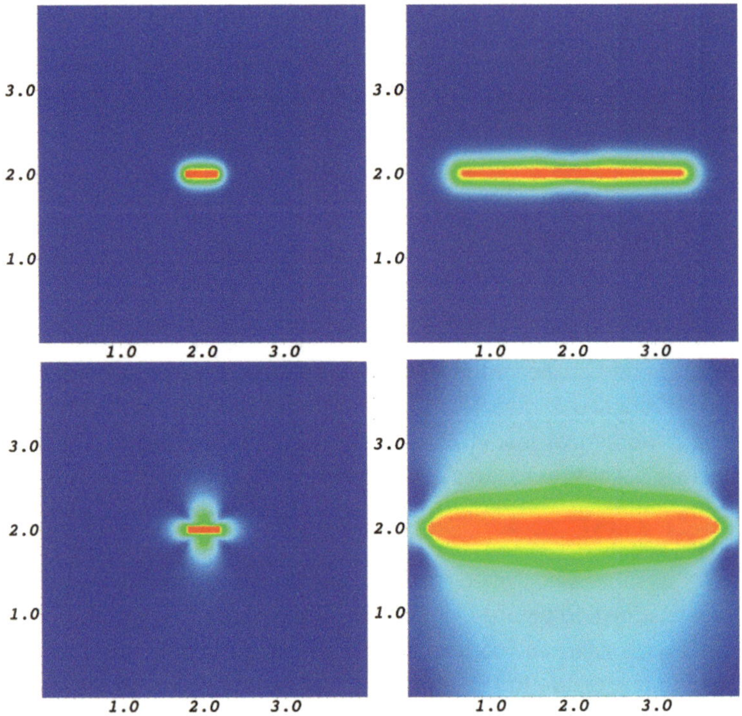

Figure 12.8: Fluid-filled fracture in a porous medium: Crack pattern (top) and pressure distribution (bottom) for $T = 0$ s and $T = 0.2$ s. The pressure ranges in the left figure from 0.36 Pa to 6776 Pa and from 791 Pa to 20 350 Pa in the right figure. [321]: Copyright @ 2015 Society for Industrial and Applied Mathematics. Reprinted with permission. All rights reserved.

12.4.1 Example 1: Constant flow

The flow is moderate and first leads to an oscillating obstacle and finally results in a stationary flow profile. According to Section 5.5.3, we took the following design decisions:

1. Proof of concept: numerical model of a fracture inside a largely deforming solid interacting with incompressible flow (see Figures 12.9, 12.11 and previously Figures 11.9 and 11.8);
2. Robustness of numerical discretization in space and time, i. e., fixing $\varepsilon > 0$ and varying the spatial discretization parameter h and the time step size k. Our findings are displayed in Figure 12.10. Most importantly, this qualitative computational convergence analysis yields stability under temporal and spatial mesh refinement. This indicates numerical robustness of the proposed model.

In [441, 443], several further challenges were extensively discussed and most of them are still current open problems.

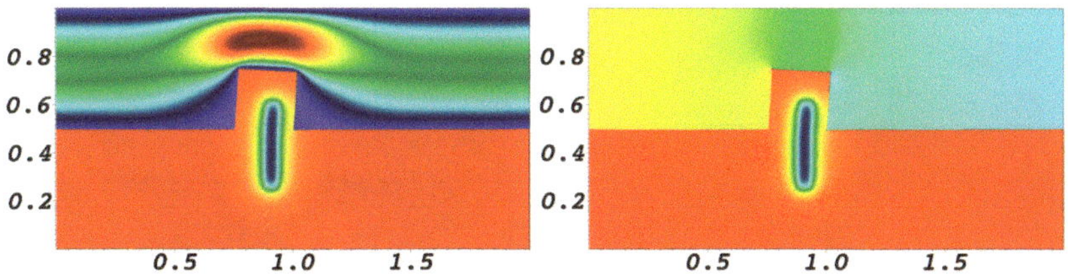

Figure 12.9: Example 1: Coupling fluid-structure interaction with phase-field fracture at $T = 50$ s. The flow field (with maximum velocity of 0.012 m/s) and the fracture location are shown. The fracture is characterized by $\hat{\varphi} = 0$ (blue). Next, the pressure field is shown. The pressure is zero (blue) at the outflow boundary, which is due to the do-nothing condition [229]. Reprinted from [441] with permission from Elsevier.

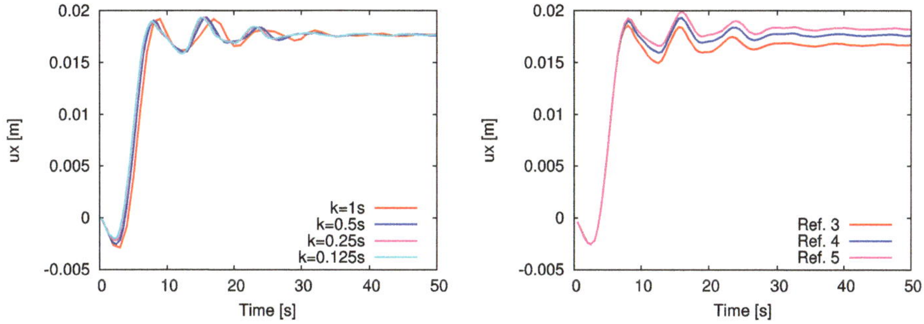

Figure 12.10: Example 1: Coupling fluid-structure interaction with phase-field fracture: Convergence tests in time (left) and space (right) for the point value $\hat{u}_x(1, 0.75)$. Reprinted from [441] with permission from Elsevier.

12.4.2 Example 2: Sinus-inflow

In this example, the inflow profile is a sinus-curve and in particular we construct the inflow in such a way that also backflow is generated; see Figure 12.11. The computations were originally done in [443] and we reevaluate the results in the following.

12.4.2.1 Inflow boundary condition

A parabolic time-dependent inflow velocity profile is given on $\widehat{\Gamma}_{in}$ by

$$v_f(0, y) = \bar{U}(y - 0.25)(y - 1), \quad \bar{U} = \left[2\sin\left(\frac{2t}{\pi} - 0.15\pi\right) + 1\right] \mathrm{m\,s}^{-1}.$$

12.4.2.2 Discussion of our findings

Our findings are shown in the Figures 12.11, 12.12, and 12.13. Brief discussions are provided in the figure captions.

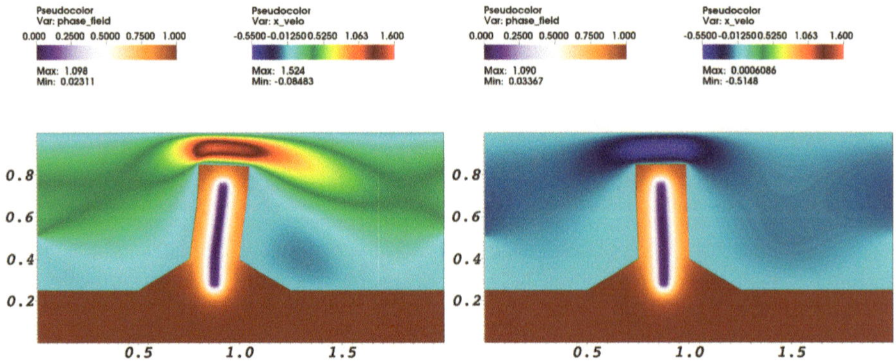

Figure 12.11: Example 2: Display of the velocity field at two different times in which the highest forward flow and the highest backward flow (negative x-velocity) are obtained. Figures reprinted with permission by de Gruyter: Thomas Wick, Coupling fluid-structure interaction with phase-field fracture: Algorithmic details; in S. Frei, B. Holm, T. Richter, T. Wick, H. Yang (Eds.) of the book Fluid-Structure Interaction: Modeling, Adaptive Discretizations and Solvers, Radon Series on Computational and Applied Mathematics, 20, de Gruyter, Berlin/Boston, 2017, page 358, Figure 9.11 [443].

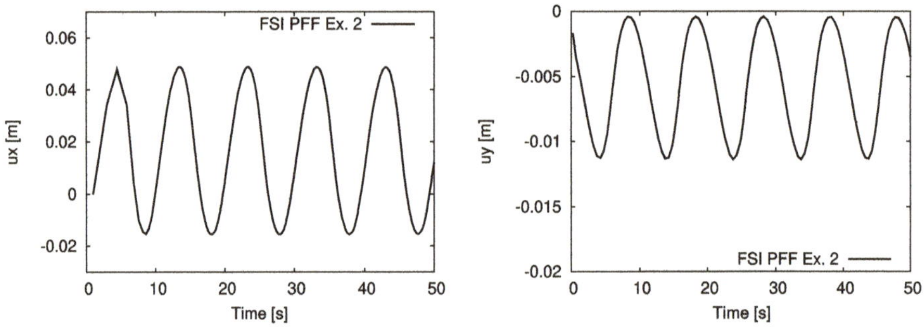

Figure 12.12: Example 2: u_x and u_y displacements measured in the right corner point $u(1, 0.85)$ of the solid obstacle. Here, we observe regular physical oscillations in both solution components. The shifted Crank–Nicolson scheme (Section 5.3) was employed as time-stepping scheme. The figures are redone with the raw data obtained in [443].

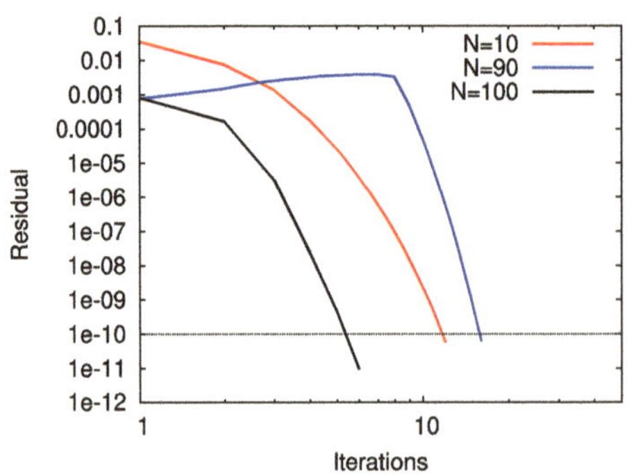

Figure 12.13: Example 2: Selected time steps displaying the convergence behavior of Newton's method. Specifically, we observe that a Newton method is required, which allows for increasing residuals, such as the modified version or an error-oriented method (Section 7.9). The figures are newly processed with the raw data obtained in [443].

13 Research software development

An important part in the presented work is sustainable software development.[1] Our implementations are based on C++.[2]

As noticed in my GAMM Leitartikel [446, pp. 4–10; specifically p. 10], there are good reasons to publish (open-source) or not publish scientific software. Such reasons are summarized in [278] and https://publiccode.eu/, and recently in [28]. Additionally, in 2012 the term Research Software Engineer (RSE) was coined in the UK. This lead to the formation of many national initiatives and associations like the German deRSE. Their first international conference deRSE19 included a workshop on sustainable software development in Germany.[3] Following the workshop, a group of scientists from various subjects got together to write [25], in which they address open challenges to further the development of sustainable research software. In my career path to date, I have met both opinions: publishing open-source software and developing in-house codes.

Several examples in this book can be reproduced either directly or with little extra work with the help of

1. [222]: T. Heister, T. Wick:
   https://github.com/tjhei/cracks
   High performance parallel computing fracture propagation with phase-field modeling, primal-dual active set solver and adaptive mesh refinement and high-performance parallel computing. The code is based on the open-source finite element package deal.II (Differential Equations Analyis Library) [36, 26].
   In the most recent deal.II version 9.2.0 https://www.dealii.org/deal92-preprint.pdf [Section 4], my doctoral researchers Katrin Mang, Jan Philipp Thiele, and Daniel Jodlbauer are officially involved. Moreover, my Bachelor students Julian Roth[4] and Julius Grundmann work currently with deal.II.
2. [207, 413]: W. Wollner, T. Wick:
   www.dopelib.net
   DOpElib: Differential Equations and Optimization Environment library, which is a toolbox providing modularized highlevel algorithms for PDE/CVIS simulations and optimization problems. The underlying finite element library is again deal.II.

For multiphysics simulations, besides DOpElib, the following open-source template has appeared to be successful:

1 http://www.thomaswick.org/gallery_engl.html

2 For C++ basics, please see http://www.cplusplus.org/ and https://www.cplusplus.com/

3 Thanks to Jan Philipp Thiele for being involved here and providing a lot of information.

4 https://github.com/mathmerizing/MultigridMaxwell

https://doi.org/10.1515/9783110497397-013

3. [438]: T. Wick:
   https://media.archnumsoft.org/10305/
   Solving monolithic fluid-structure interaction problems in arbitrary Lagrangian Eulerian coordinates with the deal.II library.
   This template was originally developed in the year 2011 for fluid-structure interaction problems as described in the accompanying article [438]. In the year 2013, the template further served for our first phase-field numerical implementation [318] and is the precursor of [221] and [222].
   Recently, problems in poroelasticity (linear Biot equations) [447] and thermo-poroelasticity [424, 426] have been solved with this template.[5] Updates of [438] to more recent deal.II versions can be found on
   http://www.thomaswick.org/gallery_engl.html
4. [183]: S. Frei, T. Richter, T. Wick:
   https://zenodo.org/record/1457758#.XqsilqZCRUR
   An implementation of a locally modified finite element method for interface problems in deal.II. Therein, an implementation of the original work [179] was done, including a careful documentation according to deal.II standards. This approach provides basic techniques for fully Eulerian solid descriptions as considered in the Sections 11.7.3 and 11.8.

Lastly, we mention the multiphysics in-house code for fracture modeling in porous media

5. [431]: M. F. Wheeler, T. Wick, S. Lee: IPACS: Integrated Phase-Field Advanced Crack Propagation Simulator. An adaptive, parallel, physics-based-discretization phase-field framework for fracture propagation in porous media.
   The governing phase-field model is based on [221, 222]. Physics-based discretizations for flow and transport in porous media are based on enriched Galerkin (EG) discretiations [400, 279, 283, 284] that are used together with fracture in [281, 280].

5 I want to thank Guido Kanschat and Wolfgang Bangerth for encouraging me to prepare [438]. It was at that time a very good exercise and it turned out to be a sustainable development as highlighted here through the given references. This might be a good place to thank various people. In the year 2007, Franz-Theo Suttmeier, developer of DEAL (deal.I; one not two!), pointed me to deal.II. Shortly after, I became involved in two important events: In 2010, I organized together with Bärbel Janssen, Timo Heister, and Martin Kronbichler the second deal.II workshop in Heidelberg. It was a great success! Indeed, we managed to have all three authors from [36] as key note speakers! Two years later, in 2012, I organized together with Bärbel Janssen, Timo Heister, and Matthias Maier the third deal.II workshop—again in Heidelberg. Despite never becoming a true deal.II-developer, but rather a user (I am always equally interested in the mixture of mathematical modeling, numerical analysis, doing proofs, and coding. Therefore, I could not spend more time on deal.II software developments). It however seems that I am one of the most productive ones; see https://www.dealii.org/publications.html. Thanks to Wolfgang Bangerth, Guido Kanschat, Timo Heister, Martin Kronbichler, Matthias Maier, and so many more, see https://www.dealii.org/authors.html, for a great, great package!

Predictor-corrector mesh refinement and parallel computing are originally based on [221].

13.1 Current developments in my group

Direct and associated members (Postoc, doctoral researchers, Master, Bachelor) of my Hannover group and my associated Austrian Linz team (through the FWF project[6] P29181) currently further use and/or develop research software: mainly using deal.II and DOpElib, but also in octave [143] and python.[7]

1. Bernhard Endtmayer (RICAM[8] Linz): Multigoal-oriented error estimation;
2. Daniel Jodlbauer (RICAM Linz): Parallel multigrid solvers for phase-field fracture and fluid-structure interaction;
3. Jan Philipp Thiele (LUH[9]): Goal-oriented space-time adaptivity for flow problems;
4. Katrin Mang (LUH): Phase-field fracture in incompressible materials;
5. Sebastian Kinnewig (LUH): Numerical modeling of Maxwell's equations;
6. Nima Noii (LUH): Nonisothermal and pressurized fractures, global-local phase-field fracture;
7. Meng Fan (LUH/China University of Petroleum): Stress splitting laws for phase-field fracture;[10]
8. Amirreza Khodadadian (LUH): Bayesian estimation for phase-field fracture;
9. Denis Khimin (LUH): Optimal control with phase-field fracture;
10. Gregor Pfau (LUH): Discontinuous Galerkin methods;
11. Julius Grundmann (LUH): Flux-corrected transport;
12. Julian Roth (LUH): Multigrid solvers for Maxwell's equations.

6 https://www.numa.uni-linz.ac.at/Research/Projects/P29181/index.shtml
7 https://www.python.org/
8 RICAM = Radon Institute of Computational and Applied Mathematics.
9 LUH = Leibniz University Hannover.
10 CSC scholarship from 2018–2019.

14 Conclusions and future interest

In this monograph, we discussed numerical concepts for multiphysics phase-field fracture problems. Multiphysics problems belong to the class of coupled equations, which are nonstationary, nonlinear, coupled partial differential equations that can be subject to variational inequality constraints referred to as CVIS.

In this text, such coupled problems are classified in great detail and state-of-the-art algorithms are derived. Examples and numerical simulations demonstrate their performance and numerous references to the literature, showing further results, are provided.

For the crack irreversibility constraint regularization, we implemented five methods: penalization, augmented Lagrangian penalization, primal-dual active set, Miehe's et al. [313] strain history function, and recently a complementarity system. In view of the computational treatment of the crack irreversibility constraint, this topic is considered to be closed for the moment. It remains to investigate further the robustness of the primal-dual active set method, which may run into cycles from time to time for certain configurations [214, 120]. A simple solution was implemented in [222], but further analysis is necessary here. The fields of rigorous numerical analysis for temporal and spatial convergence are still largely open. These notes provide some hints with the help of some numerical analysis of a simplified prototype configuration. But thorough investigations, demonstrated with carefully tailored numerical experiments, are definitely missing to date.

Basic nonlinear [444, 445, 286, 88] and linear solutions [221, 285, 222, 250] of phase-field fracture problems, including performance tests for parallel computing, have been achieved in own studies and in several national and international collaborations. However, for special topics, such as accurate functional evaluations of technical quantities of interest (goal functionals) only preliminary results exist. Indeed goal-oriented a posteriori error estimation was derived in [442], but only the primal part of the estimator was considered, and propagating fractures need further quantitative consolidation. In multiphysics environments, several goal functionals might be of interest. Here, rigorous numerical analysis and numerical concepts are currently laid out in joint studies with RICAM Linz [153, 150, 152]. Moreover, these adaptive methods allow for balancing of the discretization and iteration errors and need to be applied in future work to phase-field fracture problems.

In the field of multiphysics problems with fracture or damage, we achieved to design models and numerical simulations for porous media applications and large solid deformations. For both, qualitative convergence in space and time could be shown by a computational analysis. For the latter in particular, rigorous numerical analysis as well as the development of linear parallel solvers and preconditioners are open to date.

https://doi.org/10.1515/9783110497397-014

14.1 Open problems

In this final section, we provide a noncomplete list of some challenges, open problems, and ongoing work in phase-field fracture and multiphysics interactions:

1. Numerical verification in general; even for academic tests
2. Rigorous numerical analysis, even for "simple" phase-field fracture models; ideas may be obtained from [41, 39, 40, 463, 163, 164]
3. Linear and nonlinear solvers that are robust, efficient, and accurate with respect to the phase-field regularization and discretization parameters (for quasi-static brittle phase-field fracture itself, it seems that excellent (parallel) solvers on uniformly refined meshes have been constructed [161, 222, 250]). Moreover, employing our fluid-structure solvers developed in [251, 248], frameworks such as presented in Section 11.8 might be addressable
4. Space/time adaptivity of dynamic phase-field fracture
5. Adaptivity and parallel computing of models and numerical solvers in conjunction with discretization error control: Preliminary work for quasi-linear problems with multiple goal functionals and efficiency and reliability estimates for goal-oriented error estimation using the dual-weighted residual method has been performed in [150, 152]. The extension to nonstationary problems with variational inequalities has not been addressed to date
6. Multiscale modeling for treating large-scale problems with localized fracture/damage phenomena
7. Numerical analysis of the nonintrusive global-local phase-field approach
8. Constitutive laws, their verification and analysis
9. Better understanding of crack nucleation in elasticity (some studies exists [407, 108, 427]), for ductile fracture and porous media
10. Improvements of the accuracy of fracture width computations
11. Modeling and analysis of interface conditions on the fracture boundary in conjunction with the smeared transition zone
12. Implementation and verification of Section 11.8
13. Reconstruction algorithms to design sharp interfaces out of smeared zones
14. Optimal control, inverse problems, parameter estimation, and uncertainty quantification are largely untouched. We just started [336, 337, 259, 258, 342] and the results are promising, but also highlight current challenges, i. e., cost efficiency and interaction of various model parameters. Indeed, optimization adds another iterative loop, and therefore the development of efficient forward solutions (what this book was about) is of utter importance
15. Multiphysics phase-field fracture as high-fidelity solution within model order reduction, reduced basis methods, machine learning, and data-augmented simulations

15 The end

Finishing this book fell into the period of the 2020 “confinement/déconfinement” as spring/summer seasons came and we enjoyed being in the nature.

The family garden sessions for gaining new energy and new inspirations can be best described by these pictures combined with the following poem:

https://doi.org/10.1515/9783110497397-015

Johann Wolfgang von Goethe (1749–1832), German poet and naturalist:

Mailied

Wie herrlich leuchtet
Mir die Natur!
Wie glänzt die Sonne!
Wie lacht die Flur!

Es dringen Blüten
Aus jedem Zweig
Und tausend Stimmen
Aus dem Gesträuch,

Und Freud und Wonne
Aus jeder Brust.
O Erd, o Sonne!
O Glück, o Lust!

O Lieb, o Liebe!
So golden schön,
Wie Morgenwolken
Auf jenen Höhn!

Du segnest herrlich
Das frische Feld,
Im Blütendampfe
Die volle Welt.

O Mädchen, Mädchen,
Wie lieb ich dich!
Wie blickt dein Auge!
Wie liebst du mich!

So liebt die Lerche
Gesang und Luft,
Und Morgenblumen
Den Himmelsduft,

Wie ich dich liebe
Mit warmen Blut,
Die du mir Jugend
Und Freud und Mut

Zu neuen Liedern
Und Tänzen gibst.
Sei ewig glücklich,
Wie du mich liebst!

Bibliography

[1] A. Abdulle. On a priori error analysis of fully discrete heterogeneous multiscale FEM. *Multiscale Modeling & Simulation*, 4(2):447–459, 2005.

[2] R. A. Adams. *Sobolev Spaces*. Academic Press, 1975.

[3] M. Ainsworth and D. Kay. The approximation theory for the p-version finite element method and application to non-linear elliptic PDEs. *Numerische Mathematik*, 82(3):351–388, 1999.

[4] M. Ainsworth and J. T. Oden. *A Posteriori Error Estimation in Finite Element Analysis*. Pure and Applied Mathematics (New York). Wiley-Interscience [John Wiley & Sons], New York, 2000.

[5] J. E. Akin, T. Tezduyar, and M. Ungor. Computation of flow problems with the mixed interface-tracking/interface-capturing technique (MITICT). *Computers & Fluids*, 36:2–11, 2007.

[6] F. Aldakheel. A microscale model for concrete failure in poro-elasto-plastic media. *Theoretical and Applied Fracture Mechanics*, 107:102517, 2020.

[7] F. Aldakheel, B. Hudobivnik, A. Hussein, and P. Wriggers. Phase-field modeling of brittle fracture using an efficient virtual element scheme. *Computer Methods in Applied Mechanics and Engineering*, 341:443–466, 2018.

[8] F. Aldakheel, B. Hudobivnik, and P. Wriggers. Virtual element formulation for phase-field modeling of ductile fracture. *International Journal of Multiscale Computational Engineering*, 17(2), 2019.

[9] F. Aldakheel, N. Noii, T. Wick, and P. Wriggers. A global-local approach for hydraulic phase-field fracture in poroelastic media. *Computers & Mathematics with Applications*, accepted for publication, 2020.

[10] R. Alessi, J.-J. Marigo, C. Maurini, and S. Vidoli. Coupling damage and plasticity for a phase-field regularisation of brittle, cohesive and ductile fracture: one-dimensional examples. *International Journal of Mechanical Sciences*, 2017.

[11] G. Allaire, F. Jouve, and N. V. Goethem. Damage and fracture evolution in brittle materials by shape optimization methods. *Journal of Computational Physics*, 230(12):5010–5044, 2011.

[12] S. M. Allen and J. W. Cahn. A microscopic theory for antiphase boundary motion and its application to antiphase domain coarsening. *Acta Metallurgica*, 27(6):1085–1095, 1979.

[13] E. L. Allgower and K. Georg. *Numerical Continuation Methods: An Introduction*. Springer, 1990.

[14] T. Almani, S. Lee, M. Wheeler, and T. Wick. Multirate Coupling for Flow and Geomechanics Applied to Hydraulic Fracturing Using an Adaptive Phase-Field Technique. SPE RSC 182610-MS, Feb. 2017, Montgomery, Texas, USA, 2017.

[15] J. Alvarez-Aramberri, D. Pardo, and H. Barucq. Inversion of magnetotelluric measurements using multigoal oriented hp-adaptivity. *Procedia Computer Science*, 18:1564–1573, 2013.

[16] M. Ambati, T. Gerasimov, and L. De Lorenzis. Phase-field modeling of ductile fracture. *Computational Mechanics*, 55(5):1017–1040, 2015.

[17] M. Ambati, T. Gerasimov, and L. De Lorenzis. A review on phase-field models of brittle fracture and a new fast hybrid formulation. *Computational Mechanics*, 55(2):383–405, 2015.

[18] M. Ambati and L. D. Lorenzis. Phase-field modeling of brittle and ductile fracture in shells with isogeometric nurbs-based solid-shell elements. *Computer Methods in Applied Mechanics and Engineering*, 312:351–373, 2016. Phase Field Approaches to Fracture.

[19] L. Ambrosio and N. Dancer. *Calculus of Variations and Partial Differential Equations*. Springer, 2000.

[20] L. Ambrosio and V. Tortorelli. Approximation of functionals depending on jumps by elliptic functionals via γ-convergence. *Communications on Pure and Applied Mathematics*, 43:999–1036, 1990.

https://doi.org/10.1515/9783110497397-016

[21] L. Ambrosio and V. Tortorelli. On the approximation of free discontinuity problems. *Bollettino dell'Unione Matematica Italiana B*, 6:105–123, 1992.

[22] L. Ambrosio and E. Virga. A boundary value problem for nematic liquid crystals with variable degree of orientation. *Archive for Rational Mechanics and Analysis*, 114:335–347, 1991.

[23] H. Amor, J.-J. Marigo, and C. Maurini. Regularized formulation of the variational brittle fracture with unilateral contact: numerical experiments. *Journal of the Mechanics and Physics of Solids*, 57:1209–1229, 2009.

[24] J. Andersson and H. Mikayelyan. The asymptotics of the curvature of the free discontinuity set near the cracktip for the minimizers of the Mumford–Shah functional in the plain. A revision. arXiv:1205.5328v2, 2015.

[25] H. Anzt, F. Bach, S. Druskat, F. Löffler, A. Loewe, B. Renard, G. Seemann, A. Struck, E. Achhammer, P. Aggarwal, F. Appel, M. Bader, L. Brusch, C. Busse, G. Chourdakis, P. Dabrowski, P. Ebert, B. Flemisch, S. Friedl, B. Fritzsch, M. Funk, V. Gast, F. Goth, J. Grad, S. Hermann, F. Hohmann, S. Janosch, D. Kutra, J. Linxweiler, T. Muth, W. Peters-Kottig, F. Rack, F. Raters, S. Rave, G. Reina, M. Reißig, T. Ropinski, J. Schaarschmidt, H. Seibold, J. Thiele, B. Uekerman, S. Unger, and R. Weeber. An environment for sustainable research software in Germany and beyond: current state, open challenges, and call for action [version 1; peer review: awaiting peer review]. *F1000Research*, 9(295), 2020.

[26] D. Arndt, W. Bangerth, T. C. Clevenger, D. Davydov, M. Fehling, D. Garcia-Sanchez, G. Harper, T. Heister, L. Heltai, M. Kronbichler, R. M. Kynch, M. Maier, J.-P. Pelteret, B. Turcksin, and D. Wells. The `deal.II` library, version 9.1. *Journal of Numerical Mathematics*, 2019, accepted.

[27] D. Arndt, W. Bangerth, D. Davydov, T. Heister, L. Heltai, M. Kronbichler, M. Maier, J.-P. Pelteret, B. Turcksin, and D. Wells. The `deal.II` library, version 8.5. *Journal of Numerical Mathematics*, 2017.

[28] D. Arndt, W. Bangerth, D. Davydov, T. Heister, L. Heltai, M. Kronbichler, M. Maier, J.-P. Pelteret, B. Turcksin, and D. Wells. The deal.ii finite element library: design, features, and insights. *Computers & Mathematics with Applications*, 2020.

[29] M. Arriaga and H. Waisman. Multidimensional stability analysis of the phase-field method for fracture with a general degradation function and energy split. *Computational Mechanics*, 61:181–205, 2018.

[30] M. Artina, M. Fornasier, S. Micheletti, and S. Perotto. Anisotropic mesh adaptation for crack detection in brittle materials. *SIAM Journal on Scientific Computing*, 37(4):B633–B659, 2015.

[31] I. Babuska. The finite element method for elliptic equations with discontinuous coefficients. *Computing*, 5:207–213, 1970.

[32] I. Babuska and U. Banerjee. Stable generalized finite element method (SGFEM). *Computer Methods in Applied Mechanics and Engineering*, 201–204:91–111, 2012.

[33] I. Babuska and J. M. Melenk. The partition of unity method. *International Journal for Numerical Methods in Engineering*, 40:727–758, 1997.

[34] I. Babuska and W. C. Rheinboldt. A-posteriori error estimates for the finite element method. *International Journal for Numerical Methods in Engineering*, 12(10):1597–1615, 1978.

[35] H. Badnava, M. A. Msekh, E. Etemadi, and T. Rabczuk. An h-adaptive thermo-mechanical phase field model for fracture. *Finite Elements in Analysis and Design*, 138:31–47, 2018.

[36] W. Bangerth, R. Hartmann, and G. Kanschat. deal.II – a general purpose object oriented finite element library. *ACM Transactions on Mathematical Software*, 33(4):24/1–24/27, 2007.

[37] W. Bangerth and R. Rannacher. *Adaptive Finite Element Methods for Differential Equations*. Birkhäuser, Lectures in Mathematics, ETH Zürich, 2003.

[38] J. W. Barrett, J. F. Blowey, and H. Garcke. Finite element approximation of the Cahn–Hilliard equation with degenerate mobility. *SIAM Journal on Numerical Analysis*, 37(1):286–318, 1999.

[39] S. Bartels. A posteriori error analysis for time-dependent Ginzburg–Landau type equations. *Numerische Mathematik*, 99:557–583, 2005.

[40] S. Bartels. Error control and adaptivity for a variational model problem defined on functions of bounded variation. *Mathematics of Computation*, 84(293):1217–1240, 2015.

[41] S. Bartels, R. Müller, and C. Ortner. Robust a priori and a posteriori error analysis for the approximation of Allen–Cahn and Ginzburg–Landau equations past topological changes. *SIAM Journal on Numerical Analysis*, 49(1):110–134, 2011.

[42] S. Basava, K. Mang, M. Walloth, T. Wick, and W. Wollner. Adaptive and Pressure-Robust Discretization of Incompressible Pressure-Driven Phase-Field Fracture. SPP 1748 final report, 2020, submitted, arXiv:2006.16566.

[43] S. Basting and R. Prignitz. An interface-fitted subspace projection method for finite element simulations of particulate flows. *Computer Methods in Applied Mechanics and Engineering*, 267:133–149, 2013.

[44] S. Basting and M. Weissmann. A hybrid level set-front tracking finite element approach for fluid-structure interaction and two-phase flow applications. *Journal of Computational Physics*, 255:228–244, 2013.

[45] M. Bause, F. Radu, and U. Köcher. Space-time finite element approximation of the Biot poroelasticity system with iterative coupling. *Computer Methods in Applied Mechanics and Engineering*, 320:745–768, 2017.

[46] Y. Bazilevs, K. Takizawa, and T. Tezduyar. *Computational Fluid-Structure Interaction: Methods and Applications*. Wiley, 2013.

[47] R. Becker. An Optimal-Control Approach to a Posteriori Error Estimation for Finite Element Discretizations of the Navier–Stokes Equations. SFB 359, University of Heidelberg, 2000.

[48] R. Becker, M. Braack, D. Meidner, T. Richter, and B. Vexler. The finite element toolkit GASCOIGNE. http://www.gascoigne.uni-hd.de.

[49] R. Becker, M. Innerberger, and D. Praetorius. Optimal Convergence Rates for Goal-Oriented Fem with Quadratic Goal Functional, 2020.

[50] R. Becker, C. Johnson, and R. Rannacher. Adaptive error control for multigrid finite element methods. *Computing*, 55:271–288, 1995.

[51] R. Becker and R. Rannacher. A feed-back approach to error control in finite element methods: basic analysis and examples. *East-West Journal of Numerical Mathematics*, 4:237–264, 1996.

[52] R. Becker and R. Rannacher. An optimal control approach to a posteriori error estimation in finite element methods. *Acta Numerica*, Cambridge University Press, pages 1–102, 2001.

[53] L. Beirão da Veiga, A. Buffa, G. Sangalli, and R. Vázquez. Mathematical analysis of variational isogeometric methods. *Acta Numerica*, 23:157–287, 2014.

[54] A. Bensoussan, J. Lions, and G. Papanicolaou. *Asymptotic Analysis for Periodic Structures*. North-Holland, 1978.

[55] M. Besier. *Adaptive Finite Element Methods for Computing Nonstationary Incompressible Flows*. PhD thesis, University of Heidelberg, 2009.

[56] M. Besier and R. Rannacher. Goal-oriented space-time adaptivity in the finite element Galerkin method for the computation of nonstationary incompressible flow. *International Journal for Numerical Methods in Fluids*, 70:1139–1166, 2012.

[57] C. Bilgen, S. Homberger, and K. Weinberg. Phase-field fracture simulations of the Brazilian splitting test. *International Journal of Fracture*, 220:85–98, 2019.

[58] M. Biot. Consolidation settlement under a rectangular load distribution. *Journal of Applied Physics*, 12(5):426–430, 1941.

[59] M. Biot. General theory of three-dimensional consolidation. *Journal of Applied Physics*, 12(2):155–164, 1941.

[60] H. Blum and F.-T. Suttmeier. An adaptive finite element discretisation for a simplified Signorini problem. CALCOLO, 37(2):65–77, 1999.

[61] H. Blum and F.-T. Suttmeier. Weighted error estimates for finite element solutions of variational inequalities. *Computing*, 65(2):119–134, 2000.

[62] F. Bobaru, J. Foster, P. Geubelle, and S. Silling. *Handbook of Peridynamic Modeling*. CRC Press, 2017.

[63] H. G. Bock, E. Kostina, and J. P. Schlöder. On the role of natural level functions to achieve global convergence for damped Newton methods. In M. Powell and S. Scholtes, editors, *System Modeling and Optimization: Methods, Theory and Applications*, pages 51–74. Kluwer Academic Publishers, 2000.

[64] T. Bodnár, G. Galdi, and Š. Nečasová. *Fluid-Structure Interaction and Biomedical Applications*. Advances in Mathematical Fluid Mechanics. Springer Basel, 2014.

[65] A. Bonnet and G. David. *Cracktip is a global Mumford–Shah minimizer*. Astérisque No. 274, 2001.

[66] M. J. Borden, T. J. Hughes, C. M. Landis, and C. V. Verhoosel. A higher-order phase-field model for brittle fracture: formulation and analysis within the isogeometric analysis framework. *Computer Methods in Applied Mechanics and Engineering*, 273:100–118, 2014.

[67] M. J. Borden, C. V. Verhoosel, M. A. Scott, T. J. R. Hughes, and C. M. Landis. A phase-field description of dynamic brittle fracture. *Computer Methods in Applied Mechanics and Engineering*, 217:77–95, 2012.

[68] J. Both, M. Borregales, J. Nordbotten, K. Kumar, and F. Radu. Robust fixed stress splitting for Biot's equations in heterogeneous media. *Applied Mathematics Letters*, 68:101–108, 2017.

[69] B. Bourdin. Image segmentation with a finite element method. *Mathematical Modelling and Numerical Analysis*, 33(2):229–244, 1999.

[70] B. Bourdin. Numerical implementation of the variational formulation for quasi-static brittle fracture. *Interfaces and Free Boundaries*, 9:411–430, 2007.

[71] B. Bourdin, C. Chukwudozie, and K. Yoshioka. A variational approach to the numerical simulation of hydraulic fracturing. *SPE Journal*, Conference Paper 159154-MS, 2012.

[72] B. Bourdin, G. Francfort, and J.-J. Marigo. Numerical experiments in revisited brittle fracture. *Journal of the Mechanics and Physics of Solids*, 48(4):797–826, 2000.

[73] B. Bourdin, G. Francfort, and J.-J. Marigo. The variational approach to fracture. *Journal of Elasticity*, 91(1–3):1–148, 2008.

[74] B. Bourdin and G. A. Francfort. Past and present of variational fracture. *SIAM News*, 52(9), 2019.

[75] B. Bourdin, C. Larsen, and C. Richardson. A time-discrete model for dynamic fracture based on crack regularization. *International Journal of Fracture*, 168(2):133–143, 2011.

[76] B. Bourdin, J.-J. Marigo, C. Maurini, and P. Sicsic. Morphogenesis and propagation of complex cracks induced by thermal shocks. *Physical Review Letters*, 112:014301, Jan 2014.

[77] M. Braack and A. Ern. A posteriori control of modeling errors and discretization errors. *Multiscale Modeling & Simulation*, 1(2):221–238, 2003.

[78] M. Braack and T. Richter. Solutions of 3d Navier–Stokes benchmark problems with adaptive finite elements. *Computers & Fluids*, 35(4):372–392, 2006.

[79] D. Braess. *Finite Elemente*. Springer-Verlag Berlin Heidelberg, Berlin, Heidelberg, vierte, überarbeitete und erweiterte edition, 2007.

[80] A. Braides. *Approximation of Free-Discontinuity Problems*. Springer Berlin Heidelberg, 1998.

[81] K. Bredies and D. Lorenz. *Mathematische Bildverarbeitung*. Vieweg+Teubner, 2011.

[82] S. C. Brenner and L. R. Scott. *The Mathematical Theory of Finite Element Methods*, volume 15 in *Texts in Applied Mathematics*. Springer, New York, NY, 3th edition, 2008.

[83] I. Brevis, I. Muga, and K. G. van der Zee. Data-Driven Finite Elements Methods: Machine Learning Acceleration of Goal-Oriented Computations, 2020.
[84] F. Brezzi, W. Hager, and P. Raviart. Error estimates for the finite element solution of variational inequalities. *Numerische Mathematik*, 28:431–443, 1977.
[85] F. Brezzi, W. Hager, and P. Raviart. Error estimates for the finite element solution of variational inequalities. Part II. Mixed methods. *Numerische Mathematik*, 31:1–16, 1978.
[86] M. Bristeau, R. Glowinski, and J. Periaux. Numerical methods for the Navier–Stokes equations. *Computer Physics Reports*, 6:73–187, 1987.
[87] I. Bronshtein, K. Semendyayev, G. Musiol, and H. Muehlig. *Handbook of Mathematics*. Springer-Verlag Berlin Heidelberg, 5th edition, 2007.
[88] M. K. Brun, T. Wick, I. Berre, J. M. Nordbotten, and F. A. Radu. An iterative staggered scheme for phase field brittle fracture propagation with stabilizing parameters. *Computer Methods in Applied Mechanics and Engineering*, 361:112752, 2020.
[89] E. C. Bryant and W. Sun. A mixed-mode phase field fracture model in anisotropic rocks with consistent kinematics. *Computer Methods in Applied Mechanics and Engineering*, 342:561–584, 2018.
[90] M. Bukac, S. Canic, R. Glowinski, B. Muha, and A. Quaini. A modular, operator-splitting scheme for fluid-structure interaction problems with thick structures. *International Journal for Numerical Methods in Fluids*, 74(8):577–604, 2014.
[91] M. Bukac, I. Yotov, and P. Zunino. Dimensional model reduction for flow through fractures in poroelastic media. *Modélisation Mathématique et Analyse Numérique*, 51(4):1429–1471, 2017.
[92] H.-J. Bungartz, M. Mehl, and M. Schäfer. *Fluid-Structure Interaction II: Modelling, Simulation, Optimization*. Lecture Notes in Computational Science and Engineering. Springer, 2010.
[93] H.-J. Bungartz and M. Schäfer. *Fluid-Structure Interaction: Modelling, Simulation, Optimization*, volume 53 of *Lecture Notes in Computational Science and Engineering*. Springer, 2006.
[94] S. Burke, C. Ortner, and E. Süli. An adaptive finite element approximation of a variational model of brittle fracture. *SIAM Journal on Numerical Analysis*, 48(3):980–1012, 2010.
[95] S. Burke, C. Ortner, and E. Süli. An adaptive finite element approximation of a generalized Ambrosio–Tortorelli functional. *Mathematical Models and Methods in Applied Sciences*, 23(9):1663–1697, 2013.
[96] E. Burman, S. Claus, P. Hansbo, M. G. Larson, and A. Massing. Cutfem: discretizing geometry and partial differential equations. *International Journal for Numerical Methods in Engineering*, 104(7):472–501, 2015.
[97] C. Burstedde, L. C. Wilcox, and O. Ghattas. P4est: scalable algorithms for parallel adaptive mesh refinement on forests of octrees. *SIAM Journal on Scientific Computing*, 33(3):1103–1133, May 2011.
[98] J. W. Cahn and J. E. Hilliard. Free energy of a nonuniform system. I. Interfacial free energy. *Journal of Chemical Physics*, 28(2):258–267, 1958.
[99] X.-C. Cai and D. E. Keyes. Nonlinearly preconditioned inexact Newton algorithms. *SIAM Journal on Scientific Computing*, 24(1):183–200, 2002.
[100] T. Cajuhi, L. Sanavia, and L. De Lorenzis. Phase-field modeling of fracture in variably saturated porous media. *Computational Mechanics*, 61:299–318, 2018.
[101] G. F. Carey and J. T. Oden. *Finite Elements. Volume III. Computational Aspects*. The Texas Finite Element Series, Prentice-Hall, Inc., Englewood Cliffs, 1984.
[102] T. Carraro, C. Goll, A. Marciniak-Czochra, and A. Mikelic. Pressure jump interface law for the Stokes–Darcy coupling: confirmation by direct numerical simulations. *Journal of Fluid Mechanics*, 732:510–536, 2013.

[103] T. Carraro, C. Goll, A. Marciniak-Czochra, and A. Mikelić. Effective interface conditions for the forced infiltration of a viscous fluid into a porous medium using homogenization. *Computer Methods in Applied Mechanics and Engineering*, 292:195–220, 2015. Special Issue on Advances in Simulations of Subsurface Flow and Transport (Honoring Professor Mary F. Wheeler).

[104] C. Carstensen and R. Verfürth. Edge residuals dominate a posteriori error estimates for low order finite element methods. *SIAM Journal on Numerical Analysis*, 36(5):1571–1587, 1999.

[105] N. Castelletto, J. A. White, and H. A. Tchelepi. Accuracy and convergence properties of the fixed-stress iterative solution of two-way coupled poromechanics. *International Journal for Numerical and Analytical Methods in Geomechanics*, 39(14):1593–1618, 2015.

[106] H. D. Ceniceros, R. L. Nós, and A. M. Roma. Three-dimensional, fully adaptive simulations of phase-field fluid models. *Journal of Computational Physics*, 229(17):6135–6155, 2010.

[107] A. Chambolle. Image segmentation by variational methods: Mumford and Shah functional and the discrete approximations. *SIAM Journal on Applied Mathematics*, 55:827–863, 1995.

[108] A. Chambolle, A. Giacomini, and M. Ponsiglione. Crack initiation in brittle materials. *Archive for Rational Mechanics and Analysis*, 188:309–349, 2008.

[109] A. Chambolle and G. D. Maso. Discrete approximation of the Mumford–Shah functional in dimension two. *ESAIM: Mathematical Modelling and Numerical Analysis – Modélisation Mathématique et Analyse Numérique*, 33(4):651–672, 1999.

[110] C. Chang and M. E. Mear. A boundary element method for two dimensional linear elastic fracture analysis. *International Journal of Fracture*, 74:219–251, 1995.

[111] X. Chen, C. M. Elliot, A. Gardiner, and J. J. Zhao. Convergence of numerical solutions to the Allen–Cahn equation. *Applicable Analysis*, 69(1–2):95–108, 1998.

[112] C. Chukwudozie, B. Bourdin, and K. Yoshioka. A variational phase-field model for hydraulic fracturing in porous media. *Computer Methods in Applied Mechanics and Engineering*, 347:957–982, 2019.

[113] P. G. Ciarlet. *Lectures on Three Dimensional Elasticity*. Springer Berlin Heidelberg New York, 1983.

[114] P. G. Ciarlet. *Mathematical Elasticity. Volume 1: Three Dimensional Elasticity*. North-Holland, 1984.

[115] P. G. Ciarlet. *The Finite Element Method for Elliptic Problems*. North-Holland, Amsterdam [u. a.], 2th pr. edition, 1987.

[116] P. G. Ciarlet. *Linear and Nonlinear Functional Analysis with Applications*. SIAM, 2013.

[117] S. Conti, M. Focardi, and F. Iurlano. Existence of strong minimizers for the Griffith static fracture model in dimension two. *Annales de l'Institut Henri Poincaré C, Analyse non linéaire*, 36(2):455–474, 2019.

[118] J. A. Cottrell, T. J. R. Hughes, and Y. Bazilevs. *Isogeometric Analysis: Toward Integration of CAD and FEA*. John Wiley & Sons, Chichester, England, 2009.

[119] O. Coussy. *Poromechanics*. Wiley, 2004.

[120] F. E. Curtis, Z. Han, and D. P. Robinson. A globally convergent primal-dual active-set framework for large-scale convex quadratic optimization. *Computational Optimization and Applications*, 60(2):311–341, Mar 2015.

[121] G. dal Maso, G. A. Francfort, and R. Toader. Quasistatic crack growth in nonlinear elasticity. *Archive for Rational Mechanics and Analysis*, 176:165–225, 2005.

[122] G. dal Maso and R. Toader. A model for the quasistatic growth of brittle fractures: existence and approximation results. *Archive for Rational Mechanics and Analysis*, 162:101–135, 2002.

[123] H. Darcy. *Les Fontaines de la Ville de Dijon*. Dalmont, Paris, 1856.

[124] R. Dautray and J.-L. Lions. *Mathematical Analysis and Numerical Methods for Science and Technology*, volume 5. Springer-Verlag, Berlin, Heidelberg, 2000.

[125] T. A. Davis and I. S. Duff. An unsymmetric-pattern multifrontal method for sparse LU factorization. *SIAM Journal on Matrix Analysis and Applications*, 18(1):140–158, 1997.

[126] D. Davydov, J. Pelteret, D. Arndt, M. Kronbichler, and P. Steinmann. A matrix-free approach for finite-strain hyperelastic problems using geometric multigrid. *International Journal for Numerical Methods in Engineering*, 121(13):2874–2895, 2020.

[127] R. de Borst and C. Verhoosel. Gradient damage vs phase-field approaches for fracture: similarities and differences. *Computer Methods in Applied Mechanics and Engineering*, 312(December):78–94, 2016.

[128] R. Dean and J. Schmidt. Hydraulic-fracture predictions with a fully coupled reservoir simulator. SPE Journal 14(4):707–714, 2009.

[129] R. Dembo, S. Eisenstat, and T. Steihaug. Inexact Newton methods. *SIAM Journal on Numerical Analysis*, 19(2):400–408, 1982.

[130] P. Deuflhard. A relaxation strategy for the modified Newton method. In R. Bulirsch, W. Oettli, and J. Stoer, editors, *Optimization and Optimal Control*, pages 59–73. Springer, 1974.

[131] P. Deuflhard. *Newton Methods for Nonlinear Problems*, volume 35 of *Springer Series in Computational Mathematics*. Springer Berlin Heidelberg, 2011.

[132] L. Diening and M. Růžička. Interpolation operators in Orlicz–Sobolev spaces. *Numerische Mathematik*, 107(1):107–129, 2007.

[133] M. Dittmann, F. Aldakheel, J. Schulte, P. Wriggers, and C. Hesch. Variational phase-field formulation of non-linear ductile fracture. *Computer Methods in Applied Mechanics and Engineering*, 342:71–94, 2018.

[134] J. Donea, S. Giuliani, and J. Halleux. An arbitrary Lagrangian–Eulerian finite element method for transient dynamic fluid-structure interactions. *Computer Methods in Applied Mechanics and Engineering*, 33:689–723, 1982.

[135] J. Donea, A. Huerta, J.-P. Ponthot, and A. Rodriguez-Ferran. *Arbitrary Lagrangian–Eulerian Methods*, pages 1–25. Encyclopedia of Computational Mechanics. John Wiley and Sons, 2004.

[136] J. Donéa, P. Fasoli-Stella, and S. Giuliani. Lagrangian and Eulerian finite element techniques for transient fluid-structure interaction problems. In *Trans. 4th Int. Conf. on Structural Mechanics in Reactor Technology*, Paper B1/2, 1977.

[137] F. Duda, A. Ciarbonetti, P. Sanchez, and A. Huespe. A phase-field/gradient damage model for brittle fracture in elastic-plastic regimes. *International Journal of Plasticity*, 65:269–296, 2015.

[138] T. Dunne. An Eulerian approach to fluid-structure interaction and goal-oriented mesh adaption. *International Journal for Numerical Methods in Fluids*, 51:1017–1039, 2006.

[139] T. Dunne. *Adaptive Finite Element Approximation of Fluid-Structure Interaction Based on Eulerian and Arbitrary Lagrangian–Eulerian Variational Formulations*. PhD thesis, University of Heidelberg, 2007.

[140] T. Dunne, T. Richter, and R. Rannacher. *Numerical Simulation of Fluid-Structure Interaction Based on Monolithic Variational Formulations*, pages 1–75. Comtemporary Challenges in Mathematical Fluid Mechanics. Springer, World Scientific, Singapore, 2010.

[141] G. Duvaut and J. L. Lions. *Inequalities in Mechanics and Physics*. Springer, Berlin, Heidelberg, New York, 1976.

[142] W. Dörfler. A convergent adaptive algorithm for Poisson's equation. *SIAM Journal on Numerical Analysis*, 33(3):1106–1124, 1996.

[143] J. W. Eaton, D. Bateman, S. Hauberg, and R. Wehbring. *GNU Octave Version 3.8.1 Manual: A High-Level Interactive Language for Numerical Computations*. CreateSpace Independent Publishing Platform, 2014. ISBN 1441413006.

[144] C. Eck, H. Garcke, and P. Knabner. *Mathematische Modellierung*. Springer, 2008.

[145] H. Emmerich. *The Diffusive Interface Approach in Materials Science*. Springer, 2003.

[146] B. Endtmayer. *Multi-Goal Oriented a Posteriori Error Estimates for Nonlinear Partial Differential Equations*. PhD thesis, Johannes Kepler University Linz, in preparation, 2020.
[147] B. Endtmayer, U. Langer, I. Neitzel, T. Wick, and W. Wollner. Multigoal-oriented optimal control problems with nonlinear PDE constraints. *Computers & Mathematics with Applications*, 79(10):3001–3026, 2020.
[148] B. Endtmayer, U. Langer, J. P. Thiele, and T. Wick. Hierarchical DWR error estimates for the Navier Stokes equation: h and p enrichment, arXiv:1912.04819, 2019.
[149] B. Endtmayer, U. Langer, and T. Wick. Multiple goal-oriented error estimates applied to 3d non-linear problems. *Proceedings in Applied Mathematics and Mechanics (PAMM)*, 18(1):e201800048, 2018.
[150] B. Endtmayer, U. Langer, and T. Wick. Multigoal-oriented error estimates for non-linear problems. *Journal of Numerical Mathematics*, 27(4):215–236, 2019.
[151] B. Endtmayer, U. Langer, and T. Wick. Reliability and Efficiency of DWR-Type a Posteriori Error Estimates with Smart Sensitivity Weight Recovering. RICAM report No. 2020-18, 2020.
[152] B. Endtmayer, U. Langer, and T. Wick. Two-side a posteriori error estimates for the dual-weighted residual method. *SIAM Journal on Scientific Computing*, 42(1):A371–A394, 2020.
[153] B. Endtmayer and T. Wick. A partition-of-unity dual-weighted residual approach for multi-objective goal functional error estimation applied to elliptic problems. *Computational Methods in Applied Mathematics*, 17(2):575–599, 2017.
[154] C. Engwer, S. I. Pop, and T. Wick. Dynamic and weighted stabilizations of the l-scheme applied to a phase-field model for fracture propagation. arXiv:1912.07096, 2019.
[155] C. Engwer and L. Schumacher. A phase field approach to pressurized fractures using discontinuous Galerkin methods. *Mathematics and Computers in Simulation*, 2016.
[156] A. Ern and M. Vohralík. Adaptive inexact Newton methods with a posteriori stopping criteria for nonlinear diffusion PDEs. *SIAM Journal on Scientific Computing*, 35(4):A1761–A1791, 2013.
[157] F. Ernesti, M. Schneider, and T. Böhlke. Fast implicit solvers for phase-field fracture problems on heterogeneous microstructures. *Computer Methods in Applied Mechanics and Engineering*, 363:112793, 2020.
[158] L. C. Evans. *Partial differential equations*. American Mathematical Society, 2010.
[159] L. Failer and T. Wick. Adaptive time-step control for nonlinear fluid-structure interaction. *Journal of Computational Physics*, 366:448–477, 2018.
[160] M. Fan, Y. Jin, and T. Wick. *A Phase-Field Model for Mixed-Mode Fracture*. Institutionelles Repositorium der Leibniz Universität Hannover, Hannover, 2019, 40 S. DOI: https://doi.org/10.15488/5369.
[161] P. E. Farrell and C. Maurini. Linear and nonlinear solvers for variational phase-field models of brittle fracture. *International Journal for Numerical Methods in Engineering*, 109:648–667, 2017.
[162] M. Feischl, D. Praetorius, and K. G. van der Zee. An abstract analysis of optimal goal-oriented adaptivity. *SIAM Journal on Numerical Analysis*, 54(3):1423–1448, 2016.
[163] X. Feng and A. Prohl. Numerical analysis of the Allen–Cahn equation and approximation for mean curvature flows. *Numerische Mathematik*, 94:33–65, 2003.
[164] X. Feng and A. Prohl. Analysis of a fully discrete finite element method for the phase field model and approximation of its sharp interface limits. *Mathematics of Computation*, 73:541–567, 2004.
[165] M. Fernández and J.-F. Gerbeau. *Algorithms for fluid-structure interaction problems*, pages 307–346, volume 1 of Formaggia et al. [170], 2009.
[166] M. A. Fernandez and J. Mullaert. Displacement-velocity correction schemes for incompressible fluid-structure interaction. *Comptes Rendus Mathematique*, 349(17):1011–1015, 2011.

[167] R. P. Feynman. *The Pleasure of Finding Things Out*. Helix Books, Perseus Books, Cambridge, Massachusetts, 1999.

[168] R. P. Feynman, R. B. Leighton, and M. Sands. *The Feynman Lectures on Physics, vol. II: The New Millennium Edition: Mainly Electromagnetism and Matter*. Basic Books; New Millennium edition, 2011.

[169] L. Formaggia and F. Nobile. A stability analysis for the arbitrary Lagrangian Eulerian formulation with finite elements. *East-West Journal of Numerical Mathematics*, 7:105–132, 1999.

[170] L. Formaggia, A. Quarteroni, and A. Veneziani. *Cardiovascular Mathematics: Modeling and Simulation of the Circulatory System*. Springer-Verlag, Italia, Milano, 2009.

[171] M. Fortin and R. Glowinski. *Augmented Lagrangian Methods: Applications to the Numerical Solution of Boundary Value Problems*. Studies in Mathematics and Its Applications 15. North Holland, Amsterdam, 1983.

[172] G. Francfort. *Un résumé de la théorie variationnelle de la rupture. Séminaire Laurent Schwartz – EDP et applications*, Institut des hautes études scientifiques, 2011–2012, Exposé no. XXII, 1–11. http://slsedp.cedram.org/slsedp-bin/fitem?id=SLSEDP_2011-2012, 2011.

[173] G. Francfort, N. Le, and S. Serfaty. Critical points of Ambrosio–Tortorelli converge to critical points of Mumford–Shah in the one-dimensional Dirichlet case. *ESAIM. Control, Optimisation and Calculus of Variations*, 15:576–598, 2009.

[174] G. Francfort and J.-J. Marigo. Revisiting brittle fracture as an energy minimization problem. *Journal of the Mechanics and Physics of Solids*, 46(8):1319–1342, 1998.

[175] G. A. Francfort and C. J. Larsen. Existence and convergence for quasi-static evolution in brittle fracture. *Communications on Pure and Applied Mathematics*, 56(10):1465–1500, 2003.

[176] F. Freddi and G. Royer-Carfagni. Regularized variational theories of fracture: a unified approach. *Journal of the Mechanics and Physics of Solids*, 58(8):1154–1174, 2010.

[177] S. Frei. *Eulerian Finite Element Methods for Interface Problems and Fluid-Structure Interactions*. PhD thesis, Heidelberg University, 2016.

[178] S. Frei, B. Holm, T. Richter, T. Wick, and H. Yang. *Fluid-Structure Interactions: Fluid-Structure Interaction: Modeling, Adaptive Discretisations and Solvers*. de Gruyter, 2017.

[179] S. Frei and T. Richter. A locally modified parametric finite element method for interface problems. *SIAM Journal on Numerical Analysis*, 52(5):2315–2334, 2014.

[180] S. Frei and T. Richter. An accurate Eulerian approach for fluid-structure interaction. In S. Frei, B. Holm, T. Richter, T. Wick, and H. Yang, editors, *Fluid-Structure Interaction: Modeling, Adaptive Discretization and Solvers*, Radon Series on Computational and Applied Mathematics. Walter de Gruyter, Berlin, 2017.

[181] S. Frei and T. Richter. A second order time-stepping scheme for parabolic interface problems with moving interfaces. *Modélisation Mathématique et Analyse Numérique*, 51(4):1539–1560, 2017.

[182] S. Frei, T. Richter, and T. Wick. Long-term simulation of large deformation, mechano-chemical fluid-structure interactions in ALE and fully Eulerian coordinates. *Journal of Computational Physics*, 321:874–891, 2016.

[183] S. Frei, T. Richter, and T. Wick. An implementation of a locally modified finite element method for interface problems in deal.ii. zenodo, DOI: 10.5281/zenodo.1457758, 2018.

[184] T.-P. Fries and T. Belytschko. The extended/generalized finite element method: an overview of the method and its applications. *International Journal for Numerical Methods in Engineering*, 84:253–304, 2010.

[185] J. Frohne, T. Heister, and W. Bangerth. Efficient numerical methods for the large-scale, parallel solution of elastoplastic contact problems. *International Journal for Numerical Methods in Engineering*, 105, 07 2015.

[186] X. Gai. *A Coupled Geomechanics and Reservoir Flow Model on Parallel Computers*. PhD thesis, The University of Texas at Austin, 2004.

[187] G. Galdi and R. Rannacher. *Fundamental Trends in Fluid-Structure Interaction*. World Scientific, 2010.

[188] M. Gee, U. Küttler, and W. Wall. Truly monolithic algebraic multigrid for fluid–structure interaction. *International Journal for Numerical Methods in Engineering*, 85(8):987–1016, 2011.

[189] R. Geelen, J. Plews, M. Tupek, and J. Dolbow. An extended/generalized phase-field finite element method for crack growth with global-local enrichment. *International Journal for Numerical Methods in Engineering*, 2020.

[190] R. J. Geelen, Y. Liu, T. Hu, M. R. Tupek, and J. E. Dolbow. A phase-field formulation for dynamic cohesive fracture. *Computer Methods in Applied Mechanics and Engineering*, 348:680–711, 2019.

[191] L. Gendre, O. Allix, P. Gosselet, and F. Comte. Non-intrusive and exact global/local technique for structural problems with local plasticity. *Computational Mechanics*, 44(2):233–245, 2009.

[192] T. Gerasimov and L. D. Lorenzis. A line search assisted monolithic approach for phase-field computing of brittle fracture. *Computer Methods in Applied Mechanics and Engineering*, 312:276–303, 2016.

[193] T. Gerasimov and L. D. Lorenzis. On penalization in variational phase-field models of brittle fracture. *Computer Methods in Applied Mechanics and Engineering*, 354:990–1026, 2019.

[194] T. Gerasimov, N. Noii, O. Allix, and L. De Lorenzis. A non-intrusive global/local approach applied to phase-field modeling of brittle fracture. *Advanced Modeling and Simulation in Engineering Sciences*, 5(1):14, May 2018.

[195] C. Geuzaine and J.-F. Remacle. Gmsh: a three-dimensional finite element mesh generator with built-in pre- and post-processing facilities. *International Journal for Numerical Methods in Engineering*, 79(11):1309–1331, 2009.

[196] A. Ghassemi and G. Kumar. Changes in fracture aperture and fluid pressure due to thermal stress and silica dissolution/precipitation induced by heat extraction from subsurface rocks. *Geothermics*, 36:115–140, 2006.

[197] A. Giacomini. Ambrosio–Tortorelli approximation of quasi-static evolution of brittle fractures. *Calculus of Variations and Partial Differential Equations*, 22:129–172, 2005.

[198] E. D. Giorgi and L. Ambrosio. Un nuovo funzionale del calcolo delle variazioni. *Atti della Accademia Nazionale dei Lincei. Classe di Scienze Fisiche, Matematiche e Naturali. Rendiconti*, 82:199–210, 1988.

[199] B. Giovanardi, A. Scotti, and L. Formaggia. A hybrid xfem–phase field (xfield) method for crack propagation in brittle elastic materials. *Computer Methods in Applied Mechanics and Engineering*, 320:396–420, 2017.

[200] V. Girault, G. Pencheva, M. F. Wheeler, and T. Wildey. Domain decomposition for poroelasticity and elasticity with DG jumps and mortars. *Mathematical Models and Methods in Applied Sciences*, 21(1):169–213, 2011.

[201] V. Girault and P.-A. Raviart. *Finite Element Method for the Navier–Stokes Equations*, volume 5 in *Computer Series in Computational Mathematics*. Springer-Verlag, 1986.

[202] R. Glowinski, S. Leung, and J. Qian. Operator-splitting based fast sweeping methods for isotropic wave propagation in a moving fluid. *SIAM Journal on Scientific Computing*, 38(2):A1195–A1223, 2016.

[203] R. Glowinski and A. Marroco. Sur l'approximation per elements finite d'ordre un, et la resolution, par penalization-dualite, d'une classe de Dirichlet non-lineaires. *RAIRO R-2*, pages 41–76, 1975.

[204] R. Glowinski, T.-W. Pan, and J. Periaux. A fictitious domain method for Dirichlet problem and applications. *Computer Methods in Applied Mechanics and Engineering*, 111(3–4):283–303, 1994.

[205] R. Glowinski and J. Periaux. Numerical methods for nonlinear problems in fluid dynamics. In *Proc. Intern. Seminar on Scientific Supercomputers*. North Holland, Feb 2–6 1987.

[206] R. Glowinski and P. L. Tallec. *Augmented Lagrangian and Operator-Splitting Methods in Nonlinear Mechanics*, volume 9 in *SIAM Studies in Applied Mathematics* SIAM, Philadelphia, 1989.

[207] C. Goll, T. Wick, and W. Wollner. DOpElib: Differential equations and optimization environment; A goal oriented software library for solving PDEs and optimization problems with PDEs. *Archive of Numerical Software*, 5(2):1–14, 2017.

[208] A. A. Griffith. The phenomena of rupture and flow in solids. *Philosophical Transactions of the Royal Society of London*, 221:163–198, 1921.

[209] C. Großmann, H.-G. Roos, and M. Stynes. *Numerical Treatment of Partial Differential Equations*. Springer, 2007.

[210] K. Gustafsson, M. Lundh, and G. Soederlind. A PI stepsize control for the numerical solution of ordinary differential equations. *BIT*, 28(2):270–287, 1988.

[211] H. Guzman. *Domain Decomposition Methods in Geomechanics*. PhD thesis, The University of Texas at Austin, 2012.

[212] J. Hagoort. *Waterflood-Induced Hydraulic Fracturing*. Technical Report, PhD Dissertation, Delft, The Netherlands, 1981.

[213] W. Han. *A Posteriori Error Analysis via Duality Theory: With Applications in Modeling and Numerical Approximations*. Springer, 2005.

[214] Z. Han. *Primal-Dual Active-Set Methods for Convex Quadratic Optimization with Applications*. PhD thesis, Lehigh University, 2015.

[215] R. Hartmann. Multitarget error estimation and adaptivity in aerodynamic flow simulations. *SIAM Journal on Scientific Computing*, 31(1):708–731, 2008.

[216] R. Hartmann and P. Houston. Goal-oriented a posteriori error estimation for multiple target functionals. In T. Hou and E. Tadmor, editors, *Hyperbolic Problems: Theory, Numerics, Applications*, pages 579–588. Springer Berlin Heidelberg, 2003.

[217] F. Hausdorff. Dimension und äußeres Maß. *Mathematische Annalen*, 79(1):157–179, Mar 1918.

[218] Y. Heider and B. Markert. A phase-field modeling approach of hydraulic fracture in saturated porous media. *Mechanics Research Communications*, 80:38–46, 2017. Multi-Physics of Solids at Fracture.

[219] Y. Heider, S. Reiche, P. Siebert, and B. Markert. Modeling of hydraulic fracturing using a porous-media phase-field approach with reference to experimental data. *Engineering Fracture Mechanics*, 202:116–134, 2018.

[220] M. Heil. An efficient solver for the fully coupled solution of large-displacement fluid-structure interaction problems. *Computer Methods in Applied Mechanics and Engineering*, 193:1–23, 2004.

[221] T. Heister, M. F. Wheeler, and T. Wick. A primal-dual active set method and predictor-corrector mesh adaptivity for computing fracture propagation using a phase-field approach. *Computer Methods in Applied Mechanics and Engineering*, 290:466–495, 2015.

[222] T. Heister and T. Wick. Parallel solution, adaptivity, computational convergence, and open-source code of 2d and 3d pressurized phase-field fracture problems. *Proceedings in Applied Mathematics and Mechanics (PAMM)*, 18(1):e201800353, 2018.

[223] B. Helenbrook. Mesh deformation using the biharmonic operator. *International Journal for Numerical Methods in Engineering*, pages 1–30, 2001.

[224] M. A. Heroux, R. A. Bartlett, V. E. Howle, R. J. Hoekstra, J. J. Hu, T. G. Kolda, R. B. Lehoucq, K. R. Long, R. P. Pawlowski, E. T. Phipps, A. G. Salinger, H. K. Thornquist, R. S. Tuminaro, J. M. Willenbring, A. Williams, and K. S. Stanley. An overview of the Trilinos project. *ACM Transactions on Mathematical Software*, 31(3):397–423, 2005.

[225] C. Hesch, A. Gil, R. Ortigosa, M. Dittmann, C. Bilgen, P. Betsch, M. Franke, A. Janz, and K. Weinberg. A framework for polyconvex large strain phase-field methods to fracture. *Computer Methods in Applied Mechanics and Engineering*, 317:649–683, 2017.

[226] C. Hesch, S. Schuß, M. Dittmann, M. Franke, and K. Weinberg. Isogeometric analysis and hierarchical refinement for higher-order phase-field models. *Computer Methods in Applied Mechanics and Engineering*, 303:185–207, 2016.

[227] M. R. Hestenes. Multiplier and gradient methods. *Journal of Optimization Theory and Applications*, 4:303–320, 1969.

[228] J. G. Heywood and R. Rannacher. Finite-element approximation of the nonstationary Navier–Stokes problem part iv: error analysis for second-order time discretization. *SIAM Journal on Numerical Analysis*, 27(2):353–384, 1990.

[229] J. G. Heywood, R. Rannacher, and S. Turek. Artificial boundaries and flux and pressure conditions for the incompressible Navier–Stokes equations. *International Journal for Numerical Methods in Fluids*, 22:325–352, 1996.

[230] M. Hintermüller, R. Hoppe, and C. Löbhard. Dual-weighted goal-oriented adaptive finite elements for optimal control of elliptic variational inequalities. *ESAIM. Control, Optimisation and Calculus of Variations*, 20:524–546, 2014.

[231] M. Hintermüller, K. Ito, and K. Kunisch. The primal-dual active set strategy as a semismooth Newton method. *SIAM Journal on Optimization*, 13(3):865–888, 2002.

[232] M. Hinze, R. Pinnau, M. Ulbrich, and S. Ulbrich. *Optimization with PDE Constraints*, volume 23 in Mathematical Modelling: Theory and Applications. Springer, Dordrecht u. a., 2009.

[233] A. Hirn. Finite element approximation of singular power-law systems. *Mathematics of Computation*, 82(283):1247–1268, 2013.

[234] C. Hirt, A. Amsden, and J. Cook. An arbitrary Lagrangian–Eulerian computing method for all flow speeds. *Journal of Computational Physics*, 14:227–253, 1974.

[235] M. Hofacker and C. Miehe. Continuum phase field modeling of dynamic fracture: variational principles and staggered FE implementation. *International Journal of Fracture*, 178:113–129, 2012.

[236] G. Holzapfel. *Nonlinear Solid Mechanics: A Continuum Approach for Engineering*. John Wiley and Sons, LTD, 2000.

[237] Q. Hong and J. Kraus. Parameter-robust stability of classical three-field formulation of Biot's consolidation model. *Electronic Transactions on Numerical Analysis*, 48:202–226, 2018.

[238] J. Hron, A. Ouazzi, and S. Turek. A computational comparison of two FEM solvers for nonlinear incompressible flow. In *Lecture Notes in Computational Science and Engineering*, vol. 35, pages 87–109. Springer, 2003.

[239] J. Hron and S. Turek. *A Monolithic FEM/Multigrid Solver for ALE Formulation of Fluid Structure with Application in Biomechanics*, volume 53, pages 146–170. Springer-Verlag, 2006.

[240] J. Hron and S. Turek. *Proposal for Numerical Benchmarking of Fluid-Structure Interaction Between an Elastic Object and Laminar Incompressible Flow*, volume 53, pages 146–170. Springer-Verlag, 2006.

[241] T. Hughes. *The Finite Element Method*. Dover Publications, 2000.

[242] T. Hughes, J. Cottrell, and Y. Bazilevs. Isogeometric analysis: cad, finite elements, nurbs, exact geometry and mesh refinement. *Computer Methods in Applied Mechanics and Engineering*, 194(39–41):4135–4195, 2005.

[243] T. Hughes, W. Liu, and T. Zimmermann. Lagrangian–Eulerian finite element formulation for incompressible viscous flows. *Computer Methods in Applied Mechanics and Engineering*, 29:329–349, 1981.

[244] K. Ito and K. Kunisch. Augmented Lagrangian methods for nonsmooth, convex optimization in Hilbert spaces. *Nonlinear Analysis*, 41:591–616, 2000.

[245] K. Ito and K. Kunisch. *Lagrange Multiplier Approach to Variational Problems and Applications*, volume 15 of *Advances in Design and Control*. Society for Industrial and Applied Mathematics (SIAM), Philadelphia, PA, 2008.

[246] W. Jäger and A. Mikelić. On the interface boundary condition of Beavers, Joseph, Saffman. *SIAM Journal on Applied Mathematics*, 60(4):1111–1127, 2000.

[247] M. Jammoul, M. Wheeler, and T. Wick. A phase-field multirate scheme with stabilized iterative coupling for pressure driven fracture propagation in porous media, under review, 2020.

[248] D. Jodlbauer, U. Langer, and T. Wick. Parallel block-preconditioned monolithic solvers for fluid-structure interaction problems. *International Journal for Numerical Methods in Engineering*, 117(6):623–643, 2019.

[249] D. Jodlbauer, U. Langer, and T. Wick. Parallel matrix-free higher-order finite element solvers for phase-field fracture problems, *Mathematical and Computational Applications*, 25(3):40, 2020.

[250] D. Jodlbauer, U. Langer, and T. Wick. Matrix-free multigrid solvers for phase-field fracture problems, math.NA: arXiv:1902.08112, 2019.

[251] D. Jodlbauer and T. Wick. A monolithic FSI solver applied to the FSI 1,2,3 benchmarks. In S. Frei, B. Holm, T. Richter, T. Wick, and H. Yang, editors, *Fluid-Structure Interaction: Modeling, Adaptive Discretization and Solvers*, Radon Series on Computational and Applied Mathematics. Walter de Gruyter, Berlin, 2017.

[252] V. John and J. Rang. Adaptive time step control for the incompressible Navier–Stokes equations. *Computer Methods in Applied Mechanics and Engineering*, 199(9–12):514–524, 2010.

[253] C. Johnson. *Numerical Solution of Partial Differential Equations by the Finite Element Method*. Cambridge University Press, Cambridge, 1987.

[254] C. Johnson. Adaptive finite element methods for the obstacle problem. *Mathematical Models and Methods in Applied Sciences*, 02(04):483–487, 1992.

[255] L. Kantorovich. On Newton's method for functional equations (Russian). *Doklady Akademii Nauk SSSR*, 59:1237–1249, 1948.

[256] L. Kantorovich and G. Akhilov. *Functional Analysis in Normed Spaces*. Fizmatgiz, Moscow, 1959. German translation: Berlin, Akademie-Verlag, 1964.

[257] K. Kergrene, S. Prudhomme, L. Chamoin, and M. Laforest. A new goal-oriented formulation of the finite element method. *Computer Methods in Applied Mechanics and Engineering*, 327:256–276, 2017. Advances in Computational Mechanics and Scientific Computation – the Cutting Edge.

[258] D. Khimin. Numerische Modellierung von Optimalsteuerungsproblemen für Phasenfeld-riss-ausbreitung. Master's thesis, Leibniz University Hannover, Jun 2020.

[259] A. Khodadadian, N. Noii, M. Parvizi, M. Abbaszadeh, T. Wick, and C. Heitzinger. A Bayesian estimation method for variational phase-field fracture problems. *Computational Mechanics*, published online, Jul 2020, DOI: https://doi.org/10.1007/s00466-020-01876-4.

[260] N. Kikuchi and J. Oden. *Contact Problems in Elasticity*, Studies in Applied Mathematics. Society for Industrial and Applied Mathematics (SIAM), Philadelphia, PA, 1988.

[261] D. Kinderlehrer and G. Stampacchia. *An Introduction to Variational Inequalities and Their Applications*, Classics in Applied Mathematics. Society for Industrial and Applied Mathematics, 2000.

[262] A. Kopanicakova and R. Krause. A recursive multilevel trust region method with application to fully monolithic phase-field models of brittle fracture. *Computer Methods in Applied Mechanics and Engineering*, 360:112720, 2020.

[263] M. Kronbichler. *Computational Techniques for Coupled Flow-Transport Problems*. PhD thesis, Uppsala Universitet, 2011.

[264] C. Kuhn and R. Müller. A continuum phase field model for fracture. *Engineering Fracture Mechanics*, 77(18):3625–3634, 2010. Computational Mechanics in Fracture and Damage: a Special Issue in Honor of Prof. Gross.

[265] C. Kuhn and R. Müller. A new finite element technique for a phase field model of brittle fracture. *Journal of Theoretical and Applied Mechanics*, 49(4):1115–1133, 2011.

[266] C. Kuhn, T. Noll, and R. Müller. On phase field modeling of ductile fracture. *GAMM-Mitteilungen*, 39(1):35–54, 2016.

[267] A. Kumar, B. Bourdin, G. A. Francfort, and O. Lopez-Pamies. Revisiting nucleation in the phase-field approach to brittle fracture. In *Workshop on Experimental and Computational Fracture Mechanics, Feb 26–28, 2020, Baton Rouge, LA*, 2020.

[268] G. S. Kumar and A. Ghassemi. Numerical modeling of non-isothermal quartz dissolution/precipitation in a coupled fracture-matrix system. *Geothermics*, 34:411–439, 2005.

[269] U. Köcher, M. P. Bruchhäuser, and M. Bause. Efficient and scalable data structures and algorithms for goal-oriented adaptivity of space-time FEM codes. *SoftwareX*, 10:100239, 2019.

[270] K. Königsberger. *Analysis 1*, Springer Lehrbuch. Springer, Berlin, Heidelberg, New York, 6th edition, 2004.

[271] K. Königsberger. *Analysis 2*, Springer Lehrbuch. Springer, Berlin, Heidelberg, New York, 5th edition, 2004.

[272] O. Ladyzhenskaja, V. Solonnikov, and N. Uralceva. *Linear and Quasi-Linear Equations of Parabolic Type*, volume 23 in *Translations of Mathematical Monographs, AMS*, 1968.

[273] U. Langer and O. Steinbach, editors. *Space-Time Methods: Application to Partial Differential Equations*, volume 25 of *Radon Series on Computational and Applied Mathematics*. de Gruyter, Berlin, 2019.

[274] U. Langer and H. Yang. Robust and efficient monolithic fluid-structure-interaction solvers. *International Journal for Numerical Methods in Engineering*, 108:303–325, 2016.

[275] U. Langer and H. Yang. Recent development of robust monolithic fluid-structure interaction solvers. In S. Frei, B. Holm, T. Richter, T. Wick, and H. Yang, editors, *Fluid-Structure Interaction: Modeling, Adaptive Discretization and Solvers*, Radon Series on Computational and Applied Mathematics. Walter de Gruyter, Berlin, 2017.

[276] C. J. Larsen. Models for dynamic fracture based on Griffith's criterion. In *Proceedings of the IUTAM Symposium on Variational Concepts with Applications to the Mechanics of Materials, Bochum, Germany, September 22–26, 2008*, pages 131–140. Springer Netherlands, Dordrecht, 2010.

[277] C. J. Larsen, C. Ortner, and E. Süli. Existence of solutions to a regularized model of dynamics fracture. *Methods in Applied Sciences*, 20:1021–1048, 2010.

[278] R. le Veque. Top ten reasons to not share your code (and why you should anyway). http://faculty.washington.edu/rjl/pubs/topten/topten.pdf, Dec 2012.

[279] S. Lee, Y.-J. Lee, and M. F. Wheeler. A locally conservative enriched Galerkin approximation and efficient solver for elliptic and parabolic problems. *SIAM Journal on Scientific Computing*, 2016.

[280] S. Lee, A. Mikelić, M. Wheeler, and T. Wick. Phase-field modeling of two phase fluid filled fractures in a poroelastic medium. *Multiscale Modeling & Simulation*, 16(4):1542–1580, 2018.

[281] S. Lee, A. Mikelić, M. F. Wheeler, and T. Wick. Phase-field modeling of proppant-filled fractures in a poroelastic medium. *Computer Methods in Applied Mechanics and Engineering*, 312:509–541, 2016. Phase Field Approaches to Fracture.

[282] S. Lee, B. Min, and M. F. Wheeler. Optimal hydraulic fracturing design using the phase field model coupled with global-objective genetic algorithm. *Computational Geosciences*, 22(3), 2018.

[283] S. Lee and M. F. Wheeler. Adaptive enriched Galerkin methods for miscible displacement problems with entropy residual stabilization. *Journal of Computational Physics*, 331:19–37, 2017.

[284] S. Lee and M. F. Wheeler. Enriched Galerkin methods for two-phase flow in porous media with capillary pressure. *Journal of Computational Physics*, 367:65–86, 2018.

[285] S. Lee, M. F. Wheeler, and T. Wick. Pressure and fluid-driven fracture propagation in porous media using an adaptive finite element phase field model. *Computer Methods in Applied Mechanics and Engineering*, 305:111–132, 2016.

[286] S. Lee, M. F. Wheeler, and T. Wick. Iterative coupling of flow, geomechanics and adaptive phase-field fracture including level-set crack width approaches. *Journal of Computational and Applied Mathematics*, 314:40–60, 2017.

[287] S. Lee, M. F. Wheeler, T. Wick, and S. Srinivasan. Initialization of phase-field fracture propagation in porous media using probability maps of fracture networks. *Mechanics Research Communications*, 80:16–23, 2017. Multi-Physics of Solids at Fracture.

[288] S. Lee, H. Chul Yoon, and S. Muddamallappa. Nonlinear strain-limiting elasticity for fracture propagation with phase-field approach. arXiv:2007.11101, 2020.

[289] J. L. Lions and E. Magenes. *Problèmes aux limites non homogènes et applications*. Travaux et Recherches Mathématiques, Paris, 1968.

[290] F. List and F. A. Radu. A study on iterative methods for solving Richards' equation. *Computational Geosciences*, 20(2):341–353, 2016.

[291] W. B. Liu and J. W. Barrett. A remark on the regularity of the solutions of the p-Laplacian and its applications to their finite element approximation. *Journal of Mathematical Analysis and Applications*, 178(2):470–487, 1993.

[292] M. Luskin and R. Rannacher. On the smoothing property of the Crank–Nicolson scheme. *Applicable Analysis*, 14(2):117–135, 1982.

[293] C. M. Maes. *A Regularized Active-Set Method for Sparse Convex Quadratic Programming*. PhD thesis, Stanford University, 2010.

[294] M. Maier and R. Rannacher. A duality-based optimization approach for model adaptivity in heterogeneous multiscale problems. *Multiscale Modeling & Simulation*, 16(1):412–428, 2018.

[295] G. Mallik, M. Vohralík, and S. Yousef. Goal-oriented a posteriori error estimation for conforming and nonconforming approximations with inexact solvers. Working paper or preprint, Dec. 2018.

[296] S. Mandal, A. Ouazzi, and S. Turek. Modified Newton solver for yield stress fluids. In *Numerical Mathematics and Advanced Applications, ENUMATH 2015*. Springer, 2016.

[297] J. Mandel. Consolidation des sols (étude mathémathique). *Geotechnique*, 3:287–299, 1953.

[298] K. Mang, M. Walloth, T. Wick, and W. Wollner. Adaptive numerical simulation of a phase-field fracture model in mixed form tested on an l-shaped specimen with high Poisson ratios, arXiv:2003:09459, 2020.

[299] K. Mang, M. Walloth, T. Wick, and W. Wollner. Mesh adaptivity for quasi-static phase-field fractures based on a residual-type a posteriori error estimator. *GAMM-Mitteilungen*, e202000003, 2019.

[300] K. Mang and T. Wick. Numerical Methods for Variational Phase-Field Fracture Problems. Institutionelles Repositorium der Leibniz Universität Hannover, Hannover. DOI: https://doi.org/10.15488/5129, July 2019.
[301] K. Mang, T. Wick, and W. Wollner. A phase-field model for fractures in nearly incompressible solids. *Computational Mechanics*, 65:61–78, 2020.
[302] J.-J. Marigo, C. Maurini, and K. Pham. An overview of the modelling of fracture by gradient damage models. *Meccanica*, 51:3107–3128, 2018.
[303] B. Markert and Y. Heider. Coupled multi-field continuum methods for porous media fracture. In *Recent Trends in Computational Engineering – CE2014: Optimization, Uncertainty, Parallel Algorithms, Coupled and Complex Problems*, pages 167–180. Springer International Publishing, Cham, 2015.
[304] H.-G. Maschke and M. Kuna. A review of boundary and finite element methods in fracture mechanics. *Theoretical and Applied Fracture Mechanics*, 4(3):181–189, 1985.
[305] S. A. Mattis and B. Wohlmuth. Goal-oriented adaptive surrogate construction for stochastic inversion. *Computer Methods in Applied Mechanics and Engineering*, 339:36–60, 2018.
[306] C. Mehlmann and T. Richter. A modified global Newton solver for viscous-plastic sea ice models. *Ocean Modelling*, 116:96–107, 2017.
[307] D. Meidner, R. Rannacher, and J. Vihharev. Goal-oriented error control of the iterative solution of finite element equations. *Journal of Numerical Mathematics*, 17:143–172, 2009.
[308] D. Meidner and T. Richter. Goal-oriented error estimation for the fractional step theta scheme. *Computational Methods in Applied Mathematics*, 14(2):203–230, 2014.
[309] D. Meidner and T. Richter. A posteriori error estimation for the fractional step theta discretization of the incompressible Navier–Stokes equations. *Computer Methods in Applied Mechanics and Engineering*, 288:45–59, 2015.
[310] J. Melenk and I. Babuška. The partition of unity finite element method: basic theory and applications. *Computer Methods in Applied Mechanics and Engineering*, 139(1):289–314, 1996.
[311] A. Mesgarnejad, B. Bourdin, and M. Khonsari. Validation simulations for the variational approach to fracture. *Computer Methods in Applied Mechanics and Engineering*, 290:420–437, 2015.
[312] C. Miehe, M. Hofacker, L.-M. Schaenzel, and F. Aldakheel. Phase field modeling of fracture in multi-physics problems. Part II. Coupled brittle-to-ductile failure criteria and crack propagation in thermo-elastic–plastic solids. *Computer Methods in Applied Mechanics and Engineering*, 294:486–522, 2015.
[313] C. Miehe, M. Hofacker, and F. Welschinger. A phase field model for rate-independent crack propagation: robust algorithmic implementation based on operator splits. *Computer Methods in Applied Mechanics and Engineering*, 199:2765–2778, 2010.
[314] C. Miehe and S. Mauthe. Phase field modeling of fracture in multi-physics problems. Part III. Crack driving forces in hydro-poro-elasticity and hydraulic fracturing of fluid-saturated porous media. *Computer Methods in Applied Mechanics and Engineering*, 304:619–655, 2016.
[315] C. Miehe, S. Mauthe, and S. Teichtmeister. Minimization principles for the coupled problem of Darcy–Biot-type fluid transport in porous media linked to phase field modeling of fracture. *Journal of the Mechanics and Physics of Solids*, 82:186–217, 2015.
[316] C. Miehe, L.-M. Schaenzel, and H. Ulmer. Phase field modeling of fracture in multi-physics problems. Part I. balance of crack surface and failure criteria for brittle crack propagation in thermo-elastic solids. *Computer Methods in Applied Mechanics and Engineering*, 294:449–485, 2015.

[317] C. Miehe, F. Welschinger, and M. Hofacker. Thermodynamically consistent phase-field models of fracture: variational principles and multi-field FE implementations. *International Journal for Numerical Methods in Engineering*, 83:1273–1311, 2010.

[318] A. Mikelić, M. Wheeler, and T. Wick. *A Phase-Field Approach to the Fluid Filled Fracture Surrounded by a Poroelastic Medium*. ICES Report 13-15, Jun 2013.

[319] A. Mikelić and M. F. Wheeler. Convergence of iterative coupling for coupled flow and geomechanics. *Computational Geosciences*, 17(3):455–462, 2012.

[320] A. Mikelić and M. F. Wheeler. On the interface law between a deformable porous medium containing a viscous fluid and an elastic body. *Mathematical Models and Methods in Applied Sciences*, 22(11):32, 2012.

[321] A. Mikelić, M. F. Wheeler, and T. Wick. A phase-field method for propagating fluid-filled fractures coupled to a surrounding porous medium. *Multiscale Modeling & Simulation*, 13(1):367–398, 2015.

[322] A. Mikelić, M. F. Wheeler, and T. Wick. Phase-field modeling of a fluid-driven fracture in a poroelastic medium. *Computational Geosciences*, 19(6):1171–1195, 2015.

[323] A. Mikelić, M. F. Wheeler, and T. Wick. A quasi-static phase-field approach to pressurized fractures. *Nonlinearity*, 28(5):1371–1399, 2015.

[324] A. Mikelić, M. F. Wheeler, and T. Wick. Phase-field modeling through iterative splitting of hydraulic fractures in a poroelastic medium. *GEM – International Journal on Geomathematics*, 10(1), Jan 2019.

[325] P. Mital. The Enriched Galerkin Method for Linear Elasticity and Phase Field Fracture Propagation. Master's thesis, The University of Texas at Austin, 2015.

[326] P. Mital, T. Wick, M. Wheeler, and G. Pencheva. Discontinuous and enriched Galerkin methods for phase-field fracture propagation in elasticity. In *Numerical Mathematics and Advanced Applications, ENUMATH2015*. Springer, 2016.

[327] L. Modica and S. Mortola. Il limite nella γ-convergenza di una famiglia di funzionali ellittici. *Bollettino dell'Unione Matematica Italiana A (5)*, 14(3):526–529, 1977.

[328] N. Moes, J. Dolbow, and T. Belytschko. A finite element method for crack growth without remeshing. *International Journal for Numerical Methods in Engineering*, 46:131–150, 1999.

[329] W. Mulder, S. Osher, and J. Sethian. Computing interface motion in compressible gas dynamics. *Journal of Computational Physics*, 100(2):209–228, 1992.

[330] D. Mumford and J. Shah. Optimal approximations by piecewise smooth functions and associated variational problems. *Communications on Pure and Applied Mathematics*, 42:577–685, 1989.

[331] G. Nabh. *On High Order Methods for the Stationary Incompressible Navier–Stokes Equations*. PhD thesis, University of Heidelberg, 1998.

[332] F. Natterer. Optimale L^2-Konvergenz finite Elemente bei Variationsungleichungen. *Bonner Mathematische Schriften*, 89:1–12, 1976.

[333] M. Negri. The anisotropy introduced by the mesh in the finite element approximation of the Mumford–Shah functional. *Numerical Functional Analysis and Optimization*, 20(9–10):957–982, 1999.

[334] M. Negri. Γ-convergence for high order phase field fracture: Continuum and isogeometric formulations. *Computer Methods in Applied Mechanics and Engineering*, 362:112858, 2020.

[335] P. Neittaanmäki and S. Repin. *Reliable Methods for Computer Simulation: Error Control and Posteriori Estimates*. Elsevier, Amsterdam, 2004.

[336] I. Neitzel, T. Wick, and W. Wollner. An optimal control problem governed by a regularized phase-field fracture propagation model. *SIAM Journal on Control and Optimization*, 55(4):2271–2288, 2017.

[337] I. Neitzel, T. Wick, and W. Wollner. An optimal control problem governed by a regularized phase-field fracture propagation model. Part II: the regularization limit. *SIAM Journal on Control and Optimization*, 57(3):1672–1690, 2019.
[338] T. Nguyen, J. Yvonnet, Q.-Z. Zhu, M. Bornert, and C. Chateau. A phase-field method for computational modeling of interfacial damage interacting with crack propagation in realistic microstructures obtained by microtomography. *Computer Methods in Applied Mechanics and Engineering*, 312:567–595, 2016.
[339] L. Ni, X. Zhuang, L. Zou, and J. Huang. Phase-field modeling of hydraulic fracture network propagation in poroelastic rocks. *Computational Geosciences*, 2020.
[340] J. Nocedal and S. J. Wright. *Numerical Optimization*. Springer Series in Operations Research and Financial Engineering, 2006.
[341] N. Noii, F. Aldakheel, T. Wick, and P. Wriggers. An adaptive global-local approach for phase-field modeling of anisotropic brittle fracture. *Computer Methods in Applied Mechanics and Engineering*, 361:112744, 2020.
[342] N. Noii, A. Khodadadian, and T. Wick. Bayesian inversion for anisotropic hydraulic phase-field fracture. arXiv:2007.16038, 2020.
[343] N. Noii and T. Wick. A phase-field description for pressurized and non-isothermal propagating fractures. *Computer Methods in Applied Mechanics and Engineering*, 351:860–890, 2019.
[344] J. Oden and S. Prudhomme. Estimation of modeling error in computational mechanics. *Journal of Computational Physics*, 182(2):496–515, 2002.
[345] J. T. Oden. Adaptive multiscale predictive modelling. *Acta Numerica*, 27:353–450, 2018.
[346] S. Osher and J. Sethian. Fronts propagating with curvature-dependent speed: algorithms based on Hamiltonian–Jacobi formulations. *Journal of Computational Physics*, 79(1):12–49, 1988.
[347] E. Pan. A general boundary element analysis of 2-d linear elastic fracture mechanics. *International Journal of Fracture*, 88:41–59, 1997.
[348] D. Pardo. Multigoal-oriented adaptivity for hp-finite element methods. *Procedia Computer Science*, 1(1):1953–1961, 2010.
[349] J. Parvizian, A. Düster, and E. Rank. Finite cell method: h- and p-extension for embedded domain problems in solid mechanics. *Computational Mechanics*, pages 121–133, 2007.
[350] R. Patil, B. Mishra, and I. Singh. An adaptive multiscale phase field method for brittle fracture. *Computer Methods in Applied Mechanics and Engineering*, 329:254–288, 2018.
[351] C. Peskin. The immersed boundary method. *Acta Numerica*, 11:479–517, 2002. Cambridge University Press.
[352] K. Pham, H. Amor, J.-J. Marigo, and C. Maurini. Gradient damage models and their use to approximate brittle fracture. *International Journal of Damage Mechanics*, pages 1–36, May 2011.
[353] G. Pillo, G. Liuzzi, S. Lucidi, and L. Palagi. A truncated Newton method in an augmented Lagrangian framework for nonlinear programming. *Computational Optimization and Applications*, 45(2):311–352, 2008.
[354] I. S. Pop, F. Radu, and P. Knabner. Mixed finite elements for the Richards' equation: linearization procedure. *Journal of Computational and Applied Mathematics*, 168(1–2):365–373, 2004.
[355] M. J. D. Powell. A method for nonlinear constraints in minimization problems. In *Optimization (Sympos., Univ. Keele, Keele, 1968)*, pages 283–298. Academic Press, 1969.
[356] N. Provatas and K. Elder. *Phase-Field Methods in Materials Science and Engineering*. Wiley, 2010.
[357] A. Rademacher. *Adaptive Finite Element Methods for Nonlinear Hyperbolic Problems of Second Order*. PhD thesis, Technische Universität, Dortmund, 2009.

[358] E. Rank, M. Ruess, S. Kollmannsberger, D. Schillinger, and A. Düster. Geometric modeling, isogeometric analysis and the finite cell method. *Computer Methods in Applied Mechanics and Engineering*, 249–252:104–115, 2012. Higher Order Finite Element and Isogeometric Methods.

[359] R. Rannacher. Finite element solution of diffusion problems with irregular data. *Numerische Mathematik*, 43:309–327, 1984.

[360] R. Rannacher. *Finite Element Methods for the Incompressible Navier–Stokes Equations*. Lecture Notes, August 1999.

[361] R. Rannacher. *Numerische Methoden der Kontinuumsmechanik*, Numerische Mathematik, vol. 3. Vorlesungsskriptum, 2001.

[362] R. Rannacher. *Numerik gewöhnlicher Differentialgleichungen*. Heidelberg University Publishing, 2017.

[363] R. Rannacher. *Numerik partieller Differentialgleichungen*. Heidelberg University Publishing, 2017.

[364] R. Rannacher. *Probleme der Kontinuumsmechanik und ihre numerische Behandlung*. Heidelberg University Publishing, 2017.

[365] R. Rannacher. *Special Topics in Numerics: I. FEM for Nonlinear Problems*. Vorlesungsskriptum, 2017.

[366] R. Rannacher. On the stabilization of the Crank–Nicolson scheme for long time calculations. Preprint, August 1986.

[367] R. Rannacher and F.-T. Suttmeier. A posteriori error control in finite element methods via duality techniques: application to perfect plasticity. *Computational Mechanics*, 21(2):123–133, 1998.

[368] R. Rannacher and F.-T. Suttmeier. A posteriori error estimation and mesh adaptation for finite element models in elasto-plasticity. *Computer Methods in Applied Mechanics and Engineering*, 176(1–4):333–361, 1999.

[369] R. Rannacher and F.-T. Suttmeier. Error estimation and adaptive mesh design for FE models in elasto-plasticity. In E. Stein, editor, *Error-Controlled Adaptive FEMs in Solid Mechanics*. John Wiley, 2000.

[370] R. Rannacher and J. Vihharev. Adaptive finite element analysis of nonlinear problems: balancing of discretization and iteration errors. *Journal of Numerical Mathematics*, 21(1):23–61, 2013.

[371] S. Repin. *A Posteriori Estimates for Partial Differential Equations*, volume 4 of *Radon Series on Computational and Applied Mathematics*. Walter de Gruyter GmbH & Co. KG, Berlin, 2008.

[372] J. Rice. Mathematical analysis in the mechanics of fracture. In *Fracture: An Advanced Treatise*, pages 191–311, chapter 3. Academic Press, New York, 1968.

[373] T. Richter. A fully Eulerian formulation for fluid-structure interaction problems. *Journal of Computational Physics*, 233:227–240, 2012.

[374] T. Richter. Goal-oriented error estimation for fluid-structure interaction problems. *Computer Methods in Applied Mechanics and Engineering*, 223–224:38–42, 2012.

[375] T. Richter. A monolithic geometric multigrid solver for fluid-structure interactions in ale formulation. *International Journal for Numerical Methods in Engineering*, 2015.

[376] T. Richter. *Fluid-Structure Interactions: Models, Analysis, and Finite Elements*. Springer, 2017.

[377] T. Richter and T. Wick. Finite elements for fluid-structure interaction in ALE and fully Eulerian coordinates. *Computer Methods in Applied Mechanics and Engineering*, 199:2633–2642, 2010.

[378] T. Richter and T. Wick. Optimal control and parameter estimation for stationary fluid-structure interaction. *SIAM Journal on Scientific Computing*, 35(5):B1085–B1104, 2013.

[379] T. Richter and T. Wick. On time discretizations of fluid-structure interactions. In T. Carraro, M. Geiger, S. Körkel, and R. Rannacher, editors, *Multiple Shooting and Time Domain Decomposition Methods*, pages 377–400. Contributions in Mathematical and Computational Science, 2015.

[380] T. Richter and T. Wick. Variational localizations of the dual weighted residual estimator. *Journal of Computational and Applied Mathematics*, 279:192–208, 2015.

[381] T. Richter and T. Wick. *Einführung in die numerische Mathematik – Begriffe, Konzepte und zahlreiche Anwendungsbeispiele*. Springer, 2017.

[382] J. Rungamornrat and M. E. Mear. A weakly-singular SGBEM for analysis of cracks in 3D anisotropic media. *Computer Methods in Applied Mechanics and Engineering*, 197(49):4319–4332, 2008.

[383] J. B. Russ and H. Waisman. Topology optimization for brittle fracture resistance. *Computer Methods in Applied Mechanics and Engineering*, 347:238–263, 2019.

[384] P. Saffman. On the boundary condition at the interface of a porous medium. *Studies in Applied Mathematics*, 1:93–101, 1971.

[385] D. Santillan, R. Juanes, and L. Cueto-Felgueroso. Phase field model of fluid-driven fracture in elastic media: immersed-fracture formulation and validation with analytical solutions. *Journal of Geophysical Research. Solid Earth*, 122(4):2565–2589, 2017.

[386] J. M. Sargado, E. Keilegavlen, I. Berre, and J. M. Nordbotten. High-accuracy phase-field models for brittle fracture based on a new family of degradation functions. *Journal of the Mechanics and Physics of Solids*, 111:458–489, 2018.

[387] A. Schlüter, A. Willenbücher, C. Kuhn, and R. Müller. Phase field approximation of dynamic brittle fracture. *Computational Mechanics*, 54:1141–1161, 2014.

[388] M. Schmich and B. Vexler. Adaptivity with dynamic meshes for space-time finite element discretizations of parabolic equations. *SIAM Journal on Scientific Computing*, 30(1):369–393, 2008.

[389] A. Schroeder and A. Rademacher. Goal-oriented error control in adaptive mixed FEM for Signorini's problem. *Computer Methods in Applied Mechanics and Engineering*, 200(1–4):345–355, 2011.

[390] H. R. Schwarz. *Methode der finiten Elemente*. Teubner Studienbuecher Mathematik, 1989.

[391] M. Schäfer and S. Turek. Benchmark computations of laminar flow around a cylinder. In *Flow Simulation with High-Performance Computer II*, volume 52 of *Notes on Numerical Fluid Mechanics*. Vieweg, Braunschweig Wiesbaden, 1996.

[392] J. Shen and X. Yang. Energy stable schemes for Cahn–Hilliard phase-field model of two-phase incompressible flows. *Chinese Annals of Mathematics, Series B*, pages 743–758, 2010.

[393] R. Shen, H. Waisman, and L. Guo. Fracture of viscoelastic solids modeled with a modified phase field method. *Computer Methods in Applied Mechanics and Engineering*, 346:862–890, 2019.

[394] S. Shin and D. Juric. Modeling three-dimensional multiphase flow using a level contour reconstruction method for front tracking without connectivity. *Journal of Computational Physics*, 180(2):427–470, 2002.

[395] I. N. Sneddon. The distribution of stress in the neighbourhood of a crack in an elastic solid. *Proceedings of the Royal Society of London. Series A*, 187:229–260, 1946.

[396] I. N. Sneddon and M. Lowengrub. *Crack Problems in the Classical Theory of Elasticity*. SIAM Series in Applied Mathematics. John Wiley and Sons, Philadelphia, 1969.

[397] K. Stein, T. Tezduyar, and R. Benney. Mesh moving techniques for fluid-structure interactions with large displacements. *Journal of Applied Mechanics*, 70:58–63, 2003.

[398] C. Steinke and M. Kaliske. A phase-field crack model based on directional stress decomposition. *Computational Mechanics*, 63(5):1019–1046, 2019.

[399] M. Strobl and T. Seelig. A novel treatment of crack boundary conditions in phase field models of fracture. *Proceedings in Applied Mathematics and Mechanics (PAMM)*, 15(1):155–156, 2015.

[400] S. Sun and J. Liu. A locally conservative finite element method based on piecewise constant enrichment of the continuous Galerkin method. *SIAM Journal on Scientific Computing*, 31(4):2528–2548, 2009.

[401] M. Sussman, P. Smereka, and S. Osher. A level set approach for computing solutions to incompressible two-phase flow. *Journal of Computational Physics*, 114:146–159, 1994.

[402] F. Suttmeier. *Adaptive Finite Element Approximation of Problems in Elasto-Plasticity Theory*. PhD thesis, University of Heidelberg, 1996.

[403] F. Suttmeier. *Numerical Solution of Variational Inequalities by Adaptive Finite Elements*. Vieweg+Teubner, 2008.

[404] F.-T. Suttmeier. General approach for a posteriori error estimates for finite element solutions of variational inequalities. *Computational Mechanics*, 27(4):317–323, 2001.

[405] L. Svolos, C. A. Bronkhorst, and H. Waisman. Thermal-conductivity degradation across cracks in coupled thermo-mechanical systems modeled by the phase-field fracture method. *Journal of the Mechanics and Physics of Solids*, 137:103861, 2020.

[406] A. Z. Szeri. *Fluid Film Lubrication*. Cambridge University Press, 2nd edition, 2010.

[407] E. Tanné, T. Li, B. Bourdin, J.-J. Marigo, and C. Maurini. Crack Nucleation in Variational Phase-Field Models of Brittle Fracture. *Journal of the Mechanics and Physics of Solids*, 09 2017.

[408] S. Teichtmeister, D. Kienle, F. Aldakheel, and M.-A. Keip. Phase field modeling of fracture in anisotropic brittle solids. *International Journal of Non-Linear Mechanics*, 97:1–21, 2017.

[409] R. Temam. *Navier–Stokes Equations: Theory and Numerical Analysis*. AMS Chelsea Publishing, Providence, Rhode Island, 2001.

[410] T. Tezduyar. Finite element methods for flow problems with moving boundaries and interfaces. *Archives of Computational Methods in Engineering*, 8(2):83–130, 2001.

[411] T. Tezduyar. Interface-tracking and interface-capturing techniques for finite element computation of moving boundaries and interfaces. *Computer Methods in Applied Mechanics and Engineering*, 195:2983–3000, 2006.

[412] T. Tezduyar, K. Takizawa, C. Moorman, S. Wright, and J. Christopher. Space-time finite element computation of complex fluid-structure interaction. *International Journal for Numerical Methods in Fluids*, 64:1201–1218, 2010.

[413] The Differential Equation and Optimization Environment. DOpElib. http://www.dopelib.net.

[414] I. Tolstoy. *Acoustic, Elasticity, and Thermodynamics of Porous Media, Twenty-One Papers by M. A. Biot*. Acoustical Society of America, New York, 1992.

[415] I. Toulopoulos and T. Wick. Numerical methods for power-law diffusion problems. *SIAM Journal on Scientific Computing*, 39(3):A681–A710, 2017.

[416] D. Tran, A. T. Settari, and L. Nghiem. Predicting growth and decay of hydraulic-fracture width in porous media subjected to isothermal and nonisothermal flow. *SPE Journal*, pages 781–794, 2013.

[417] R. Trémolières, J. Lions, and R. Glowinski. *Analyse numérique des inéquations variationnelles*. Dunod, 1976.

[418] R. Trémolières, J. Lions, and R. Glowinski. *Numerical Analysis of Variational Inequalities*. Studies in Mathematics and Its Applications. Elsevier Science, 2011.

[419] F. Tröltzsch. *Optimale Steuerung partieller Differentialgleichungen – Theorie, Verfahren und Anwendungen*. Vieweg und Teubner, Wiesbaden, 2nd edition, 2009.

[420] S. Turek. *Efficient Solvers for Incompressible Flow Problems*. Springer-Verlag, 1999.

[421] S. Turek, L. Rivkind, J. Hron, and R. Glowinski. Numerical Analysis of a New Time-Stepping θ-Scheme for Incompressible Flow Simulations. Technical report, TU Dortmund and University of Houston, 2005. Dedicated to David Gottlieb on the occasion of his 60th anniversary.

[422] H. Ulmer, M. Hofacker, and C. Miehe. Phase field modeling of brittle and ductile fracture. *Proceedings in Applied Mathematics and Mechanics (PAMM)*, 13:533–536, 2013.

[423] E. van Brummelen, S. Zhuk, and G. van Zwieten. Worst-case multi-objective error estimation and adaptivity. *Computer Methods in Applied Mechanics and Engineering*, 313:723–743, 2017.

[424] C. van Duijn, A. Mikelić, M. F. Wheeler, and T. Wick. Thermoporoelasticity via homogenization: modeling and formal two-scale expansions. *International Journal of Engineering Science*, 138:1–25, 2019.

[425] C. van Duijn, A. Mikelić, and T. Wick. A monolithic phase-field model of a fluid-driven fracture in a nonlinear poroelastic medium. *Mathematics and Mechanics of Solids*, 24(5):1530–1555, 2019.

[426] C. J. van Duijn, A. Mikelić, and T. Wick. Mathematical theory and simulations of thermoporoelasticity. *Computer Methods in Applied Mechanics and Engineering*, 366:113048, 2020.

[427] N. van Goethem and A. Novotny. Crack nucleation sensitivity analysis. *Mathematical Methods in the Applied Sciences*, 33(16), 2010.

[428] R. Verfürth. *A Review of A Posteriori Error Estimation and Adaptive Mesh-Refinement Techniques*. Wiley/Teubner, New York, Stuttgart, 1996.

[429] D. Werner. *Funktionalanalysis*. Springer, 2004.

[430] M. Wheeler, T. Wick, and W. Wollner. An augmented-Lagangrian method for the phase-field approach for pressurized fractures. *Computer Methods in Applied Mechanics and Engineering*, 271:69–85, 2014.

[431] M. F. Wheeler, T. Wick, and S. Lee. IPACS: integrated phase-field advanced crack propagation simulator. An adaptive, parallel, physics-based-discretization phase-field framework for fracture propagation in porous media. *Computer Methods in Applied Mechanics and Engineering*, 367:113124, 2020.

[432] D. Wick. *Simulation sowie experimentelle Untersuchung von Materialschädigungen bei Umformprozessen*. PhD thesis, Universität Siegen, Siegen, 2019.

[433] D. Wick, T. Wick, H. R. Hellmig, and H.-J. Christ. Numerical simulations of crack propagation in screws with phase-field modeling. *Computational Materials Science*, 109:367–379, 2015.

[434] T. Wick. *Adaptive Finite Element Simulation of Fluid-Structure Interaction with Application to Heart-Valve Dynamics*. PhD thesis, University of Heidelberg, 2011.

[435] T. Wick. Fluid-structure interactions using different mesh motion techniques. *Computers & Structures*, 89(13–14):1456–1467, 2011.

[436] T. Wick. Coupling of fully Eulerian and arbitrary Lagrangian–Eulerian methods for fluid-structure interaction computations. *Computational Mechanics*, 52(5):1113–1124, 2013.

[437] T. Wick. Fully Eulerian fluid-structure interaction for time-dependent problems. *Computer Methods in Applied Mechanics and Engineering*, 255:14–26, 2013.

[438] T. Wick. Solving monolithic fluid-structure interaction problems in arbitrary Lagrangian Eulerian coordinates with the deal.II library. *Archive of Numerical Software*, 1:1–19, 2013.

[439] T. Wick. Flapping and contact FSI computations with the fluid-solid interface-tracking/interface-capturing technique and mesh adaptivity. *Computational Mechanics*, 53(1):29–43, 2014.

[440] T. Wick. *Modeling, Discretization, Optimization, and Simulation of Fluid-Structure Interaction*. Lecture Notes at Heidelberg University, TU Munich, and JKU Linz available on https://www-m17.ma.tum.de/Lehrstuhl/LehreSoSe15NMFSIEn, 2015.

[441] T. Wick. Coupling fluid-structure interaction with phase-field fracture. *Journal of Computational Physics*, 327:67–96, 2016.

[442] T. Wick. Goal functional evaluations for phase-field fracture using PU-based DWR mesh adaptivity. *Computational Mechanics*, 57(6):1017–1035, 2016.

[443] T. Wick. Coupling fluid-structure interaction with phase-field fracture: algorithmic details. In S. Frei, B. Holm, T. Richter, T. Wick, and H. Yang, editors, *Fluid-Structure Interaction: Modeling, Adaptive Discretization and Solvers, Radon Series on Computational and Applied Mathematics*. Walter de Gruyter, Berlin, 2017.

[444] T. Wick. An error-oriented Newton/inexact augmented Lagrangian approach for fully monolithic phase-field fracture propagation. *SIAM Journal on Scientific Computing*, 39(4):B589–B617, 2017.

[445] T. Wick. Modified Newton methods for solving fully monolithic phase-field quasi-static brittle fracture propagation. *Computer Methods in Applied Mechanics and Engineering*, 325:577–611, 2017.

[446] T. Wick. Zielorientierte Numerik für Multiphysiksimulationen. *GAMM-Rundbrief*, 2:4–10, 2018.

[447] T. Wick. *Numerical Methods for Partial Differential Equations*. Institutionelles Repositorium der Leibniz Universität Hannover, Hannover. DOI: https://doi.org/10.15488/9248, January 2020.

[448] T. Wick, S. Lee, and M. Wheeler. *3D Phase-Field for Pressurized Fracture Propagation in Heterogeneous Media*. ECCOMAS and IACM Coupled Problems Proc., May 2015 at San Servolo, Venice, Italy, 2015.

[449] J. Schröder, T. Wick, S. Reese, P. Wriggers, R. Müller, S. Kollmannsberger, M. Kästner, A. Schwarz, M. Igelbüscher, N. Viebahn, H. Bayat, S. Wulfinghoff, K. Mang, E. Rank, T. Bog, D. D'Angella, M. Elhaddad, P. Hennig, A. Düster, W. Garhuom, S. Hubrich, M. Walloth, W. Wollner, C. Kuhn, and T. Heister. A selection of benchmark problems in solid mechanics and applied mathematics. *Archives of Computational Methods in Engineering*, accepted for publication, Aug 2020.

[450] T. Wick, G. Singh, and M. Wheeler. Fluid-filled fracture propagation using a phase-field approach and coupling to a reservoir simulator. *SPE Journal*, 21(03):981–999, 2016.

[451] T. Wick and W. Wollner. On the differentiability of fluid-structure interaction problems with respect to the problem data. *Journal of Mathematical Fluid Mechanics*, 21(3):34, Jun 2019.

[452] T. Wick and W. Wollner. Optimization with nonstationary, nonlinear monolithic fluid-structure interaction. *International Journal for Numerical Methods in Engineering*, 2020.

[453] Z. A. Wilson and C. M. Landis. Phase-field modeling of hydraulic fracture. *Journal of the Mechanics and Physics of Solids*, 96:264–290, 2016.

[454] B. Winkler. *Traglastuntersuchungen von unbewehrten und bewehrten Betonstrukturen auf der Grundlage eines objektiven Werkstoffgesetzes fuer Beton*. PhD thesis, University of Innsbruck, 2001.

[455] J. Wloka. *Partial Differential Equations*. Cambridge University Press, 1987.

[456] B. Wohlmuth. Variationally consistent discretization schemes and numerical algorithms for contact problems. *Acta Numerica*, 20:569–734, 2011.

[457] C. Wu, J. Fang, S. Zhou, Z. Zhang, G. Sun, G. P. Steven, and Q. Li. Level-set topology optimization for maximizing fracture resistance of brittle materials using phase-field fracture model. *International Journal for Numerical Methods in Engineering*, 121(13):2929–2945, 2020.

[458] J.-Y. Wu. A unified phase-field theory for the mechanics of damage and quasi-brittle failure. *Journal of the Mechanics and Physics of Solids*, 103:72–99, 2017.

[459] J.-Y. Wu, V. P. Nguyen, C. Thanh Nguyen, D. Sutula, S. Bordas, and S. Sinaie. Phase field modelling of fracture. *Advances in Applied Mechanics*, 53: https://www.sciencedirect.com/science/article/pii/S0065215619300134, 09 2019.

[460] K. Yoshioka, D. Naumov, and O. Kolditz. On crack opening computation in variational phase-field models for fracture. *Computer Methods in Applied Mechanics and Engineering*, 369:113210, 2020.
[461] A. Zehnder. *Fracture Mechanics*. Springer-Verlag, 2012.
[462] F. Zhang, W. Huang, X. Li, and S. Zhang. Moving mesh finite element simulation for phase-field modeling of brittle fracture and convergence of Newton's iteration. *Journal of Computational Physics*, 356:127–149, 2018.
[463] J. Zhang and Q. Du. Numerical studies of discrete approximations to the Allen–Cahn equation in the sharp interface limit. *SIAM Journal on Scientific Computing*, 31(4):3042–3063, 2009.
[464] X. Zhang, S. W. Sloan, C. Vignes, and D. Sheng. A modification of the phase-field model for mixed mode crack propagation in rock-like materials. *Computer Methods in Applied Mechanics and Engineering*, 322:123–136, 2017.
[465] S. Zhou, X. Zhuang, and T. Rabczuk. A phase-field modeling approach of fracture propagation in poroelastic media. *Engineering Geology*, 240:189–203, 2018.
[466] S. Zhou, X. Zhuang, and T. Rabczuk. Phase-field modeling of fluid-driven dynamic cracking in porous media. *Computer Methods in Applied Mechanics and Engineering*, 350:169–198, 2019.
[467] S. Zhou, X. Zhuang, and T. Rabczuk. Phase-field modeling of fluid-driven dynamic cracking in porous media. *Computer Methods in Applied Mechanics and Engineering*, 350:169–198, 2019.
[468] V. Ziaei-Rad and Y. Shen. Massive parallelization of the phase field formulation for crack propagation with time adaptivity. *Computer Methods in Applied Mechanics and Engineering*, 312:224–253, 2016. Phase Field Approaches to Fracture.

Index

Radon Series on Computational and Applied Mathematics

Volume 27
Multivariate Algorithms and Information-Based Complexity
Fred J. Hickernell, Peter Kritzer (Eds.), 2020
ISBN 978-3-11-063311-5, e-ISBN 978-3-11-063546-1

Volume 26
Discrepancy Theory
Dmitriy Bilyk, Josef Dick, Friedrich Pillichshammer (Eds.), 2020
ISBN 978-3-11-065115-7, e-ISBN 978-3-11-065258-1

Volume 25
Space-Time Methods. Applications to Partial Differential Equations
Ulrich Langer, Olaf Steinbach (Eds.), 2019
ISBN 978-3-11-054787-0, e-ISBN 978-3-11-054848-8

Volume 24
Maxwell's Equations. Analysis and Numerics
Ulrich Langer, Dirk Pauly, Sergey I. Repin (Eds.), 2019
ISBN 978-3-11-054264-6, e-ISBN 978-3-11-054361-2

Volume 23
Combinatorics and Finite Fields. Difference Sets, Polynomials, Pseudorandomness and Applications
Kai-Uwe Schmidt, Arne Winterhof (Eds.), 2019
ISBN 978-3-11-064179-0, e-ISBN 978-3-11-064209-4

Volume 22
The Radon Transform. The First 100 Years and Beyond
Ronny Ramlau, Otmar Scherzer (Eds.), 2019
ISBN 978-3-11-055941-5, e-ISBN 978-3-11-056085-5

Volume 21
Hamilton–Jacobi–Bellman Equations. Numerical Methods and Applications in Optimal Control
Dante Kalise, Karl Kunisch, Zhiping Rao (Eds.), 2018
ISBN 978-3-11-054263-9, e-ISBN 978-3-11-054359-9

www.degruyter.com

Bei Fragen zur Produktsicherheit wenden Sie sich bitte an:
If you have any questions regarding product safety,
please contact:

Walter de Gruyter GmbH
Genthiner Straße 13
10785 Berlin
productsafety@degruyterbrill.com